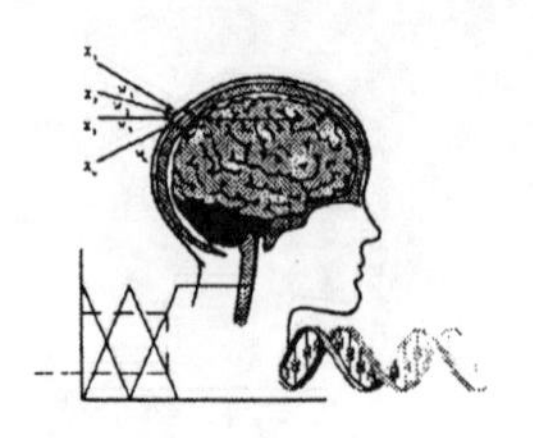

The CRC Press

International Series on Computational Intelligence

Series Editor
L.C. Jain, Ph.D.

L.C. Jain, R.P. Johnson, Y. Takefuji, and L.A. Zadeh
Knowledge-Based Intelligent Techniques in Industry

L.C. Jain and C.W. de Silva
Intelligent Adaptive Control: Industrial Applications

L.C. Jain and N.M. Martin
Fusion of Neural Networks, Fuzzy Systems, and Genetic Algorithms: Industrial Applications

H.N. Teodorescu, A. Kandel, and L.C. Jain
Fuzzy and Neuro-Fuzzy Systems in Medicine

C.L. Karr and L.M. Freeman
Industrial Applications of Genetic Algorithms

L.C. Jain and Beatrice Lazzerini
Knowledge-Based Intelligent Techniques in Character Recognition

L.C. Jain and V. Vemuri
Industrial Applications of Neural Networks

FUZZY and NEURO-FUZZY SYSTEMS in MEDICINE

Edited by
Horia-Nicolai Teodorescu
Abraham Kandel
Lakhmi C. Jain

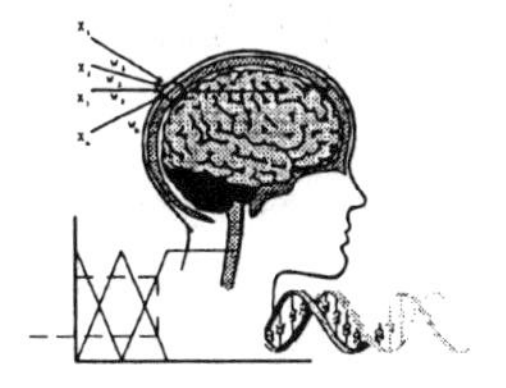

CRC Press
Boca Raton London New York Washington, D.C.

Library of Congress Cataloging-in-Publication Data

Fuzzy and neuro-fuzzy systems in medicine / H.N. Teodorescu, A. Kandel, L.C. Jain, editors.
p. cm. -- (International series on computational intelligence)
Includes bibliographical references and index.
ISBN 0-8493-9806-1 (alk. paper)
1. Fuzzy systems in medicine. 2. Neural networks (Neurobiology)
I. Teodorescu, Horia-Nicolai. II. Kandel, Abraham. III. Jain, L. C. IV. Series.
R859.7.F89F89 1998
610'.285'63—dc21 98-36332
CIP

International Standard Book Number 0-8493-9806-1
Library of Congress Card Number 98-36332
Printed in the United States of America 1 2 3 4 5 6 7 8 9 0
Printed on acid-free paper

Preface

This volume is entirely devoted to the applications of fuzzy and neuro-fuzzy systems in medicine. In this volume, we thoroughly review the state of the art, explain the problems to be addressed, and show the ways problems are solved.

Well-balanced chapters were selected, covering several major fields of application in medicine and bio-medical engineering. The volume introduces the reader to medical applications of fuzzy and neuro-fuzzy systems. The presentation progresses from fundamentals to applications. The applications go from modeling and signal analysis to medical knowledge processing systems and equipment.

A special emphasis is placed on clinical applications. In this respect, the volume addresses, beyond the group of fuzzy engineering, an audience from medicine and bio-medical engineering. Researchers and students in various fields of medicine, from radiology to neurology and from dentistry to cardiology will find this essential reading for applying the most recent advances in technical methods and equipment. In the field of engineering, the volume addresses biomedical engineers, computer scientists, and electrical engineers who are interested in medical applications.

The book is composed of three sections. The first section deals with the fundamentals and signal processing. The second section deals with knowledge processing and expert systems in clinical applications. The third section presents control issues and hardware implementations.

Part 1

The structure of the first section is as follows. The first chapter is introductory in nature. The second chapter deals with a modeling problem in neuro-physiology; namely, the foundation of fuzzy logic is seen as *wired* in the human brain by the way the neurons operate. Signal analysis and classification are discussed in Chapters 3 to 5. Chapter 3 addresses monodimensional signals (EEG), while Chapters 4 and 5 are devoted to medical imagistics and the use of neuro-fuzzy systems in this field.

In **Chapter 1**, the applications of fuzzy systems technology and neuro-fuzzy systems technology in medicine and biology are viewed from a historical perspective. Presentation of the first bibliography of the early years, analysis of past evolution, current development of the field and its perspectives are given. This chapter also offers the first ever bibliography in the field of medical applications for the first 15 years after the origination of fuzzy systems theory. Moreover, an analysis of the state of the art is presented as a background.

In **Chapter 2**, the authors discuss, from a neuroscience perspective, how fuzzy logic may originate in the brain. This may be the first time the intimate neurophysiological nature of fuzzy logic is explained and related to the biological underlying mechanics in the human brain. This paradigm may relate neural networks and fuzzy logic in a fusing mechanism that better explain both fields. After providing a neurophysiologic basis for the fuzzy logic, the authors explain in particular the color vision process as a neuro-fuzzy process in the brain.

In **Chapter 3**, the authors investigate the analysis and classification of signals using fuzzy, neuro-fuzzy, and wavelet methods. Unsupervised fuzzy clustering of electroencephalogram (EEG) signals is applied to sleep-stage scoring and to prediction of epilepsy. After briefly presenting the fundamentals of electroencephalographic activity and the main characteristics of EEG signals, the authors introduce various methods of EEG signal analysis; namely, time-frequency analysis, multiscale decomposition by the fast wavelet transform, multichannel model based, and decomposition by matching pursuit. Subsequently, various methods of classification are introduced, such as the unsupervised optimal fuzzy clustering algorithm, the weighted fuzzy k-means algorithm. The clustering validity criteria are discussed. The applications presented are sleep stage scoring, forecasting epilepsy, and classification of evoked- and event-related potentials by waveform.

In **Chapter 4**, the authors present the use of wavelet analysis in combination with neuro-fuzzy systems in the contouring gated SPECT images of ventricles. After an introduction to G-SPECT images, the authors present the principles of a method to contouring the myocardial cavities based on a combination of techniques. The analysis of the signal by wavelet transform is followed by neural network based image segmentation and fuzzy logic based recognition of the regions of the ventricles. A detailed analysis of the methods used to generate membership functions and rules, based on several techniques, is presented. Both experiments with phantoms and clinical test cases are presented in detail and the results are contrasted. Several implementation issues are discussed, as well as future research and possible developments.

Chapter 5 also deals with image analysis and processing by fuzzy methods. The authors present a detailed application based on a knowledge-based system that combines fuzzy techniques, multispectral analysis, and image processing algorithms. The unsupervised system discussed here is capable of automatically segmenting and labeling a specific tumor volume from transaxial MR image over a period of time when the tumor is treated. The fuzzy approach to edge detection helps solve the problem of detecting edge in tumor with diffuse contours, along with fuzzy c-means (FCM) clustering algorithm used in the initial stage for pathology detection. These are excellent examples of fuzzy techniques used to solve the problems with varying content of imprecision.

Part 2

The second section of the volume is devoted to knowledge processing and integration in decision and expert systems for preclinical or clinical use. Three of the chapters in this section deal with applications related to cardiology, two of them with ischemia diagnosis. By including two chapters with the same purpose – development of expert systems for cardiac ischemia diagnosis – but written by two different research groups, we offer the reader a good example and understanding of the variety of possible approaches. The presented approaches are of comparable complexity and have similar performances. Both have a foundation in fuzzy logic. Yet, one approach (Zahan, Michael, and Nikolakeas) emphasizes neural-related techniques in achieving the goals, while Rotshtein (and his co-workers in applying the system in medical practice) emphasizes classical nonlinear optimization and genetic algorithms. While comparable, the approaches point to different optimization tools and are based on different theoretical foundations. Similarly, Pilz and Engler deal with an expert system used in cardiology, but with different requirements, as the envisaged application is for critical conditions as encountered in intensive care units. The system is designed to offer a fast response and relies more heavily on knowledge directly derived from medical experts, with few adjustments. Although the system operation is based on fuzzy logic, it is developed closer to the basic knowledge of a computer engineer.

Chapter 6 integrates aspects related to fuzzy signal analysis and knowledge processing in a medical system, thus connecting the first and second sections of the volume. After introducing the main concepts, principles and methods related to screening, the authors present the select function used in the decision step, and show how disease specific knowledge can be defined and integrated into a system for screening the health state. A breast cancer case study is presented to exemplify the fundamental concepts and details of tuning the system performance. The integrated system is presented and discussed in the last section of the chapter.

An application in the field of dentistry is presented in **Chapter 7**. The author, an expert in occlusion analysis in pediatric dentistry, applies the concepts of fuzzy logic to the description and analysis of occlusal harmony and to the correctness of the masticatory function, mainly to the assessment of dental developmental age. The dental developmental age is used for planning the treatment and the evaluation of oral function in relation to maxillo-facial growth. The design and operation of a dental expert system for the evaluation of the dental developmental age is presented. The system is based on fuzzy modeling in the inferring part, on the principal component analysis in the extraction of relevant information from the data, and on genetic algorithms in optimizing the fuzzy system. The results generated by the fuzzy system models are contrasted with the results obtained using direct visual inspection by pediatric dentists who had special expertise in dental age assessment. The system helps decrease patient exposure to X-rays, among other benefits.

Chapter 8 presents a dedicated system for a major medical diagnosis problem, namely myocardial ischemia diagnosis. The authors emphasize the hierarchical structure of the system (DIFUS – Hierarchical Diagnosis Fuzzy System) and on knowledge organization inside the system. The multimethod myocardial ischemia diagnosis system presented by the authors uses fuzzy score-based tests in a compact representation and knowledge on medical patterns to derive a diagnosis. The system accomplishes, to a large degree, the attributes of a good expert system. It ensures the explicit separation between knowledge and inference mechanisms, knowledge-based dynamism, the interaction with the user in a natural language, the reasoning in uncertain conditions and using incomplete knowledge, and the understanding of its own actions. The system allows the implementation of certain operations with words, has a high adaptability and, hence, it offers a convenient tool for the development of different specialized applications. Experiments demonstrated a good level of diagnosis accuracy compared both to classical diagnosis (based on bivalent logic and on classical score computing) and to conventional diagnoses.

In **Chapter 9**, a typical expert system based on fuzzy knowledge processing used in intensive care units is presented. The system treats a large amount of data and is designed to provide answers on the state of the patient, primarily described by physiological conditions like "heart failure" or "shock," and to evaluate the patient's progress during several hours or days. The problems of the system interfacing with external programs and presentation of the results are also discussed.

In developing fuzzy decision systems and fuzzy expert systems, the designers face the problems of "tuning" the rules and the membership functions to achieve the optimal results. In **Chapter 10**, a method is presented for designing and tuning fuzzy rules for medical diagnoses. The tuning includes a rough and a fine tuning stage. The rough stage consists of fuzzy rules generation by a medical expert and selection of fuzzy membership functions by the method of paired comparisons. The fine tuning stage consists of finding the parameters of the membership functions and the weights of the rules based on training data for a specific optimization problem. The optimization method allows tuning of fuzzy rules with both continuous and discrete consequents. The effectiveness of the tuning models is demonstrated by several examples of a fuzzy expert system for differential diagnosis of ischemia heart disease.

Part 3

Chapter 11 presents a good example of how knowledge processing, decision-making, and control strategies are combined with control methods in medical equipment for a critical application: hemodynamic control in anesthesia. Several supervisory controls are used to cope with the complexity of the control problem where uncertainty of the controlled system is high, and the system is changing its parameters during the process.

Chapter 12 is an extensively documented overview of current technological problems and trends in the field of neural and fuzzy hardware implementation. The chapter includes two parts. In the first part, the authors present the main requirements for medical equipment and discuss several applications. Problems related to hardware minimization are detailed. In the second part of this chapter, the contributors analyze the possibilities for the integration of both the neural and fuzzy techniques. These techniques are analyzed in relation to medical instrumentation. In the first section, the preliminary design phase that implies software design and off-line experimenting of the concepts being implemented in integrated circuits are emphasized. The required properties for embedded medical systems are enumerated as a basis for the constraints that the integrated circuits must challenge. The authors investigate the main constraints: autonomy, reliability, and precision of computation. The strong link of these topics with architectural and design matters is very well illustrated. Digital as well as analogue implementations are reviewed, and benefits and limitations from either field are specified. In the third section, they describe the main hardware families of neural networks and fuzzy systems (analog, digital, and hybrid). The principles and performances of two neuro-fuzzy applications are briefly presented in the last part of the chapter.

The contributors to this volume are computer scientists, engineers, and physicians. Almost all chapters were written by teams of physicians, bio-medical engineers, computer scientists, and electrical engineers. This ensures that both the medical audience and the engineering audience will find this volume interesting and useful.

The Editors will be pleased to receive any comments from the readers and scholars that use this volume in their classes.

Acknowledgments

This volume could not appear without the kind support of many people and of several organizations.

First, the Editors thank the contributors for their commitment, hard work and patience during the preparation of the chapters.

The Editors are grateful to the staff of CRC Press, Florida, U.S.A., for their help and advice during the preparation of this volume and for their commitment to the project. We are particularly thankful to Ms. Josephine Gilmore, Ms. M. Williams, Mr. J. Papke, Ms. L.M. Franco, and to the team of editors and correctors at CRC Press. Their kind and professional guidance, and their support helped this volume be completed to the needed standards.

The Editors are grateful to all the reviewers of chapters in this volume, namely (in alphabetic order): Dragos Arotaritei, James Black, Lucian Boiculesei, Claudia-Cristina Bonciu, Cristian Bonciu, Adrian Brazulianu, Scott Dick, Radu Dogaru, Adriana Dumitras, Ionel Gheorghe, Florin Grigoras, R.K. Jain, Doron Leca, Steli Loznen, Daniel Moses, Constantin Posa, Wladimir Rodriguez, Adam Schenker, Emil Sofron, Qian Song, and Adrian Stoica. The Editors are also grateful to the contributors to this volume for their work in cross-reviewing the chapters. Without the efforts of the authors and reviewers, this volume could not have come to fruition.

The first Editor thanks his students for essential contributions in shaping his ideas related to the field over 15 years, as a teacher of bio-medical engineering (medical electronics). He acknowledges the support of the Institute for Information Sciences of the Romanian Academy.

Horia-Nicolai Teodorescu
Abraham Kandel
Tampa, Florida

Lakhmi C. Jain
Adelaide, South Australia

April 1998

About the Editors

Horia-Nicolai L. Teodorescu has served as a professor in several universities, University of South Florida (currently), Swiss National Institute of Technology, Lausanne, Switzerland, and Technical University of Iasi, Iasi, Romania. Dr. Teodorescu received an M.S. degree and the Doctoral degree in Electrical Engineering, in 1975 and 1981, respectively. He served as professor and founding director of the Center for Fuzzy Systems and Approximate Reasoning at Technical University of Iasi, Iasi, Romania, from 1990. He was an invited or visiting professor in Japan (1992, 1993, 1994), Switzerland (1994, 1995, 1996), and Spain (1993, 1996).

Dr. Teodorescu has written more than 200 papers, authored, co-authored, edited or co-edited more than 20 volumes, and holds 19 patents. He won several gold and silver medals for his inventions at various invention exhibitions. He authored many papers on biomedical engineering and applications of fuzzy and neuro-fuzzy systems to medical engineering. He also holds several patents in the field of biomedical engineering. He won several grants for research on applying fuzzy systems in biomedical applications. He is a Senior Member of the IEEE, and holds several honorary titles, including "Eminent Scientist" of Fuzzy Logic Systems Institute, Japan, and he was awarded the Honorary Medal of the Higher Economic School in Barcelona, Spain. He has been a *correspondent member* of the Romanian Academy, since 1993.

Dr. Teodorescu is one of the Chief Editors of *Fuzzy Systems & A.I.– Reports and Letters, International Journal for Chaos Theory and Applications, Iasi Polytechnic Magazine* and *Magazine for Fuzzy Systems* and a Co-Director of *Fuzzy Economic Review*. He is a member of the editorial boards of *Fuzzy Sets and Systems, The Journal of Grey Systems, BUSEFAL – Bulletin for Studies and Exchange of Fuzziness and Its Applications*, *Journal of Information Sciences of Moldavia, Review for Inventions*, and *Journal of AEDEM*. He served as a chairman of the scientific committees at several international conferences and was a member of the scientific committees of more than 40 international conferences.

Abraham Kandel received a B.Sc. from the Technion – Israel Institute of Technology and a M.S. from the University of California, both in Electrical Engineering and a Ph.D. in Electrical Engineering and Computer Science from the University of New Mexico. Dr. Kandel, a Professor and the Endowed Eminent Scholar in Computer Science and Engineering, is the Chairman of the Department of Computer Science and Engineering at the University of South Florida. Previously, he was Professor and Founding Chairman of the Computer Science Department at Florida State University as well as the Director of the Institute of Expert Systems and Robotics at FSU and the Director of the State University System Center for Artificial Intelligence at FSU.

He is Editor of the Fuzzy Track – IEEE MICRO, an Associate Editor of IEEE Transaction on Systems, Man, and Cybernetics, and a member of the editorial board of the international journals *Fuzzy Sets and Systems, Information Sciences, Expert Systems, Engineering Applications of Artificial Intelligence, The Journal of Grey Systems, Control Engineering Practice, Fuzzy Systems – Reports and Letters, IEEE Transactions on Fuzzy Systems, Book Series on Studies in Fuzzy Decision and Control, Applied Computing Review Journal, Journal of Neural Network World, The Journal of Fuzzy Mathematics*, and *BUSEFAL – Bulletin for Studies and Exchange of Fuzziness and its Applications.*

Dr. Kandel has published over 350 research papers for numerous professional publications in Computer Science and Engineering. He is co-author or co-editor of numerous volumes in the field of fuzzy logic. Dr. Kandel is a Fellow of the IEEE, a Fellow of the New York Academy of Sciences, a Fellow of AAAS, as well as a member of the ACM, NAFIPS, IFSA, ASEE, and Sigma-Xi.

Dr. Kandel has been awarded the College of Engineering Outstanding Researcher Award, USF 1993-94, Sigma-Xi Outstanding Faculty Researcher Award, 1995, The Theodore and Venette-Askounes Ashford Distinguished Scholar Award, USF, 1995, MOISIL International Foundation Gold Medal for Lifetime Achievements, 1996, and the Distinguished Researcher Award, USF, 1997.

Lakhmi C. Jain is a founding director of the Knowledge-Based Intelligent Engineering Systems (KES) center, located in the Faculty of Information Technology, Adelaide, Australia. Professor Jain received his B.E. (Hons), M.E., and Ph.D. degrees in Electronic Engineering and is a fellow of the Institution of Engineers Australia. He has received a number of awards for his papers including The 1996 Sir Thomas Ward Memorial Medal, and the 1995 best paper award by the Electronics Association of South Australia (for a paper in the Proceedings published by IEEE Computer Society Press U.S.A.). Professor Jain holds a joint patent (with Mr. Udina) on the skylight Light Intensity Data Logger.

He is one of the Editors-in-Chief of the International Journal of Knowledge-Based Intelligent Engineering Systems and serves as an Associate Editor of the IEEE Transactions on Industrial Electronics. Professor Jain was the Technical Chair of the ETD2000 International Conference in 1995, Publications Chair of the Australian and New Zealand Conference on Intelligent Information Systems in 1996 and a Conference Chair of the KES'97 International Conference in 1997. He was appointed as Vice-President (International Liaison) by the Electronics Association of South Australia. He has authored/co-authored over 100 books, papers, and invited research papers in his field. He is the Editor-in-Chief of the International Book Series in Computational Intelligence, CRC Press. U.S.A. His interests focus on the use of novel techniques such as knowledge-based systems, artificial neural networks, fuzzy systems, and genetic algorithms in engineering systems, and, in particular, the application of these techniques to solving practical engineering problems.

Contributors

Catalin V. Buhusi (M.Sc. in Computer Engineering at Technical University of Iasi in 1990; M.A. in Psychology at Duke University in 1998; currently working on his doctoral dissertation on neural networks models of timing and attention at Duke University) has worked on fuzzy and neural systems since 1990. He introduced a class of dynamic fuzzy systems (fuzzy systems with variable number of rules) and studied the neural and evolutionary methods of synthesis for dynamic fuzzy systems.
Address: Department of Neurobiology, Duke University, Durham, NC 27708, U.S.A. and Institute for Computer Science, Romanian Academy, Iasi 6600, Romania

Mircea I. Chelaru (Ph.D. in Electrical Engineering at Technical University of Iasi in 1994; postdoctoral study in experimental psychology at Duke University since 1997) has worked on neural networks and fuzzy systems since 1989. He is currently working in animal time learning and spatial navigation.
Address: Department of Neurobiology, Duke University, Durham, NC 27708, U.S.A. and Institute for Computer Science, Romanian Academy, Iasi 6600, Romania. E-mail: mircea@psych.duke.edu

Matthew Clark received the B.S. degree in 1992 in Computer Engineering from the University of South Florida in Tampa, Florida, the M.S. degree in Computer Science in 1994, and his Ph.D. from the same university in 1998. He has authored several journal papers and book chapters in medical imaging. His research interests include intelligent systems and the use of artificial intelligence techniques in pattern recognition and image processing applications.
Address: Computer Science and Engineering Department, University of South Florida, 4202 E. Fowler Avenue, Tampa, Florida 33620-5350, U.S.A., Fax: 001-813-974-5456, E-mail: clark@babbage.csee.usf.edu

André Constantinesco was born in Lisbon, Portugal. He graduated from Faculté des Sciences, Université Louis Pasteur Strasbourg, France and received the Ph.D. degree in Physics in 1976. He also graduated from Faculté de Médecine, Strasbourg and received the M.D. degree in 1977. Since 1980 he has been Professor of Biophysics and Nuclear Medicine and became head of the Nuclear Medicine Department of the University Hospital of Hautepierre, Strasbourg in 1997. His research interests include image processing in nuclear medicine and low – field MRI.
Address: Laboratoire de Biophysique et Médecine Nucléaire, Hôpital de Hautepierre, 1 Avenue Molière, 67098 Strasbourg, France. E-mail : luis@andromede.u-strasbg.fr.

Also: Laboratoire des Sciences de l'Image, de l'Informatique et de la Télédétection UPRES-A CNRS 7005, Bd. Sebastien Brant, 67400 Illkirch, France.

Patrik Eklund received his PhD in mathematics at Abo Akademi University in 1986, and since 1995 has been a professor in computer science at Umea University. His research interests include foundations, AI, health informatics, distributed systems, and industrial information systems. He actively develops partnerships within both industry and health care.
Address: Umea University, Department of Computing Science, Umeå University, Se-90187, Umeå Sweden. E-mail: Phone +46-90-7869914, fax +46-90-166126, peklund@cs.umu.se, URL: http://www.cs.umu.se/~peklund

Lothar Engelmann was born in Dohna, Germany in 1944. He received his M.D. (1969) and his lecturing qualification (1982) on hemodynamic oriented differential therapy of patients with acute myocardial infarct, all from the University of Leipzig. He is a Professor in the Medical Department at the University of Leipzig and Head of the Intensive Care Unit of the Center of Internal Medicine.
Address: Universität Leipzig, Zentrum für Innere Medizin, Ph.-Rosenthal-Str. 27a, D-04103 Leipzig, Germany. E-mail: engl@server3.medizin.uni-leipzig.de

Robert P. Erickson (Ph.D. in Psychology at Brown University in 1958; postdoctoral study in neurophysiology at the University of Washington in Seattle) is a Professor of Experimental Psychology, and Neurobiology at Duke University. He has worked on a concept of neural organization based on neural activity since 1961. This concept was pointed out as an expression of "fuzzy logic" in 1993 by Max Woodbury.
Address: Department of Psychology: Experimental, Duke University, Durham, NC 27708, U.S.A.. Fax: fax 660-5726; E-mail: eric@psych.duke.edu

Fredrik Georgsson is a Ph.D. student in the Computing Science Department at Umea University, where he also earned his M.Sc. His research interest is medical imaging, texture analysis, invariant representation, and topology. Much of his work is done in cooperation with the University Hospital in Umea.
Address: Umea University, Department of Computing Science, Umeå University, Se-90187, Umeå Sweden. Phone: +46 90 786 67 94; fax: +46 90 786 61 26. E-mail: fredrikg@cs.umu

Amir B. Geva was born in Haifa, Israel in 1959. He received his B.Sc. (1981) in Computer Engineering and his M.Sc. (1987) on optimal fuzzy partition of sleep EEG and D.Sc. on spatio-temporal source estimation of evoked potentials, in biomedical Engineering, all from the Technion–Israel Institute of Technology, Haifa. His research interests include: Biomedical signal (EEG, heart rate variability, blood flow sound) processing and system modeling, algorithms and computer systems, pattern recognition, fuzzy clustering, neural networks, wavelets analysis, time series prediction, the bioelectric inverse problem, functional brain imaging and modeling.
Address: Electrical and Computer Engineering Department, Ben Gurion University of the Negev, Be'er Sheva, Israel. E-mail: geva@ee.bgu.ac.il

Dmitry B. Goldgof received the M.S. from the Rensselaer Polytechnic Institute in 1985 and his Ph.D. from the University of Illinois in 1989. He is currently an Associate Professor in the Department of Computer Science and Engineering at the University of South Florida in Tampa. Professor Goldgof's research interests include motion analysis of rigid and nonrigid objects, computer vision, image processing and its biomedical applications, and pattern recognition. Dr. Goldgof has published over 90 journal and conference publications, 4 book chapters, and has one book currently in press. He is a senior member of IEEE, member of the Editorial Board of Pattern Recognition, and an Associate Editor for IEEE Transactions on Image Processing. Dr. Goldgof is currently General Chair for IEEE Conference on Computer Vision and Pattern Recognition (CVPR'98).
Address: Dept. of Computer Science and Engineering, University of South Florida, 4202 E. Fowler Ave, ENB118, Tampa, FL 33620, (+)(813) 974-4055, fax (813) 974-5456, goldgof@csee.usf.edu; http://marathon.csee.usf.edu/~goldgof/

Larry O. Hall is a professor of Computer Science and Engineering at the University of South Florida and has written a number of papers on AI.
Address: Computer Science and Engineering Department, University of South Florida, 4202 E. Fowler Avenue, Tampa, Florida 33620-5350, U.S.A., Fax: 001-813-974-5456, E-mail: clark@babbage.csee.usf.edu

Claudio M. Held received his Lic.Eng.Sc. (EE) and his M.S. in biomedical engineering, from the Universidad de Chile, and Ph.D. in biomedical engineering from Rensselaer Polytechnic Institute, Troy, NY. He has worked on the design of a multiple drug delivery control system, and has previously worked on knowledge-based electrocardiogram interpretation. He has been an advisor for the Clinical Hospital and participated in designing tools for medical decision-making in the Electrical Engineering Department in biomedical engineering at the Universidad de Chile.
Address: Department of Bioengineering, University of Pennsylvania, Philadelphia, PA 19104, Fax: 215-222-5192.

Ernest Hirsch is a Full Professor for computer sciences at the University of Strasbourg (ENSPS). Dr. Hirsch is Head of the research unit on analysis, interpretation, and diffusion of images of the Laboratoire des Sciences de l'Image, de l'Informatique et de la Télédétection (UPRES-A CNRS 7005). His research interests are mainly in real-time image analysis and computer vision, parallel and/or distributed system architectures, 3D reconstruction (human body and manufactured parts), and knowledge-based systems. Currently, among other activities and with an emphasis on knowledge-based approaches, he investigates the extension of methods already available for other fields of application to medical image processing.
Address: Laboratoire des Sciences de l'Image, de l'Informatique et de la Télédétection UPRES-A CNRS 7005, Bd. Sebastien Brant, 67400 Illkirch, France.

Johnnie W. Huang received his B.S. degree in Biomedical Engineering from the Johns Hopkins University, Baltimore, MD, and the M.S. degree in Biomedical Engineering from the Rensselaer Polytechnic Institute, Troy, NY. His research interests

include multiple drug delivery system design, cardiovascular dynamics modeling, cardiac output measurement, and depth of anesthesia monitoring and control.
Address: Department of Bioengineering, University of Pennsylvania, Philadelphia, PA 19104, Tel: 215-898-8310, Fax: 215-222-5192. E-mail: johnnie@seas.upenn.edu

Dan H. Kerem was born in 1941 in Jerusalem, Israel. He received the B.Sc. in Biology (1961) and the M.Sc. in Physiology (1963) on blood acidity effects on the EEG of the isolated brain, from the Hebrew University in Jerusalem, and his Ph.D. in Diving Physiology (1971) on the tolerance of the seal's brain to hypoxia, from UCSD, La Jolla, California. Retired Head of Research at the Israeli Naval Medical Institute, his longtime interest in forecasting brain oxygen toxicity spurred his recent involvement in the use of time-frequency analysis and fuzzy clustering on EEG and heart rate variability, for the early prediction of brain state transitions.
Address: Israeli Naval, Israeli Naval Medical Institute, IDF Medical Corps, Haifa, Israel. Phone: 972-4-8247740, Fax: 972-4-8262638, E-mail: dankerem@research.haifa.ac.il

Christian Ch. Michael received a M.S. degree in Electronics and Telecommunications from the Polytechnic University of Bucharest, Romania (1992) and the Ph.D. degree in Biomedical Engineering from the National Technical University of Athens, (NTUA), Greece (1997). Currently he is a Senior Researcher at the Institute of Computer and Communications Systems – NTUA. His research interests include the inverse problem in electrocardiology, biosignal acquisition and processing, electric source imaging, medical instrumentation, and fuzzy logic for medical diagnosis.
Address: National Technical University of Athens, 9 Heroon Polytechniou, 15773 Zografou , Athens, Greece.

Daniel Mlynek is a professor in the Electrical Engineering Department, Swiss National Institute of Technology, Lausanne. He is involved in promoting fuzzy logic especially in designing fuzzy and neuro-chips and other intelligent systems. He also promotes and organizes interdisciplinary courses.
Address: Integrated Systems Center (C3i), Swiss Federal Institute of Technology (EPFL), Department of Electrical Engineering, CH – 1015 Lausanne, Switzerland, Tel: +41 21 693 3370, Fax: +41 21 693 4663, Daniel.Mlynek@epfl.ch

Reed Murtagh is an M.D. and experienced Radiologist with the USF radiology department.
Address: Radiology Department, University of South Florida, 4202 E. Fowler Avenue, Tampa, Florida 33620-5350, U.S.A., Fax: 001-813-974-5456.

Stefanos P. Nikolakeas, M.D., graduated from Athens University Medical School in 1986 and received his specialty in cardiology in 1994. His Ph.D. thesis refers to linear analysis of ECG in coronary disease. Since 1992, he has been practicing clinical cardiology and has been participating in several research programs concerning fuzzy logic and nonlinear applications in cardiology as a research associate to the National Technical University of Athens and the Onassis Cardiology Center.

Address: National Technical University of Athens, 9 Heroon Polytechniou, 15773 Zografou , Athens, Greece.

Masao Ozaki, D.D.S., Ph.D. is an Associate Professor at Fukuoka College of Health Science, Fukuoka, Japan. He is a member of the International Bio-Medical Fuzzy Systems Association, Japan. He is interested in the use of artificial intelligence in dentistry and has co-authored several papers in the field.
Address: Fukuoka College of Health Science, Fukuoka Dental College, Dept. of Pediatric Dentistry, 2-15-1 Tamura, Sawara-ku Fukuoka, 814-01, Japan. Phone: 81-92-801-0411, Fax: 81-92-801-4909, E-mail: mozaki@college.fdcnet.ac.jp

Luis Patino was born in Mexico City, Mexico. He received the Electronic Engineering diploma from Universidad La Salle, Mexico City, in 1992. He received the M.S. degree in control systems from Université Technologique de Compiègne, France, in 1994. He is currently working toward his Ph.D. in Sciences at the Université Louis Pasteur, Strasbourg, France, scheduled for November 1998. His research interests include pattern recognition, edge detection, medical image processing, wavelets, and neuro-fuzzy based systems.
Address: Laboratoire de Biophysique et Médecine Nucléaire, Hôpital de Hautepierre, 1 Avenue Molière, 67098 Strasbourg, France. E-mail : luis@andromede.u-strasbg.fr

Uwe Pilz was born in Leipzig, Germany in 1958. He received his Ph.D. (1991) in Electrical Engineering and his lecturing qualification (1995) on numerical simulations of the electrical field and mass transfer in electroplating, all from the Technical University of Ilmenau, Germany. Research interests include the development of patient data management systems, knowledge processing in medicine, and medical databases.
Address: Universität Leipzig, Zentrum für Innere Medizin, Ph.-Rosenthal-Str. 27a, D-04103 Leipzig, Germany; snail: Kieler Str. 63, 04357 Leipzig, Germany. Tel: (+49 341) 97 12714 (office); (+49 341) 60 10 383; Fax: (+49 341) 26 15 456. E-mail: pilz@server3.medizin.uni-leipzig.de

Alexander Rotshtein, Ph.D., Dr. Sc., Prof., is Chairman of Department of Computer-Based Information & Management Systems, State Technical University, Vinnitsa, Ukraine. He graduated from the Faculty of Rodio-Physics at Nizhniy Novgorod (Gorkiy) State University (Russia) in 1972. He received the Ph.D. in System Analyses and Computer Science from Riga Technical University (Latvia) in 1979 and Doctor of Science from Moscow Aircraft Institute in 1989. He is the author of three books: *Reliability Design of Man-Machine Systems, Fuzzy Logic-Based Medical Diagnosis, Fuzzy Reliability Analyses of Algorithmic Processes.* His current research interests include intellectual methodology of identification with application to real life problems.
Address: Vinnitsa State Technical University, Department of Computer-Based Information and Management Systems, Khmelnitsky Shosse 95, Vinnitsa 286021, UKRAINE. Phone: (+)(0432)440-157, Fax: (+)(0432)465-772. E-mail: alex@rap.rts.vinnica.ua

Rob J. Roy received his B.S.E.E. degree from Cooper Union, New York, NY, the M.S.E.E. degree from Columbia University, New, York, NY, the D.Eng.Sc. degree from Rensselaer Polytechnic Institute, Troy, NY, and the M.D. degree from Albany Medical College, Albany, NY. He has been Professor of Electrical Engineering and Chairman of Biomedical Engineering at Rensselaer Polytechnic Institute and is now Active Professor Emeritus. He is currently Professor of Anesthesia and Attending Anesthesiologist at Albany Medical Center. His research interests are in biological signal processing and adaptive control systems. He has published extensively in the area of pattern recognition, control systems, radar signal processing, process identification, cardiac output measurement, closed circuit anesthesia delivery systems, multiple drug delivery systems, and depth of anesthesia monitoring. Dr. Roy was recently chosen as one of the Best Doctors in America: Northeast Region.
Address: Department of Bioengineering, University of Pennsylvania, Philadelphia, PA 19104, Fax: 215-222-5192, E-mail: royr@rpi.edu

Alexandre Schmid received the Engineer degree in Micro-Engineering from the Swiss Federal Institute of Technology (EPFL) in 1994. He is currently working toward the Ph.D. degree at the Integrated Systems Center of EPFL. His research interests include artificial neural network integration, fast digital processing, and CAD for digital systems.
Address: Integrated Systems Center (C3i), Swiss Federal Institute of Technology (EPFL), Department of Electrical Engineering, CH – 1015 Lausanne, Switzerland, Tel.: +41 21 693 3370, Fax: +41 21 693 4663, E-mail: Alexandre.Schmid@epfl.ch

Martin Sibiger, M.D. is the Dean of the USF College of Medicine and an executive vice president of the University.
Address: College of Medicine, University of South Florida, 4202 E. Fowler Avenue, Tampa, Florida 33620-5350, U.S.A.

Robert Velthuizen is an associate in Research in the USF radiology department.
Address: Radiology Department, University of South Florida, 4202 E. Fowler Avenue, Tampa, Florida 33620-5350, U.S.A.

Sorina Zahan received the M.Sc. degree in Electronics and Communications Engineering and the Ph.D. degree in Electronics, both from the Technical University of Cluj-Napoca, Romania. She joined the same university in 1991, where she currently is an associate professor within the Communications Department. Her major research interests are applications of intelligent techniques to medical diagnosis and telecommunications. She is the author or co-author of three books and 35 papers in these fields.
Address: Technical University of Cluj-Napoca, Communications Department, 26, G. Baritiu str., 3400 Cluj-Napoca, Romania, Phone: +40-64-193161; Fax: +40-64-191689. E-mail: Sorina.Zahan@com.utcluj.ro

A Guide for Using This Volume in the Classroom

This book was written with the intent to serve both as:

- a tool for researchers, and
- reading for graduate or postgraduate courses.

As a tool for researchers, the volume offers a representative selection of significant medical applications where fuzzy and neuro-fuzzy systems, in conjunction with other methods, are used. All applications are explained in detail and the principles, methods, and results are comprehensively presented and documented. In addition, the volume offers the researcher a major source of references in the field (about 500 references, including those in the first 15 years of the field development).

As a reading for a course on advanced methods in medical sciences and biomedical engineering, the volume offers well-organized and detailed chapters on carefully selected topics, and extensive appendices. The volume can be used as a text for a course for postgraduate students by making a choice of chapters and sections, according to the course topic and audience. We would suggest the following possible selections:

- for a course for students in medicine: Chapters 1, 2, 3, 4, or 5 (according to the main field of interest for imagistics applications), Chapters 6, 7 (for applications in dentistry), Chapter 8, and Chapter 11;
- for a course for students in bio-medical engineering: the whole volume, possibly skipping Chapter 12 if the major interest is software, or Chapters 2 to 6, 8, 9, 11, and 12, if the emphasis is not on software;
- for a course on advanced applications of computer science and mainly of artificial intelligence techniques in medicine, Chapters 3, 4, 5, 6, 8, 9, Chapter 10 as an exercise, and possibly Chapter 11.

To use this book both as a research tool and a textbook in advanced courses, several appendices were added, in addition to this Guide: a detailed presentation of the volume in the Preface, a comprehensive Index of symbols, and an Index of terms.

Moreover, several chapters include indices of terms or acronyms.

The reader has access to a free demonstration software pending to Chapter 10. The software allows the reader to experiment with the system and may be used in class as well, if this volume is used in teaching.

This book is intended to be the first in a series on Intelligent Technologies in Human-Related Sciences. We hope the next volumes will add useful material for assistance in class.

Horia-Nicolai Teodorescu
Abraham Kandel
Lakhmi Jain

Index of Symbols

The notations below are extensively used in this volume and in the literature.

Sets of numbers and properties

N (bold N): set of natural numbers
R (bold R), or $\mathbb{R}$, also denoted sometimes by $\Re$: set of real numbers
Z or $\mathbb{I}$, the set of integers
$\mathbb{C}$ set of complex numbers

Order relations on the set of real numbers **R** : $<$, $\leq$, $=$, $>$, $\geq$:
$<$ less than
$\leq$ less or equal to
$>$ greater than
$\geq$ greater than or equal
$\neq$ not equal (different)

Sums: $a_1 + a_2 + a_3 + \cdots + a_k + \cdots + a_n = \sum_{k=1..n} a_k$ or $\sum_{k=1..n} a_k$ or: $\sum_{k=1}^{n} a_k$

$\prod_{k=1..n} a_k$ product $\prod_{k=1..n} a_k = a_1 a_2 \cdots a_k \cdots a_n$ (product limits are written as for sums)

$\equiv$ identical (for expressions)

Sets

A, B, C, ..., U,... usual notation for sets
a, b,..., x, y usually denote elements
{ : } or { | } set notation; are used to represent sets (by giving the elements or their properties on the right side of : or |)
$\in$ belongs to
$\notin$ does not belong to
$\cup$ union of sets
$\cap$ intersection of sets
$\supset$ includes: $A \supset B$ reads A includes (strictly) B
$\supseteq$ includes (not strictly, not necessary as a proper subset)
$\not\subset$ not included in
$\subset$ included (strictly, as a proper subset) in; subset of

$\subseteq$ included in
$\neg$ complement (complementation of a set with respect to some universal set). For example $\neg A$ denotes the complement of the set A with respect to the universal set
$\overline{A}$ another notation for the complement of the set A
$\varnothing$ empty set
U the "universal" set
$\mathcal{P}$(A) or $\wp$(A) power set of set A (the set of all subsets of A, including the empty set and the set A itself)
2^A same as $\wp(A)$, i.e., power set of set A
$\bigcup_{k=1}^{\infty} A_k$ union of an infinite family of sets $\{A_k \mid k = 1, 2, \ldots\}$
$\bigcap_{k=1}^{\infty} A_k$ intersection of an infinite family of sets

Fuzzy sets

$\tilde{A}, \tilde{B}, \ldots$ etc. reads: the fuzzy sets A, B etc. (examples). The tilde (˜ symbol) can be omitted if there is no possible confusion that A is interpreted as a classic, crisp set
$\mu_{\tilde{A}}(\cdot)$ $\mu_A(.)$ reads: the membership functions of the fuzzy set A, $A \subset \mathbf{R}$, $\mu : A \to [0,1]$
$I_\alpha(\tilde{x}) = [u_\alpha, v_\alpha]$ α-cut (interval corresponding to the α-cut of the fuzzy number or fuzzy set $\tilde{x}$

Fuzzy numbers

$\tilde{a}, \tilde{x}$ $(a, x \in \mathbf{R})$ fuzzy numbers (in general)
$\tilde{a} = (a_1, a_2, a_3)$ $(a_1, a_2, a_3 \in \mathbf{R})$ triangular fuzzy numbers
$\tilde{a} = (a_1, a_2, a_3, a_4)$ $(a_1, a_2, a_3, a_4 \in \mathbf{R})$ trapezoidal fuzzy numbers

Logic symbols

$\wedge$, & logic connective "and"
$\vee$ logic connective "or"
$\neg$ and $\neg$(.), or $\overline{(.)}$: logic negation; the point inside brackets stand for a logical expression
$\Leftrightarrow$, or $\equiv$ logically equivalent
$\forall$ for every
$\exists$ there exists
$\exists!$ there exists a unique
$\Rightarrow$ or $\rightarrow$ implies (if ... then); logic implication (from the left side to the right side)
$\Leftarrow$ or $\leftarrow$ implies (if ... then); logic implication (from the right side to the left side)
: such as; that satisfies the condition
... is used to denote a recursion (elements in a set, words in a language, etc.)

Algebraic notations

(a, b) or <a, b> couples (of numbers or of variables)

$\bar{a} = (a_1, \cdots, a_k, \cdots, a_n)$, $\bar{b} = (b_1, \cdots, b_k, \cdots, b_n)$ $\quad a_k, b_k \in \mathbf{R}$

$\bar{a} = \mathbf{a} = (a_1, a_2, \cdots, a_n)$, $\bar{x} = \mathbf{x} = (x_1, x_2, \cdots, x_n)$ vectors (generally denoted by small letters)

$\left.\begin{array}{c}[a_{ij}]\\ \mathbf{A}\end{array}\right\}$ denote matrices

Matrices can be explained by:

$$\mathbf{A} = \begin{bmatrix} a_{11} & & \\ & & \\ & & a_{nm} \end{bmatrix}$$

Matrix product and sum: **AB, A+B**
Matrices transpose: $\mathbf{A}^{\mathrm{T}}$

Analysis notations

$\bar{x}$ upper limit (also average value, in statistics)

$\underline{x}$ lower limit

$|\ |$ absolute value

Norm and squared norm (including Euclidean norm, obtained by scalar product), respectively: $\|\cdot\|$, or: $|.|$. The last is to be used only for Euclidean norm.

$\max\{a, b\}$ maximum value between a and b

$\max_k\{a_k\}$ maximum value between all values a_k

$f : \mathbf{R} \to \mathbf{R}$, $\mathbf{R} \xrightarrow{f} \mathbf{R}$, $f(\cdot) : \mathbf{R} \to \mathbf{R}$ functions (of real variable and real valued)

$\eta(x) = \begin{cases} 0 & \text{for } x \geq 0 \\ 1 & \text{for } x < 0 \end{cases}$ a function that has values according to the case

$\varphi_{c,\sigma} : \mathbf{R} \to (0,1]$, $\varphi_{c,\sigma}(x) = e^{-\frac{(x-c)^2}{2\sigma^2}}$ Gaussian function of center c and spreading σ

$E^2 = \frac{1}{2} \cdot \sum_{k=1}^{p} (d_k - y_k)^2$ Euclidean distance or error between the (desired) vector $\bar{d} = (d_1, d_2, \cdots d_n)$ and the vector $\bar{y} = (y_1, y_2, \cdots y_n)$. Also denoted by e^2, ε^2

ΔV increment of the variable V

Δt time interval; time increment (t denoting time)

$\dfrac{\partial f}{\partial x}$ partial derivative of the function $f = f(x, y,...)$ with respect to the variable x

$\dfrac{\partial f}{\partial x(t)}$ partial derivative of the composed function $f = f(x(t), y(t),...)$ with respect to the function $x(t)$ that is a variable of function f

$\mathscr{F}$ Fourier transform

Statistics

$\{x_n \mid n = 1,2,..., k,...n\} = \{x_1, x_2, x_3, \cdots, x_k, \cdots, x_n\}$ finite time series

$\{x_n \mid n = 1,2,..., k,...n,...\} = \{x_1, x_2, \cdots, x_k, \cdots, x_n, \cdots\}$ infinite time series

$\overline{x} = \dfrac{1}{n}\sum_{k=1}^{n} x_k,\ x_k \in \mathbf{R}$ average of the values $x_1, x_2, x_3, \cdots, x_k, \cdots, x_n$

σ_x^2 the variance of x; $\sigma_x^2 = \left(X - \overline{X}\right)^2$. When the variable is identified in some other way, the index x is missing

C_{ij} the covariance of two stochastic variables X_i, X_j, defined by:

$$C_{ij} = \frac{1}{N}\sum_{k=1}^{N}\left(X_i - \overline{X}_i\right)\left(X_j - \overline{X}_j\right)$$

C_{ii} the self-correlation is: $C_{ii} = \dfrac{1}{N}\sum_{k=1}^{N}\left(X_i - \overline{X}_i\right)^2 = \sigma_i^2$

R_{ij} the correlation coefficient is defined by: $R_{ij} = \dfrac{S_{ij}}{\sqrt{S_{ii}S_{jj}}}$

(Compiled by H.N. Teodorescu)

Contents

Part 2. Neuro-Fuzzy Knowledge Processing

Chapter 12.
Neuro-Fuzzy Hardware in Medical Applications

12 A. System Requirements for Fuzzy and Neuro-Fuzzy Hardware in Medical Equipment
Horia-Nicolai Teodorescu, Abraham Kandel, and Daniel Mlynek **341**

12. B. Neural Networks and Fuzzy-Based Integrated Circuit and System Solutions Applied to the Biomedical Field
Alexandre Schmid and Daniel Mlynek **361**

Part 1.

Fundamentals and Neuro-Fuzzy Signal Processing

Chapter 1

Fuzzy Logic and Neuro-Fuzzy Systems in Medicine and Bio-Medical Engineering: A Historical Perspective

Horia-Nicolai L. Teodorescu, Abraham Kandel, and Lakhmi C. Jain

Fuzzy logic has provided a basis for representing uncertain and imprecise knowledge and formed a basis for human reasoning representation in medical and bio-medical engineering applications. Artificial neural networks (ANNs) have been successfully used to mimic biological information processing mechanisms, in a limited sense. The fusion of ANNs and fuzzy logic offsets the demerits of one paradigm by the merits of the other. The tremendous advances in the theory of ANNs and fuzzy logic have resulted in the flourishing applications of these techniques in medicine and biomedical engineering.

In this chapter, the applications of fuzzy and neuro-fuzzy systems technology in medicine and biology are reviewed from a historical perspective. In addition to the presentation of the first bibliography of the early years, an analysis of past evolution and the current development of the field and its perspectives is presented.

1. THE FIRST PERIOD: THE INFANCY

Fuzzy logic is useful for representing uncertain and imprecise knowledge. Fuzzy logic provides an approximate but effective means of describing the behavior of systems that are complex, ill-defined, or not easily analyzed.

Zadeh argues that the attempts to automate various types of activities, from assembling hardware to medical diagnosis, have been impeded by the gap between the way humans reason and the way computers are programmed. He proposed fuzzy logic to bridge this gap. Indeed, fuzzy logic techniques have been successfully used in a number of applications including medicine and bio-medical engineering.

0-8493-9806-1/98/$0.00+$.50

On the other hand, ANNs mimic the biological information processing mechanism, in a limited sense. They use simple processing elements which are analogous to biological neurons interconnected in a typical way. Their capabilities, such as generalization, parallelism, distributed memory, and learning ability, make them useful in a wide variety of applications including medicine and biomedical engineering.

The first applications of fuzzy logic in the field of medical sciences date back to the early stages of fuzzy logic, during the years 1965-1975. In 1968, L.A. Zadeh presented the first paper on the possibility of developing applications of fuzzy sets in biology. In 1969, the paper was published in an edited volume.

A few papers were published before 1980 on applications of fuzzy systems in medicine. These were the pioneering years, with scientists, applying fuzzy logic, were looking for applications and for basic developments, trying to build bridges developed in concepts between the new field and the established ones.

Because fuzzy logic is related to the human way of thinking, it is not surprising that the first reported applications addressed the assessing of symptoms and the modeling of medical reasoning. Consequently, an early class of applications was in psychology and the modeling of the medical diagnostic process. From these early papers, after 1975, some of the first sound applications in the diagnostic (decision) support systems and the expert systems for medicine emerged. The papers published in the first decade of fuzzy systems theory also established the foundations for more advanced applications, such as classification (first paper by E.T. Lee, 1975), prosthetics (Saridis, 1975), image processing in medicine, therapy, etc.

The evolution of the number of papers, per year of publication, in biology, medicine, and psychology is shown in Figure 1, and a review of the early papers with publication information is presented in Table 1.

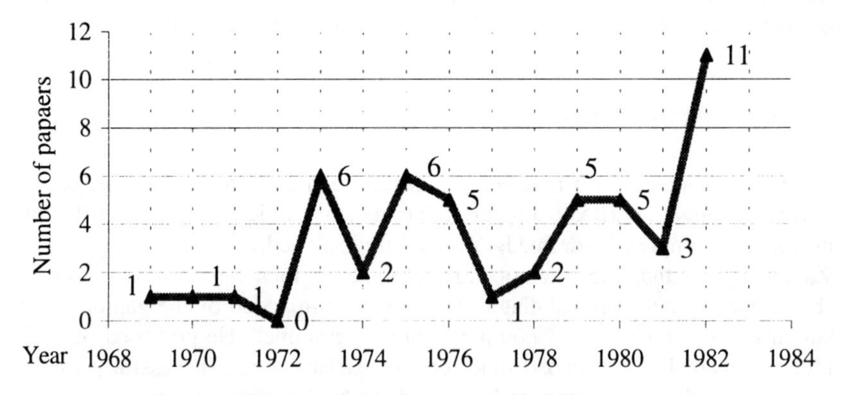

Figure 1. Number of papers on fuzzy logic in medicine and psychology (according to Kulkarni & Karwowski, 1986).

TABLE 1. Papers on Applications of Fuzzy Logic to Medicine (M) and Biology (B)

Year	Author(s)	Publication, volume, pages	Field
1969	Zadeh, L.A.	In: Biocybernetics of the Central Nervous System. Little Brown & Co., Boston, Mass., U.S.A., 199-212	M, B
1970	Lee S.C. and Lee E.T.	Proc. 4th Princeton Conf. on Information Science and Systems	B
1971	Fujisake, H.	Proc. Symp. On Fuzziness in Systems and its Processing, Professional Group of SICE	M
	Adey, W. R.	NTIS Report AD-735-178, 1971 and AFOSR TR 72-0072 Report 1971	M
1972	Adey, W.T.	Int. J. Neurology, *3*, 271-284	B
	Larsen L.E. et al.	Brain Research, *40*, 319-343	B
1973	Kalmanson, D. and Stegall, F.	La Nouvelle Presse Medicale, *41*, 2757-2760	M
	Kalmanson, D. and Stegall, F.	Amer. J. Cardio., *35*, 30-34	B
	Malvache, N., Milbred G. and Vidal, P.	Report, Contract DRME # 71-251, Paris, France	M
1974	Lee, E.T.	US-Japan Seminar on Fuzzy Sets and Their Applications, Berkeley, CA	B
	Lee, S.C. and Lee, E.T .	J. Cybernetics, *4*, 83-103	B
		Report, A.T.P.-C.N.R.S. #1K05, Paris	B
	Malvache, N. and Vidal, P.		M
	Sanchez, E.	Ph.D. Thesis, Faculte de Medicine de Marseille, France, July	M
	Woodbury, M.A. and Clive, J.	J. Cybernetics, *4*, No. 3, 111-121	
1975	Albin, M.	Ph.D. Thesis, Dept. Math., Univ. of California, Berkeley, Ca	M
			B
	Arbib, M.A.	Int. J. Man-Machine Studies, 7, 279-295	
			B
	Butnariu, D.	3rd Int. Congress of Cybernetics and Systems, Bucharest, Romania, Aug.	

	Lee, E.T.	IEEE Trans. Syst. Man Cybern., SMC-5, 629-632	M
	Lee, S.C. and Lee, E.T.	Mathematical Biosciences, vol. 23, 1975	
	Malvache, N.	Ph.D. Thesis, Lille, France, April	M, B
	Sanchez, E.	Special Interest Discussion Group, 6th IFAC World Congress, Boston, Mass., U.S.A., Aug.	M
	Saridis, G.N.	Special Interest Discussion Session on Fuzzy Automata and Decision Processes, 6th IFAC Congress, Boston, MA, U.S.A., Aug., 1975	M
	Shortliffe, E.H. and Buchanan, B.G.	Math. Biosciences, *23*, 351-379	M
	Wechsler, H.	Proc. 1975 Int. Symp. Multiple-Valued Logic, IEEE 75CH0959-7C, May	M
1976	Bezdek, J.C.	Proc. National Computer Conf. AFIPS Press, Montvale, NJ, June 1976	M
	Sanchez, E. and Sambuc, R.	IRIA Medical Data Processing Symposium, Taylor & Francis, Toulouse, France	M
	Shortliffe, E.H.	Computer-based Medical Consultation: MYCIN. Elsevier, N.Y.	M

The total number of papers in the field published before 1976 was 23. The yearly distribution of papers on applications in biology and medicine is shown in Figure 2. (The data here are mainly based on the bibliographies by Gaines and Kohout, Kandel and Kulkarni and Karwowski.)

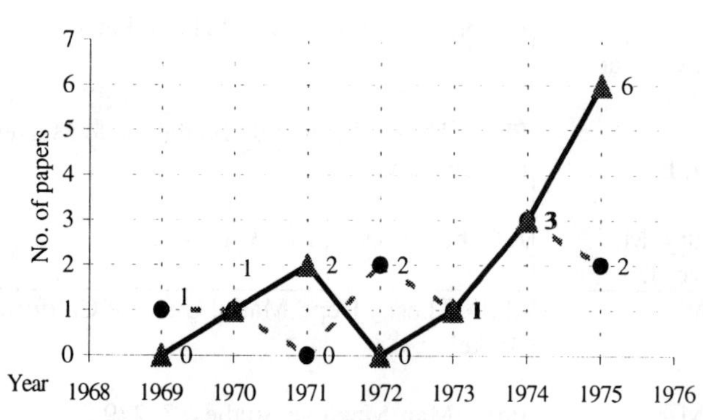

Figure 2. Number of papers on applications in medicine (triangle) and biology (circle)

The early papers published on fuzzy logic and systems in medical applications had a relatively low impact on the medical community. This was due to a lack of understanding of the fundamental concepts of fuzzy logic by the majority of physicians, and because these papers were printed in non-medical publications (or were not published at all in journals or books).

The statistics show that among the 23 papers published before 1976:

- only four (by Adley, by Larsen, and two by Kalmason) were published in medical journals;
- two were doctoral theses from medical universities (Sanchez, 1974 and Malvache, 1975, both in France);
- three appeared in mathematical- or cybernetics-oriented publications (Zadeh, 1969; Woodbury, 1974; Shortlife and Buchannan, 1975); and
- the other articles were published in engineering-oriented journals and volumes, or were not published (papers communicated in symposia).

The distribution of papers according to the type of publication is shown in Figure 3.

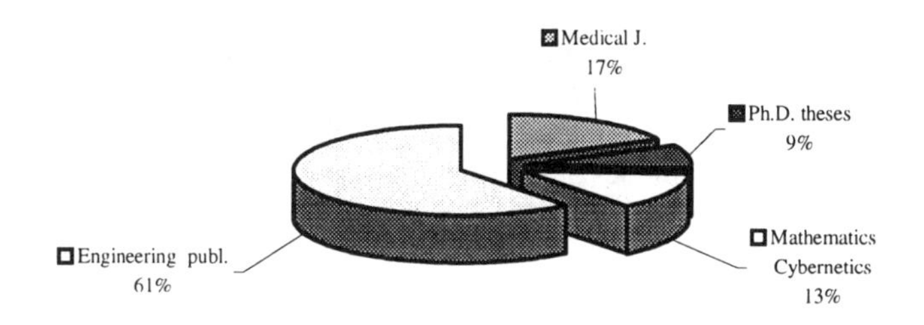

Figure 3. Distribution of papers written before 1975, according to the type of publication.

In the statistics are included the papers on applications in medicine and biology, published before 1975, moreover some papers published until 1981. The papers are classified according to the domain of the medical application. The statistics of topics show the balanced spreading of the papers over various medical fields.

- **First papers on fuzzy logic in biology**

Zadeh, L.A.: Biological applications of the theory of fuzzy sets and systems. In: Proctor, L.D. (Ed.), *Biocybernetics of the Central Nervous System*, Little Brown & Co., Boston, MA, U.S.A., 199-212, 1969

- **Early papers on fuzzy logic in medicine: basic concepts, medical reasoning, and diagnostic**

Fujisake, H.: Proc. Symp. On fuzziness in systems and its processing. Professional Group of SICE, 1971

Woodbury, M.A. and Clive J.: Clinical pure types as a fuzzy partition. *J. Cybernetics, 4*, No. 3, 111-121, 1974

Sanchez, E.: Ph.D. Thesis, Faculte de Medicine de Marseille, France, July 1974

Albin, M.: Fuzzy sets and their application to medical diagnosis. Ph.D. Thesis, Dept. Math., Univ. of California, Berkeley, CA, U.S.A., 1975

Sanchez, E.: Solutions in composite fuzzy relation equations. Application to medical diagnosis in Browerian logic. Special Interest Discussion Group, 6th IFAC World Congress, Boston, MA, U.S.A., Aug. 1975

Wechsler, H.: Applications of fuzzy logic to medical diagnosis. *Proc. 1975 Int. Symp. Multiple-Valued Logic*, IEEE 75CH0959-7C, May 1975

Sanchez, E.: Solutions in composite fuzzy relation equations: application to medical diagnosis in Brouwerian logic. In: Gupta, M.M., Saridis, G.N., Gaines, B.R., Eds.: *Fuzzy Automata and Decision Processes.* North Holland, New York, 1977, pp. 221-234

Woodbury, M.A., Clive, J., and Garson, A.: Mathematical typology: A grade of membership technique for obtaining disease definition. *Computer and Biomed. Red., 11*, 277-298, 1978

Esogbue, A.: Fuzzy sets and the modeling of physician decision making processes. I. The initial interview information gathering session. *Fuzzy Sets and Systems*, 2, 279-291, 1979

Sambuc, R. et al.: Fuzzy functions application to medical diagnosis. In *MEDINFO 80 Proc. 3rd World Conf. on Medical Information*, Tokyo, Japan, Oct. 1980, Vol. 2, pp. 784-788. North-Holland Pub., Amsterdam, The Netherlands

Adlassing, K.P.: A fuzzy logic model of computer-assisted medical diagnosis. *Methods Inf. Med.*, 19, 141-148, 1980

Regarding this field of application, one may emphasize that medical diagnosis is more than just diagnostic: it is oriented toward therapy and surgery. Hence, the medical diagnosis process includes a mandatory decision-making process. In this respect, the importance of the first paper giving an insight into fuzzy decision making cannot be overestimated. This paper is: Bellman, R.E. and Zadeh, L.A.: Decision making in a fuzzy environment. Management Science, 17 (1970), 151-169.

Some of the first papers on topics related to fundamentals of medical reasoning and diagnosis addressed the vagueness in the medical concepts – a topic that is still of interest today. Also of continued interest is the modeling of medical reasoning and decision making. It appears today that more powerful theoretical tools are needed to deal with these topics, complementing the fuzzy logic theory.

- **Early papers on fuzzy logic in preventive and occupational medicine**

Adey, W. R.: Neurophysiological estimates of human performance capabilities in aerospace systems. NTIS Report AD-735-178, 1971 and AFOSR TR 72-0072 Report 1971

Kohout, J.L.: Application of multi-valued logic to the study of human movement disorders. *Proc. 6th IEEE Int. Symp. Multi-valued Logic*, Logan, UT, U.S.A., 224-232, May 1976

Chrostek, W.: Health hazard evaluation determination. Report HHE 79-77-605, Towncenter Association Building Rockville, MD, U.S.A. and NTIS Report PB 80-14718

Sekita, Y. and Tabata, Y.: Health status index model using a fuzzy approach. *European J. Oper. Res.*, *3*, 40-49, 1979

- **Early papers on fuzzy logic in cardiology**

Kalmanson, D. and Stegall, F.: Recherche cardio-vasculaire et theorie des ensembles flous, *La Nouvelle Presse Medicale*, *41*, 2757-2760, 1973

Kalmanson, D. and Stegall, F.: Cardiovascular investigation and fuzzy concepts. *Amer. J. Cardio.*, *35*, pp. 30-34, 1975

Joly H. et al.: Application of fuzzy set theory to the evaluation of cardiac functions. In *MEDINFO 80 Proc. 3rd World Conf. on Medical Information*, Tokyo, Japan, Oct. 1980, vol. 1, pp. 91-95. North-Holland Pub., Amsterdam, The Netherlands

- **Early papers on fuzzy logic in pattern classification with application to medicine**

Lee, E.T.: Shape-oriented chromosome classification. *IEEE Trans. Syst., Man, Cybern.*, SMC-5, 629-632, 1975

- **Early papers on fuzzy logic in expert systems with application to medicine**

Shortliffe, E.H. and Buchanan, B.G.: A model of inexact reasoning in medicine. *Math. Biosciences*, *23*, 351-379, 1974

Shortliffe, E.H.: Computer-based Medical Consultation: MYCIN. Elsevier, New York, 1976

- **Early papers on fuzzy logic in prosthetics**

Saridis, G.N.: Fuzzy decision-making in prosthetic devices and other applications. in: Special interest discussion session on fuzzy automata and decision processes, 6th IFAC Congress, Boston, MA, U.S.A., Aug., 1975

Saridis, G.N. and Stephanou, H.E.: Fuzzy decision-making in prosthetic devices. In: Gupta, M.M., Saridis, G.N., Gaines, B.R., Eds., *Fuzzy Automata and Decision Processes*. North Holland, New York, pp. 387-402, 1977

- **Early paper on image processing**

The first paper reporting the use of fuzzy logic in medical image processing appeared in 1986. We quote it here because of the importance of the field:

Pathak, A. and Pal, S.K.: Fuzzy Grammars in Syntactic Recognition of Skeletal Maturity from X-Rays. *IEEE Trans. Syst., Man, Cybrn.*, vol. SMC-16, no. 5, pp. 657-667, Sept./Oct. 1986 (also in Bezdek, J., Pal, S.K., Ed., *Fuzzy Models for Pattern Recognition*, IEEE Press, New York, pp. 311-320, 1992)

2. FURTHER DEVELOPMENTS AND BACKGROUND

The first journal in the field, exclusively devoted to fuzzy sets and systems, was *Fuzzy Sets and Systems* (IFSA publication). It was published since 1978. This journal plays an essential part in forging the fuzzy systems research literature and spreading research results.

The first international conference entirely focusing on fuzzy systems was held in 1985 (the First IFSA Congress). The series of IFSA congresses was instrumental in the development of this field.

It seems that a strong research focus on fuzzy systems – unfortunately only partly known on the international scale – was developed in China. During the decade 1975-1985, several groups of Chinese scientists applied fuzzy systems to model diagnostic processes and acupuncture-related therapy. Some references can be found in: Zaifu, S., Hongyuan, S., Yanjie L., and Fu, W: Application of fuzzy set theory in health care systems. In: Karwowski, Waldemar and Mital, Anil (Eds.): *Applications of Fuzzy Set Theory in Human Factors*, Elsevier, 1986, pp. 379-391.

In 1986, the first book on fuzzy systems relating to human factors was published, namely: Karwowski, Waldemar and Mital, Anil (Eds.): *Applications of Fuzzy Set Theory in Human Factors*, Elsevier, 1986.

During the decade 1980-1990, numerous groups performed significant research using fuzzy methods in biology and medical applications. They are too numerous to be quoted here. Almost all medical fields were addressed, although a preference was given to topics such as knowledge bases, decision support systems, expert systems, and other tools for computer-assisted medical diagnosis.

In 1989, a group of Japanese scientists established the first learned society in the field: Japanese Bio-Medical Fuzzy Systems Association (BMFSA). In 1991, this Society started to have regular overseas members. The annual symposia organized by BMFSA (starting 1991) were the first to focus entirely on applications of fuzzy systems in medicine. The first journal in the field – and still the only one devoted to fuzzy systems in medicine is the Journal of BMFSA. Initially published in Japanese, with a few abstracts in English, this journal is now also published in English (one issue per year).

The field of psychology deserves a special mention. This field has seen a very active development, starting in 1975. Psychological interpretation of vagueness in the frame of experimental psychology (vagueness in memory processes, language, perception, pattern recognition, behavior, interpretation of membership functions) as well as interpretation of psychological facts by fuzzy theory, were of concern. Some of the early and significant papers and volumes are:

Oden, G.C.: Integration of fuzzy logical information. *J. of Experimental Psychology: Human Perception and Performance*, **4**, pp. 565-575, 1977

Oden, G.C.: A fuzzy model of letter identification. *J. of Experimental Psychology: Human Perception and Performance*, **5**, pp. 336-352, 1979

Hersh, H.M. and Caramazza, A.: A fuzzy set approach to modifiers and vagueness in natural language. *J. Experimental Psychology: General*, **105**, pp. 254-276, 1976

Horvath, M.J.: An identification procedure in learning disability through fuzzy set modelling of verbal theory. Unpublished doctoral thesis. Tucson, The University of Arizona, U.S.A., 1978

Horvath, M.J., Kass, C.E., and Ferrell, W.R.: An example of the use of fuzzy concepts in modeling learning disability. *American Educational Research Journal*, **17**, pp. 309-324, 1980

The first edited volumes in this field – or significantly related to it – are:

Karwowski, W. and Mital, A. (Eds.): *Applications of Fuzzy Set Theory in Human Factors*, Elsevier, pp. 395-446, 1986

Smithson, M.: *Fuzzy Set Analysis for Behavioral and Social Sciences.* New York, Springer, 1987

Zentenyi, Thomas (Ed.): *Fuzzy Sets in Psychology*, North-Holland, 1988.

The last volume includes a good review of the field before 1988 and represents a consistent reference on early developments.

Journals and Conferences

Some of the most important journals publishing papers on applications of fuzzy systems in medicine and neuro-fuzzy systems during the first two decades of development were: *IEEE Trans. Systems Man Cybernetics, Fuzzy Sets and Systems, IEEE Trans. Pattern Analysis Machine Intelligence, Int. J. Man-Mach. Studies, J. Cybernetics, Int. J. Approx. Reasoning (starting 1987), J. of Japanese Association for Fuzzy Systems*, etc.

Some of the main conferences where papers on applications of fuzzy logic in medicine and biology were presented are:

- Int. Congress of Cybernetics and Systems
- MEDINFO Conferences
- IFAC (International Federation for Automatic Control) conferences and World Congress (mainly on identification and control applications)
- Starting 1985, biannually, IFSA Congress (according to the rules, changes the continent every two years; first in Spain, 1985, then in Japan, 1987, in U.S.A. in 1989, etc.)
- NAFIPS Conferences (U.S.A.)
- Starting 1988, IIZUKA Conference series (biannually; Japan); it was one of the first to devote sections to medical applications.
- In Europe, the AMSE series of symposia "Fuzzy Systems" (starting 1988) also included many papers on medical applications.
- After 1990: IEEE conference series on neural networks and fuzzy systems and IEEE conference series on bio-medical engineering, EUFIT conference series (Aachen, Germany), BMFSA Symposia (Japan) etc.

The proceedings of these conferences are an essential source of references during the first two decades of development of fuzzy systems applications in medicine.

3. Neuro-Fuzzy Systems and Their Applications in Medicine and Biology

The domain of neural networks is as much a section of the fields of theoretical biology, neurology, and behavioral science, as it is a domain of automata theory and of engineering in general. Consequently, this field deserves a separate investigation.

The paper by Wee and Fu (1969) was the first that deals with neuro-fuzzy systems, mainly in relation to applications in pattern classification and control.

The field of neuro-fuzzy systems saw early development with the papers by Lee and Lee published in 1970, 1974, and 1975, Adey (1971 and 1972), Butnariu (1975). The bridges between the two fields naturally grew and neuro-fuzzy systems are now a strong branch of artificial intelligence, with many papers and numerous sound and well-documented applications in medicine and biology. Lee and Lee (1975) introduced the fuzzy weight concept in neural networks. They also introduced the concept of *fuzzy threshold.* The concept of fuzzy neural networks was heavily related to the discrete automaton concept in the early papers published on the subject.

During the period 1980-1995, several essential developments were introduced, including the fuzzy perceptron (Keller), the MIN/MAX-neurons of Gupta, the RBF fuzzy systems, algebraic fuzzy neurons, chaotic fuzzy neurons, and the networks of chaotic fuzzy neurons. The first integration of a fuzzy neuron into silicon is due to Yamakawa (1989). An account of neuro-fuzzy systems references can be found in: Teodorescu, H.N., Yamakawa, T.: Neuro-fuzzy systems: Hybrid configurations. In M.J. Patyra, D. Mlynek (Eds.): *Fuzzy Logic. Implementation and Applications.* Wiley & Teubner, 1996.

- **Early papers on fuzzy neurons and fuzzy logic in neural networks**

Wee, W.G. and Fu, K.S.: A formulation of fuzzy automata and its application as a model of learning systems. *IEEE Trans. Syst. Sci.and Cybern.*, vol. SSC-5, (1969) pp. 215-223 (Reprinted in: Bezdek, J., Pal, S.K., Eds., *Fuzzy Models for Pattern Recognition*, IEEE Press, New York, pp. 448-467, 1992)

Lee S.C. and Lee E.T., Fuzzy Neurons and Automata. *Proc. 4th Princeton Conf. on Information Science and Systems*, 1970, pp. 381-385

Adey, W. R.: Neurophysiological estimates of human performance capabilities in aerospace systems. NTIS Report AD-735-178, 1971 and AFOSR TR 72-0072 Report 1971

Adey, W. R.: Organization of brain tissue: is the brain a noisy processor? *Int. J. Neurology*, 3, 271-284, 1972

Lee S.C. and Lee E.T., Fuzzy sets and neural networks. *J. Cybernetics, 4*, 83-103, 1974

Lee, S.C. and Lee, E.T.: Fuzzy neural networks. *Mathematical Biosciences*, vol. 23, pp. 151-177, 1975. (Also reprinted in: Bezdek, J., Pal, S.K.,Eds., *Fuzzy Models for Pattern Recognition*, IEEE Press, New York, 1992, pp. 448-467)

Butnariu, D.: L-fuzzy automata description of a neural model. Proceedings of the 3rd Int. Congress of Cybernetics and Systems, Bucharest, Romania, Aug. 1975

Chorayan, D.D.: Formalization of a neuronal ensemble in terms of the theory of fuzzy sets. *Biofizika, 26*, pp. 876-878, 1981

Papers on fuzzy neurons published after 1980

The following list includes some contributions on neuro-fuzzy systems that have resulted in significant progress or introduced new concepts.

Yamakawa, T.: Patent Application, Japanese, *A Fuzzy Neuron*, TOKUGANHEI No. 1-33690, May, 1989

Yamakawa T. and Tomoda, S.: A fuzzy neuron and its application to pattern recognition. *Proc. 3rd IFSA Congress*, J. Bezdek (Ed.), Seattle, U.S.A., pp. 30-38, 1989

Yamakawa, T.: Patent Application, Japan, *A Nonlinear-Synapse Neuron*, TOKUGANHEI No. 4-132897, May, 1992

Kuncicky, D. and Kandel, A.: A fuzzy interpretation of neural networks. *Proc. 3rd IFSA Congress*, J. Bezdek (Ed.), Seattle, U.S.A., pp. 113-116, 1989

Keller, J.M. and Hunt, D.J.: Incorporating fuzzy membership functions into the perceptron algorithm. *IEEE Trans. Pattern Analysis and Machine Intelligence*, vol. PAMI-7, no. 6, pp. 693-699, Nov. 1985. (Also reprinted in: Bezdek, J., Pal, S.K.: *Fuzzy Models for Pattern Recognition*, IEEE Press, New York, pp. 468-474, 1992)

Teodorescu, H.N.: Chaotic networks of fuzzy systems. *Proceedings of "Fuzzy Systems and Signals," AMSE Symposium*, London, pp. 27-31, 1993

Buckley, J. and Hayashi, Y.: Hybrid neural nets can be fuzzy controllers and fuzzy expert systems. *Fuzzy Sets and Systems, 60*, 135-142, 1993

Gupta, M.M. and Rao, D.H.: On the principles of fuzzy neural networks. *Fuzzy Sets and Systems*, **61**, 1-18, 1994

Teodorescu H.N.: Chaotic Non-linear Systems, Fuzzy Systems and Neural Networks. Plenary lecture, *Proceedings 3rd International Conference on Fuzzy Logic, Neural Nets and Soft Computing*, Iizuka, Japan, August 1-7, pp. 17-28, 1994

4. Genetic Algorithms, Fuzzy Logic, and Neuro-Fuzzy Systems

Evolutionary computing is the name given to a collection of algorithms based on the evolution of a population toward an optimal solution of a certain problem. Three types of evolutionary computing techniques have been widely reported: genetic algorithms (GAs), Genetic Programming (GP), and Evolutionary Programming (EP). GAs were applied to the optimization of fuzzy systems and neuro-fuzzy systems in the early 1990s (for instance: Karr and Chunk: Applying genetics to fuzzy logic. *AI Expert*, vol. 6, nr. 3, pp. 38-43, 1991). GAs are widely used and reported in the literature (for instance: Jang, J.-S.R., Sun, C.T., and Mizutani, E.: Neuro-Fuzzy and Soft Computing, Prentice Hall, 1997; Van Rooij A., Jain, L.C., Johnson, R.P.: Neural Network Training Using Genetic Algorithms, World Scientific Publ. Co., Singapore, 1996).

Evolutionary computing techniques, in conjunction with neural networks and fuzzy logic, have been used successfully in medicine and bio-medical engineering in various applications, mainly related to control and pattern recognition.

5. Bibliographies

Several early bibliographies related to fuzzy systems and combinations of fuzzy logic and neural networks were published by Dekeref, Kandel and Davis, Kulkarni and Karwowski, Tong, and Zimmermann. Some of these excellent bibliographies are, however, somewhat biased or incomplete. Gaines and Kohout provide a good and very well-organized general bibliography up to 1975. Their list of published papers includes 763 quotations and the reviewed literature significantly includes journals not published in English. A large number of papers in this last bibliography come from publications from the East European countries and Asia and many conference papers are also quoted. According to this bibliography, the number of papers published during 1965 to mid 1976 in the fields of biology, medicine, and psychology is 10, 13, and 27, respectively.

A good source of references - and one of the excellent books in the field – is also the volume: Zimmermann, H.J.: Fuzzy Set Theory and Its Applications. Kluwer-Nijhoff Publishing, Boston-Dordrecht-Lancaster, 1985. The first extensive bibliography (covering all fields of fuzzy logic) was compiled by Kandel and Yager (1979). A continuing bibliography was done for more than a decade, since 1985, by D. Dubois and H. Prade, first in the BUSEFAL Journal (University of Toulouse, France) and then in the journal Fuzzy Sets and Systems.

List of early Bibliographies (up to 1980)

Dekeref, J.: A bibliography of fuzzy sets. *J. Comput. and Appl. Math.*, 1, 205-212, 1975

Gaines, B.R., and Kohout L.J.: The fuzzy decade: A bibliography of fuzzy systems and closely related topics. In: Gupta, M.M., Saridis, G.N., Gaines, B.R., Eds. *Fuzzy Automata and Decision Processes*. North Holland, New York, pp. 403-490, 1977

Kandel, A. and Davis, H.A.: The first fuzzy decade: Bibliography on fuzzy sets and their applications. CSR 140 Computer Science Dept., New Mexico Institute of Mining and Technology, Socorro, NM, U.S.A., Apr. 1976

Kandel, A. and Yager R.R.: A 1979 Bibliography on fuzzy sets, their applications, and related topics. In: Gupta M.M., Ragade, R.K, and Yager R.R., Eds., *Advances in Fuzzy Set Theory and Applications*. North-Holland, Amsterdam, 1979

Kulkarni, J. and Karwowski, W.: Research Guide to applications of fuzzy set theory in human factors. In: Karwowski, Waldemar, and Mital, Anil, Eds., *Applications of Fuzzy Set Theory in Human Factors*, Elsevier, Amsterdam, pp. 395-446, 1986

Tong, R.M.: An annotated bibliography of fuzzy control. Technical Memo TM 6002-1, Advanced Information and Decision Systems, Mountain View, CA, U.S.A., 1983

Zimmermann, H.J.: Bibliography: Theory and applications of fuzzy sets. RWTH Aachen, Germany, Oct. 1975

The field as reflected by MEDLINE Search Database

The MEDLINE database is considered one of the most important in the medical field. Thus, the way the papers reporting on fuzzy logic and neuro-fuzzy systems in medical applications is reflected in such a database is essential in assessing the development level of the field. Browsing this database, one finds a total of 105 papers published from 1990 to 1995 and a total of 215 published from 1990 to 1997. The growth of the number of papers is shown in Figure 4.

Contrasted to the period 1965-1975, the increase between the years 1990 and 1995 is 52 times (105 papers vs. 4). Moreover, the increase in 1997 was almost 5 times 1990 (46 papers vs. 10).

The area of neuro-fuzzy systems has only one paper during the period 1990-1997. In contrast, the number of papers reporting on the use of neural networks in medical applications was 1844 papers published from 1990 to 1997. This clearly indicates that the penetration of neuro-fuzzy systems in medicine is still to come.

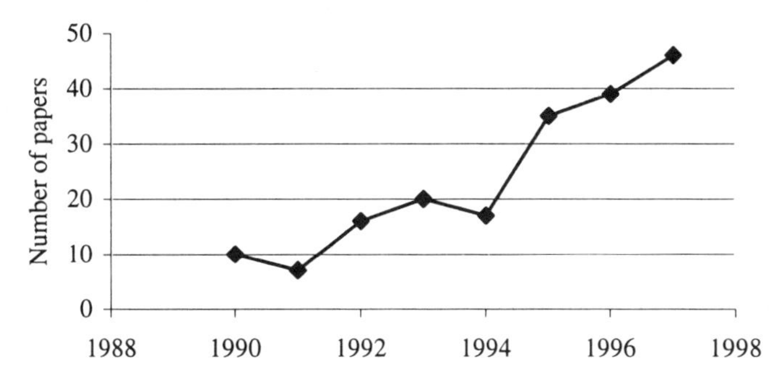

Figure 4. The number of papers on applications of fuzzy systems in medicine, published in medical journals and quoted in MEDLINE® database

We should emphasize, however, that MEDLINE reviews primarily medical journal papers. A much greater number of papers are published in conference proceedings (maybe 3 to 10 times as many).

6. Conclusions and Predictions

It is difficult to predict the future, but considering the classic shape representing the life of the scientific disciplines (see Figure 5), the field of medical applications of fuzzy systems in medicine has left behind the initial period of slow and moderate growth and is now in the phase of strong growth. One can expect that this phase will last until around 2010, then will reach the plateau for about 10 to 20 years (depending on the new discoveries).

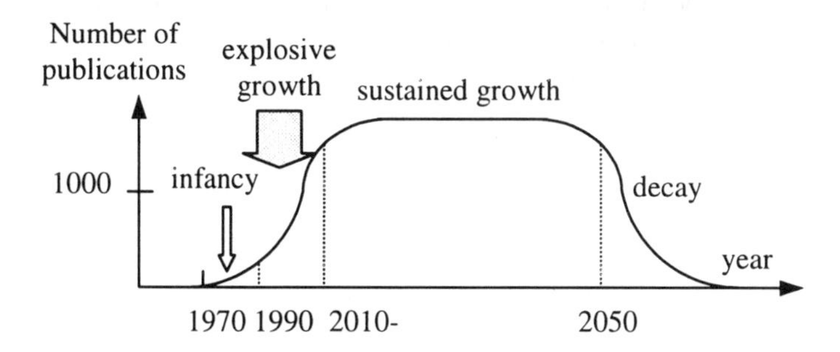

Figure 5. Predictable development of the applications of fuzzy systems in medicine and biology.

The field of applications of neural networks is in a similar stage of development.

On the other hand, the fields of neuro-fuzzy systems and of the interdisciplinary field of chaotic fuzzy and neural systems are at their emerging stage. A sustained growth will probably be reached by 2010.

It is not yet possible to foresee if all these domains will become major fields, with thousands or tens of thousands of publications. One could estimate the total number of papers on fuzzy systems in medicine, published before 1998, to be 1000 to 1500 (journal papers and conference papers as well, in a ratio of about 1 to 3...5). If regional or national conferences are considered, this number is probably around 3000...5000. Assuming the present trend will last for at least ten more years, and extrapolating present data, the number of journal papers will be about 1000-2000 per year by 2010. This figure already represents a major field. Compare, for instance, the number of papers published on topics related to "electrocardiogram/ECG." This topic is quoted for the period 1990-1997 with 886 titles (on MEDLINE database). In addition, the number of papers published on occupational medicine, an important field of medicine, with 1193 titles during the same period. Probably, "fuzzy systems" and "neural networks" papers will never reach the number of "computer"-related papers – more than 81,000 in the period 1990-1997. Nevertheless, "fuzzy systems in medical applications" will undoubtedly be a significant technical discipline of bio-medical engineering, and an essential tool in medicine in the years to come – and before 2010.

Because both the fields of neural networks and fuzzy systems (as applied to medicine) are growing very fast, one can expect the newer domain of neuro-fuzzy systems in medicine to have a bright future.

It is useful to fuse ANNs, fuzzy logic, and evolutionary computing to offset the demerits of one paradigm by the merits of another. Some of these paradigms are used as: ANNs in designing fuzzy systems, fuzzy systems for designing ANNs, evolutionary computing for designing fuzzy systems, and evolutionary computing for training and automatically generating the architecture of ANNs. Future developments will use fused paradigms to achieve improved results in bio-medical applications, based on neuro-fuzzy and GAs – fuzzy – neuro-fuzzy systems. A number of chapters in this book exemplify the use of these paradigms in medicine and biomedical engineering.

Acknowledgment. The first author thanks the support of the Institute for Information Science and a grant from FNS-Switzerland during the long period of bibliographic research and preparation of this chapter.

The material in this chapter is, in part, based on an article submitted to *Fuzzy Systems and A.I.*, to appear 1998.

Chapter 2

The Brain as a Fuzzy Machine: A Modeling Problem

Robert P. Erickson, Mircea I. Chelaru, and Catalin V. Buhusi

This chapter discusses a natural expression of fuzzy logic in the function of the brain. We start (Section 1) with a historical account of the fuzzy perspective in neuroscience, and support this perspective with recent neurophysiological data (Section 2). Next, we use the activity of taste neurons as an example of a biological fuzzy membership function, and we suggest that this conceptualization can be successfully used as an organizational tool in understanding neural function in general. Following this, we demonstrate that the dynamics of taste neurons can indeed be modeled by a fuzzy Sugeno model (Section 3). Since the brain is a biological neural network, we also address (Section 4) the possible implementation of a fuzzy machine using artificial neural networks. We conclude (Section 5) with a discussion of some implications of this view for neuroscience and fuzzy modeling.

The intent of this chapter is to integrate real aspects of neural organization into intelligent control systems and, conversely, to use the fuzzy view for a better understanding of the brain and its behavior. It appears to us that the application of fuzzy logic to describe the function of the brain is at least as reasonable as the application of neural networks. Furthermore, this approach has the added advantages which are inherent in fuzzy theory of representing a great deal of information with a limited number of neural elements, along with other benefits to be described.

The chapter is addressed not only to students of psychology and neurobiology, but also to students of statistics, computer science, and engineering. *Psychologists and neurobiologists* regard fuzzy theory as a powerful organizational tool in explaining the function of the brain (seen as a biological neural network of huge complexity). Some *statistical* tools used include fuzzy approaches to multivariate data analysis that give insight into the organization of neural processes by clustering the data into meaningful and statistically significant classes (e.g., by our colleague Max Woodbury of Duke University [1]). And the multivariate statistical tool of Multidimensional Scaling

0-8493-9806-1/98/$0.00+$.50

(MDS) is shown [2] to use fuzzy concepts and, in this way, it has been used to disclose neural organization in situations where the organization is otherwise not evident. On the other hand, *computer scientists and engineers* are interested in the approach between neural networks and fuzzy systems, and in neuro-fuzzy systems, in an effort to integrate them into intelligent control systems, and to use neural methods for automatic fuzzy system synthesis. We think that each area has much to gain from collaboration with the other.

1. THE FUZZY APPROACH IN NEUROBIOLOGY: A HISTORICAL PERSPECTIVE

The best place to begin is at the beginning, 1802. One of the most prescient, insightful and broadly heuristic contributions to neuroscience is contained in a single sentence written by Sir Thomas Young [3]. Concerning the color vision, long before the cellular nature or physiology of the nervous system was known, Young wrote:

> "*Now, as it is almost impossible to conceive each sensitive point of the retina to contain an infinite number of particles, each capable of vibrating in perfect unison with every possible undulation, it becomes necessary to suppose the number limited, for instance, to the three principal colours, red, yellow, and blue [...] and that each of the particles is capable of being put in motion less or more forcilbly, by undulations differing less or more from a perfect unison.*"

In this comment, he states what is perhaps the major problem in neural coding, and suggests a solution. The problem is the handling of large amounts of information with limited amounts of neural resources. The solution is fuzzy set theory. Young was perhaps the first fuzzy theorist.

Helmholtz made Young's theory clear in his 1860 illustration [4] of Young's concept (Figure 1). These three curves show the amount of activity evoked in each receptor type by each wavelength. The wavelength at the center of each curve excites the maximum amount of neural activity, and the adjacent wavelengths evoke progressively less activities. Our use of the term "fuzzy set" can be defined by the reference to Helmholtz' drawing. The three hypothesized receptor types are fuzzy sets in the sense that different wavelengths have various *grades of membership* (GoMs) within each of them. Also, each wavelength is not exclusive to only one set, but has GoMs in all three.

The two major aspects of Young's hypotheses are illustrated in Figure 2 [5]. In panel A three idealized receptor types are drawn following Figure 1: the degree to which four different wavelengths activate the receptors (GoMs) is shown. Panels B and C, both derived from panel A, show the two important aspects of the model: the fuzzy sets and the neural code, respectively. Panel B shows the definition of the neural fuzzy sets, with each wavelength having a GoM in each fuzzy receptor type. Panel C shows how the neural code derives from the fuzzy sets in panel B. The code is further discussed in Section 2.3.

2. THE GENERALITY OF YOUNG'S HYPOTHESIS

2.1. Simple stimuli

Was Young correct to assume that neurons behave as fuzzy sets? Young's hypothesis about the curves for the color receptors has been repeatedly shown to be true, but this circumstance could be unique to color.

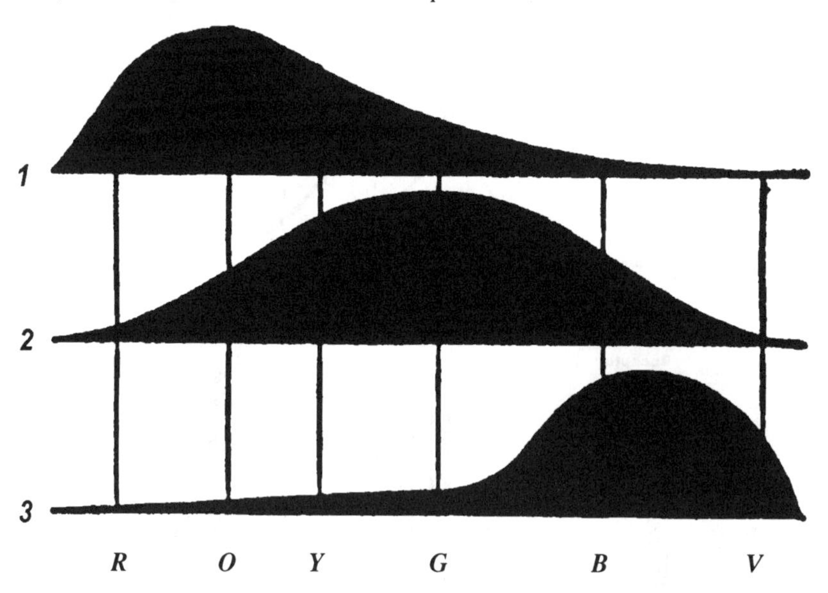

Figure 1. Young's theory [3] as drawn by Helmholtz [4]. These three curves show the amount of activity evoked in each of three visual receptor types by each wavelength. The ordinate values indicate the amount of excitation produced in the three receptor types by the various wavelengths on the abscissa. Maximum excitation in each is evoked by one wavelength: adjacent wavelengths evoke progressively less activity. Redrawn from [4].

It turns out that Young has described all sensory systems [6]. The reason is that Young's economic problem exists everywhere in the brain: although we have many neurons, there simply aren't enough to function as crisp sets – for each neuron to have a specific function (e.g., to encode red squares) distinct and disjunct from that of every other neuron (e.g., to encode blue Volvos). There is too much information by far to be encoded. As a first approximation, the sensitivity functions of all individual neurons in all sensory systems are bell-shaped, as Young hypothesized [6].

To emphasize this generality, these sensitivity functions in all sensory systems have been referred to as "Neural Response Functions" (NRFs) [7]. Some NRFs are as broadly tuned as the underlying parameters allow, or nearly so. For example, in vestibular sensitivity (head tilt), the NRFs cover a full azimuth range (360 degrees), having their peaks of sensitivity at different head angles [8]. Cortical visual neurons, which are sensitive to straight lines, each respond to lines of broadly different orientations [9] (Figure 3), and respond maximally at one angle (Young's curves describe these very well). Auditory neurons can be similarly characterized across frequency, being most sensitive at one frequency, and dropping off gracefully toward lower and higher frequencies [10]. In these cases there are few neural resources to represent many stimuli. As with color, the few neurons available must have fuzzy sets (NRFs) that are as broad as possible to cover all stimuli.

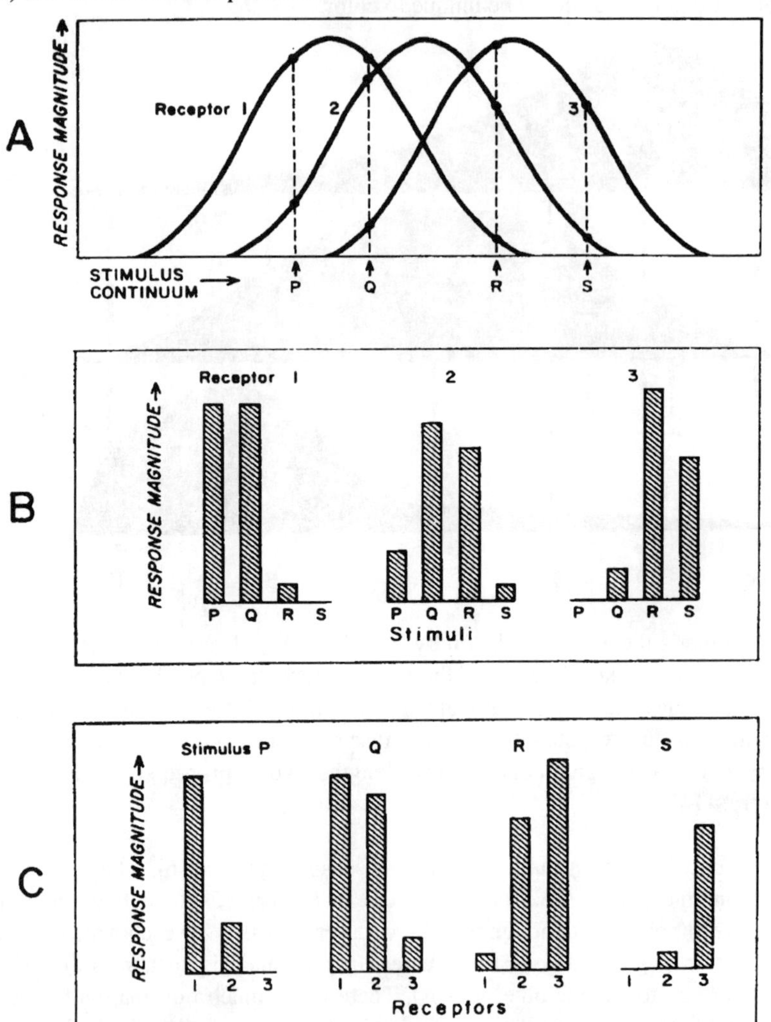

Figure 2. The way in which Young's hypothesis provides for a great amount of information, and forms an illustration of fuzzy sets and neural codes. In panel A, representing any sensory continuum, three idealized receptor types (1, 2, 3) are drawn following Young's theory (Figure 1). Panel B shows how these receptor types function as GoMs: i.e., the degree to which four different stimuli (P, Q, R, S) activate each of these receptor types. Panel C shows the neural codes for P, Q, R, and S. From [5].

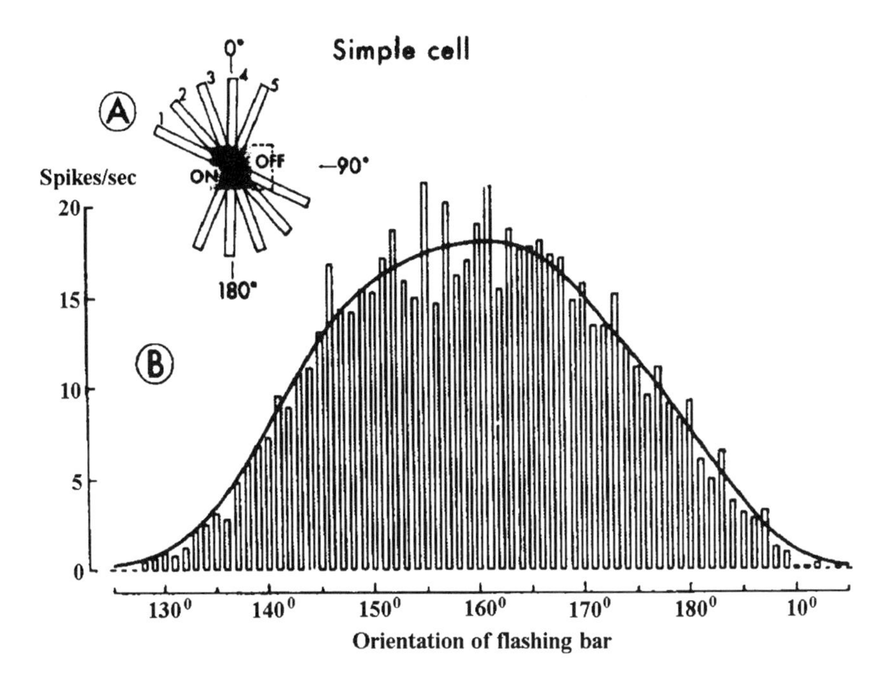

Figure 3. The sensitivity of a visual "simple cell" in cat cortex. Such cells respond best to straight lines. However, their fuzzy tuning across line orientations is similar to Young's curves for color, suggesting common coding mechanisms. From [9] with permission. Notice similarity to Figure 1 (curve 2) and Figure 2 (Panel A).

In other cases the NRFs are narrow with respect to the extent of the underlying continuum. This is true when the supply of neurons is large, such as in the broad spatial sheets across the retina or skin [6]. The reason for the reduction in NRF width with the increase in neural resources may be intuitively obvious; many neurons are activated even with narrowly tuned receptors. This important phenomenon of NRF breadth will be discussed later as an issue in the control of adequate neural mass (Section 5.1.1.).

In some cases the form of the NRF curve is not bell-shaped, although it is still simple. For example, within the brain the NRFs for color are modified from Young's bell-shape [11] to several different horizontal "S" shapes [12], which are a bell with one end turned up: their peaks occur at different wavelengths. This still conforms to Young's "fuzzy set" idea of a few broadly tuned NRFs: the fuzzy sets should be organized both to be simple for the brain to construct, and to provide a simple and sensible perceptual organization. But to meet these demands, they need not be bell-shaped. As a second example, appreciation of a limb's position (kinesthesis) is based on two overlapping fuzzy sets which have their maxima at the extremes of joint angle – flexion or extension [13].

NRFs occur beyond sensory systems. Broad, bell-shaped NRFs are seen in neurons involved in the production of movement. Georgopoulos [14] showed that cortical neurons involved in arm movement in the monkey have a maximum rate of firing for one direction of movement, with the amount of activity falling off in a bell shape as the movement differed from the maximum. The NRF maxima appear at different directions for the various neurons. These NRFs match Young's curves quite well. And as Young's, they are "maximally" broad, each neuron responding for movements in all directions.

2.2. Neural organization of cryptic events: from tastes to faces

2.2.1. The general approach

In the examples given above, the baseline parameters are clear: wavelength for color, frequency for audition, head tilt for vestibular sensitivity, and location on a topographic sheet for space across the visual field or across the skin. Can all central nervous system (CNS) neurons be described this way? For some of them, the issue is unresolved. The major problem for those neurons is that we do not have baselines along which NRFs could be drawn. For example, in the currently important field of the encoding of faces, neurons may respond well to simple figures, as a straight pink line, or to some aspects of a face, or to faces oriented in various directions; but individual neurons can never be described as crisply responsive to only one feature, or one face, and no other [15]. In other words, the responses of these neurons appear to belong to broad, fuzzy sets in that they respond broadly and to varying degrees (GoMs) to an entire stimulus complex, or to its parts. But there is no concise organizing principle for this phenomenon as we have for the simpler stimuli (wavelength, etc.).

In addressing this issue of event complexity, we might assume that the brain functions in the most simple and clear possible way. Once the brain has developed an effective mechanism (i.e., fuzzy logic) for one task, it would probably be conserved as much as possible across all tasks; this would also improve the possibility of compatibility of the various systems (sensory, memory, motor, etc.) with each other. Moreover, the need of simple organization is probably even greater for information-dense events, such as the perception of faces, than for simpler stimuli, such as colors.

The encoding of any neural event, such as the representation of a face or a memory, can be approached in a simple organized way along a baseline parameter even when that parameter is unknown. In addition, the responses of the neurons appear to be simply organized fuzzy NRFs. For example, in the simpler realm of the perception of tastes, and the complex realm of face perception, there are no known baseline parameters along which the stimuli and NRFs could be arranged. With the realization that our perceptions are not random, one could anticipate some neural organization. We will first demonstrate this for taste, and then propose a solution for complex issues.

2.2.2. Solutions possible: taste

There are plentiful data on taste (and olfaction) indicating that individual neurons each respond to many different stimuli, and in different degrees to each [16]. This suggests that these neurons may indeed be acting as fuzzy sets. In addition, it would seem that if the responses are based on an underlying organization, that organization could be determined from these responses. This turns out to be true.

We [17], [7] and many others (see [2] for review) have submitted the responses of individual taste neurons to an MDS analysis which places these neurons, and the stimuli, in a space designed to show the proximity of each neuron or stimulus to every other neuron or stimulus [2]. In this space, the neural responses to the various stimuli can be roughly described as broad bell-shaped NRFs [7], [18]. That is, stimuli evoking large responses in a given neuron are placed close to the center of its NRF, and stimuli producing progressively smaller responses are placed farther away. The shape and breadth of these NRFs are reminiscent of Young's curves.

Here is a strange effect: we can disclose the organization of a sensory system, its neurons and stimuli, without knowledge of what the organizing dimensions are! For example, we treated the responses of visual receptors to various wavelengths of light [7], [18]. We did not use the wavelength continuum as an organizing tool, as if this knowledge was not available to us. Analysis of the responses of these receptors to various wavelengths resulted in a rather accurate color circle as defined by the arrangement of the neurons and stimuli. If we had no knowledge of the wavelength parameter, perhaps some physicist could now be led to it from the arrangement provided by this analysis. It may be possible that a chemist could identify the dimensions for the chemical senses in this way.

It seems reasonable that the success of these solutions requires that the organization be based on gracefully overlapping, fuzzy NRFs arranged along a continuum. It would be difficult to imagine a simpler organization. This requirement should hold for other proposed solutions, such as for faces.

2.2.3. Solutions possible: faces

Now let us return to the analysis of neurons sensitive to complex stimuli: to faces, or to neurons involved in memory, language, affectivity, etc. It would seem reasonable for the brain to be strongly organized and stingily economic at this point. As with taste, it may be assumed that there is some underlying economical simplicity used by the nervous system in these complex tasks. If so, analysis of the responses of these "complexity"-related neurons should disclose it. Such a solution has not been attempted in fields other than the field of taste. We feel that neural modeling and neurobiology would both benefit from more collaboration.

The problem and its solution are buried in the recipe for unraveling these complex events. First, in searching for the code for "faces," we have already made the implicit assumption that words such as "faces" will provide the proper organizational principles for the brain. The search is on for the face (attention, word meaning, memory, etc.) areas, and neuroscientists have certainly found neural responses related to these categories. Without detailing the problems encountered, it should be pointed out that these neural areas are evidently not private for our chosen functions, and the functions spill over into other "areas." This is the problem: we search for our most important common nouns and, with this approach, the brain is not exactly yielding up its secrets in an uncomplicated way. The solution, as described above, is to simply use all effective stimuli in a standard manner (e.g., not just faces), and let the solution show itself in the analysis of the data. Otherwise our crisp language may let the brain keep its secrets.

The artificiality of our language – which directs our research – cannot be overemphasized. It would seem that in the real world we do not function exactly in terms of our words, but approximately in terms of broad generalities. We would probably function perfectly well without language, except for getting through college. As the non-verbal animals, whose brains are slight variants of our own, we would successfully gather food, seek comfort and protection, procreate, etc. We cannot totally trust our words to guide our understanding in general, and especially not of the brain.

2.3. Neural Codes

Assume for the argument, now, that the information is carried by neurons with fuzzy set characteristics as described in Section 2.1. There the characteristics of individual neurons were discussed. But now, what would be the neural appearance, the code for, a particular event?

First, since by definition, each fuzzy neuron responds to a variety of stimuli, the neural response to a given stimulus will be spread out across a population of neurons: this was shown in Figure 2C. The rubric describing this phenomenon is the "Across-Fiber Pattern" theory [5], [19], and, as a first approximation, is relevant to all sensory, motor and "hidden layer" neurons.

The code for each stimulus is given in the distinct profile of grades of membership across the population of receptors. That is, stimulus *P* activates receptor type 1 strongly (Figure 2C) followed by receptor type 2, and with no response in receptor 3. Stimulus *Q* produces large responses in receptors 1 and 2, and a small response to receptor 3. The important point is that each stimulus has a *distinct* pattern of responses across the receptor types, different from that caused by every other stimulus. This fulfills the basic logical requirement for coding, that each discriminably different stimulus sets up a distinct neural situation.

That this is the most parsimonious neural code may not be immediately obvious. It is simple and economically powerful in that only three receptors can encode a wide range of different stimuli, in fact a continuum. One set might be enough except for the fact that beyond wavelength, stimulus quantity or intensity (and saturation) must also be encoded – and encoding more than one parameter would be impossible for a single set. For stronger stimuli, the curves are simply considered to increase in amplitude; this is discussed in connection with Figure 13. Saturation is a direct function of how similar the GoMs are across the fuzzy sets. This model accurately describes the activity of real neurons. In Section 4, we will show that this behavior can be easily matched by artificial neurons.

We can determine whether this "pattern of activity across a population of neurons" is indeed a neural code in the same way that we know whether we have decoded the dots and dashes of the Morse code. If we can determine from the output what the sender's message was, then we can say that we understand the code. In the case of taste stimuli, we know that the patterns for NH_4Cl and KCl are very similar – further, we know that their tastes are very similar. However, the pattern for NaCl is very different from both of these, as is its taste [5]. This kind of relationship between the similarities of the patterns of amounts of activities across neurons and the similarities of the tastes of various stimuli has been confirmed many times in the last 35 years (see [19] for review). The assumption, which has proven itself, is that all we know depends on *differences* – here differences in the across-fiber patterns and their *degrees* of difference [20], [21]; e.g., if all the world were red, if nothing was different from red, then we would have no knowledge of red, or of the concept of color at all. As soon as we experience a different color, we can experience red and understand what the color is.

From these data in taste, we believe that neural codes are carried in "across-fiber patterns," since they show us what the stimuli are in terms of the degrees of their neural differences from each other – differences being the underlying principle for "knowing." The requirement for this is simple overlapping fuzzy-set neurons.

3. FUZZY MODELS FOR TASTE

3.1. Grades of membership in fuzzy sets

Modeling the responses of the taste neurons to different substances is an interesting task due to the nonexistence of a quantitative measure of the tastes. We cannot describe notions like "sweet," "salty," "bitter," or "acid" with numerical quantities. However, Erickson et al. [22] successfully described this organization in taste with a fuzzy statistical procedure in terms of a few "pure types." This is a fuzzy procedure in that each neuron and stimulus has a GoM in each pure taste type.

Erickson et al. [22] based their analysis on data collected by Doetsch [17], that described the responses of rat taste neurons to various taste substances; the real-time responses were recorded in 30 consecutive 0.1 sec. bins over 3 seconds. It was proposed [22] that the taste responses could be described in terms of the interactions between the stimuli and the neurons. Both neurons and stimuli would have GoMs in a few sets called "General Interactions" which are believed to mediate between the stimuli and the neurons. The analysis indicated that three General Interactions were sufficient to account for the data. Thus, the degree to which each neuron contained these receptors could be indicated by n_1, n_2, and n_3. The degree to which each stimulus activated these receptors could be indicated by a_1, a_2, and a_3. It was found that the sum S of the impulses of a taste neuron, over a fixed time-interval (e.g., 3 seconds), could be computed with the equation:

$$S = n_1 a_1 + n_2 a_2 + n_3 a_3 \tag{1}$$

In this way it was shown [22-24] that a set of numbers n_1, n_2, and n_3 specific to each neuron, and a_1, a_2, and a_3, specific to each substance, can be found such that

Equation (1) is satisfied for all neurons and for all substances ($r^2 = 0.97$, see [22] for details). For example, taste neuron #18821 was described by the numbers {1.03, 1.37, 5.4}, and HCl was characterized by the numbers {17.91, 24.65, 81.06}. Then, according to Equation (1), this neuron should respond to HCl with $S = 489.94$ impulses/3 seconds.

It was also shown [22] that the idiosyncratic time course of response of each neuron to each stimulus could also be successfully modeled by a combination of three temporal "pure types" [23-24]. The statistical method of selecting the GoM numbers is described in detail elsewhere [1].

Equation (1) shows that, for the same substance, the various taste neurons have different responses. This leads to the idea that the taste system could have three receptor mechanisms, characterized by a "receptor cell" having the weights vector $\{n_1, n_2, n_3\}$, that is activated by a "stimulus" vector $\{a_1, a_2, a_3\}$.

Usually, the membership function values in the fuzzy sets are positive numbers, limited to unity. However, the amount of response of these neurons can be up to 500 impulses in three seconds, although the usual rate is much lower. Neurons respond with brief electrical impulses, each from about 0.2 to 1.0 msec in duration. The typical measure of amount of response of a neuron is in terms of #impulses/time. If the numbers a_1 to a_3 and n_1 to n_3 were limited to unity, the value of the taste response S would be limited to three impulses/3 seconds. To view this in conventional terms, these real numbers could be divided by a constant so that they are each unity or less. Figure 4 from [23] provides an illustration of the GoMs of the neurons and stimuli with the total of the memberships for each neuron or stimulus equal to 1.0.

In the above proposal, the notion of "fuzzy set" is used to mean that a variety of objects (here stimuli and neurons) could belong with different GoMs to the same set (called "General Interactions" in [22-24]), and an object could belong with different GoMs to a number of different sets. The "Fuzzy Model," to follow, uses the term "fuzzy" in a more usual sense. Therein, we demonstrate that the dynamics of taste neurons can indeed be modeled by a fuzzy Sugeno model. And, since the brain is a biological neural network, we will also address the possible implementation of a fuzzy machine using artificial neural networks (Section 4).

3.2. A fuzzy model

In the following, we prove that a dynamic model of taste neurons, based on Sugeno's [25] fuzzy system, could lead to a relation similar to (1) for the computation of the total number of neuron impulses. In addition, the system can be trained to approximate the time-response of the taste neurons, for a finite time interval.

3.2.1. The model

The model has two inputs: x_1, the "substance" input and x_2, the "time" input. The output y approximates the number of neuron impulses in equal and consecutive time intervals (Figure 5).

A substance activating input x_1 is described by three numbers that are the GoMs to three discrete fuzzy sets, $\tilde{A}_1$, $\tilde{A}_2$, and $\tilde{A}_3$ (Figure 6). The numbers a_{i1}, a_{i2} and a_{i3} – having values in the interval [0,1] – identify the substance i, sent to the input x_1.

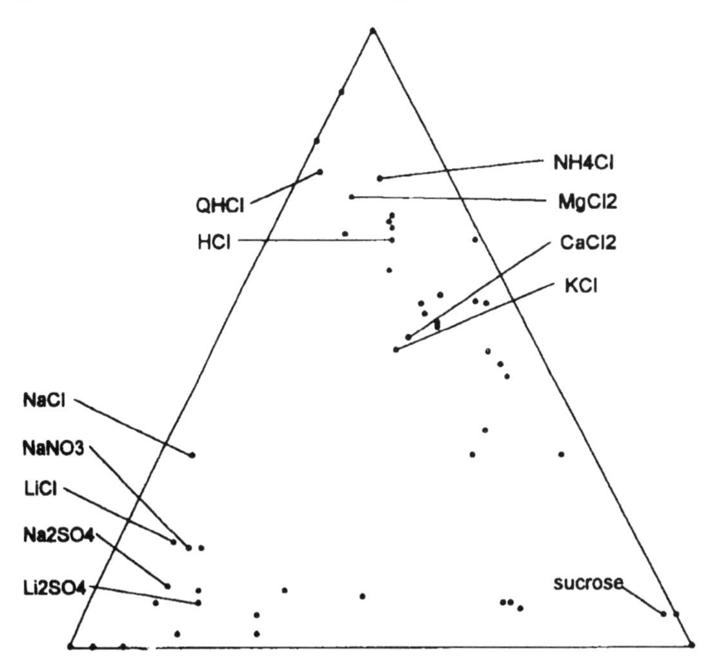

Figure 4. A map of the GoMs for the stimuli and neurons in three pure types. The GoMs for each neuron (dots) and stimulus in each pure type are given by the distance of each point from each of the three sides of the triangle. For example, sucrose, in the lower right of the triangle, has a large membership in the pure type represented by the side opposite (left), and very small memberships on the adjacent sides. In each case, these three distances sum to a constant, which may be considered to be 1.0. From [23], with permission.

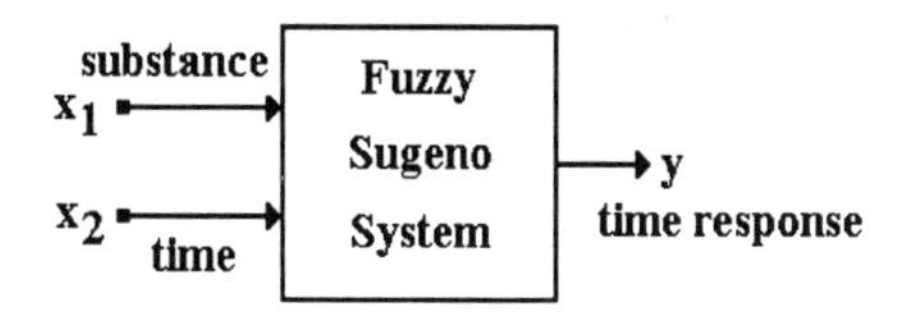

Figure 5. The structure of the fuzzy model for the dynamics of taste neurons. $\mathbf{x_1}$: input substance; $\mathbf{x_2}$: time; output variable is the neural response (number of spikes).

Input x_2 receives crisp integer values in a fixed interval [0, 30], since the time interval of 3 seconds, used for modeling, is divided into 30 bins of 0.1 seconds. During this time interval, the output y of the model should approximate the taste neuron's dynamics. In the following, we consider the time as a linguistic variable having four linguistic values: $\tilde{B}_1$ (very small), $\tilde{B}_2$ (small), $\tilde{B}_3$ (medium), and $\tilde{B}_4$ (large) with continuous membership functions. We shall denote by μ_1 to μ_4 the membership functions of the sets $\tilde{B}_1$ to $\tilde{B}_4$. The functions μ_1 to μ_4 are defined over the interval of integer numbers [0,30] and take values in the interval of real numbers [0, 1] (Figure 7).

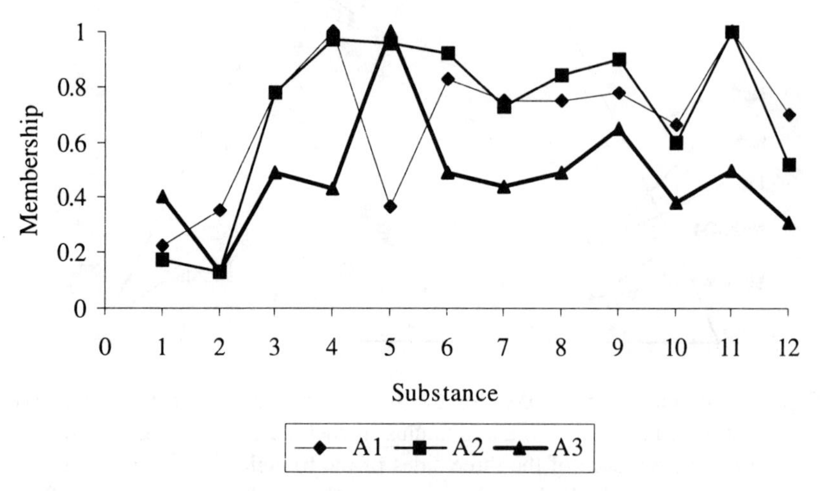

Figure 6. Discrete membership functions for the substance input.

The fuzzy system has twelve rules of the form:

$$\text{(rule \#r)} \quad \textit{If } x_1 \textit{ is } \tilde{A}_r \textit{ AND } x_2 \textit{ is } \tilde{B}_r \textit{ then } y_r = w_r,\ r = 1,2,\ldots,12 \tag{2}$$

where $\tilde{A}_r$ is one of the sets $\tilde{A}_1$ to $\tilde{A}_3$, $\tilde{B}_r$ is one, the sets $\tilde{B}_1$ to $\tilde{B}_4$, y_r is the crisp rule output, and w_r is a real number.

The output y of the system is computed as:

$$y = \sum_{r=1}^{12} t_r y_r = \sum_{r=1}^{12} t_r w_r \tag{3}$$

where t_r represents the truth degree (TD) of the rule premise *(If x_1 is $\tilde{A}_r$ AND x_2 is $\tilde{B}_r$)*, $t_r \in [0,1]$.

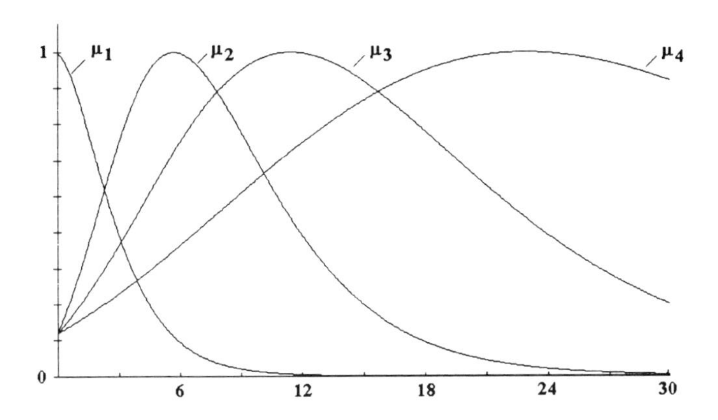

Figure 7. Continuous membership function for the time input (see text).

If the input x_1 is the substance number i and the time moment, received by x_2, is u, the TD t_r of the rule r, from the equation (3), is computed as

$$t_r = a_{ir}\mu_r(u), \qquad r = 1,2,...,12 \tag{4}$$

where a_{ir} is the GoM of the substance i to the set $\tilde{A}_r$, and $\mu_r(u)$ is the GoM of the time moment u to the fuzzy set $\tilde{B}_r$.

Using Equations (3) and (4), the output $y(u,i)$ of the system, for the substance i, at the time u, is given by

$$y(u,i) = \sum_{r=1}^{12} a_{ir} w_r \mu_r(u) \tag{5}$$

In the following, we will place the system weights w_r into a 3×4 matrix **W**:

$$W = \begin{bmatrix} w_{11} & w_{12} & w_{13} & w_{14} \\ w_{21} & w_{22} & w_{23} & w_{24} \\ w_{31} & w_{32} & w_{33} & w_{34} \end{bmatrix}$$

The matrix **W** corresponds to the rules matrix of the fuzzy system:

	$\tilde{B}_1$	$\tilde{B}_2$	$\tilde{B}_3$	$\tilde{B}_4$
$\tilde{A}_1$	w_{11}	w_{12}	w_{13}	w_{14}
$\tilde{A}_2$	w_{21}	w_{22}	w_{23}	w_{24}
$\tilde{A}_3$	w_{31}	w_{32}	w_{33}	w_{34}

By using the matrix **W,** the equation (5) could be written as

$$y(u,i)=\sum_{k=1}^{3}\sum_{j=1}^{4} a_{ik} w_{kj} \mu_j(u) \tag{6}$$

The output y of the fuzzy model approximates the number of neuron impulses in equal and consecutive time intervals. Thus, the total number $S(i)$ of the neuron impulses, for the substance i, is the sum of the outputs $y(u,i)$ over the time interval [0,T].

$$S(i)=\sum_{u\in[0,T]} y(u,i) \tag{7}$$

In the following we will prove that the sum $S(i)$ has a form similar to equation (1), and the terms of the sum have similar meanings. By using equation (6), the sum $S(i)$ can be written as

$$S(i)=\sum_{u\in[0,T]} y(u,i)=\sum_{k=1}^{3} a_{ik}\sum_{j=1}^{4} w_{kj}\sum_{u\in[0,T]}\mu_j(u)=\sum_{k=1}^{3} a_{ik} n_k \tag{8}$$

The numbers n_k, defined as

$$n_k=\sum_{j=1}^{4} w_{kj}\sum_{u\in[0,T]}\mu_j(u) \qquad k=1,2,3 \tag{9}$$

do not depend directly on the input substance. A given taste neuron is defined through the set $\{n_1, n_2, n_3\}$, since every taste neuron model is defined by a specific matrix **W** of weights. Meanwhile, the sums of the functions μ_1 to μ_4 over [0, T] are the same for all the neurons.

Thus, the sum $S(i)$ can be written as

$$S(i)=a_{i1}n_1+a_{i2}n_2+a_{i3}n_3 \tag{10}$$

where the discrete set $\{a_{i1}, a_{i2}, a_{i3}\}$ defines the substance i and the discrete set $\{n_1,n_2,n_3\}$ defines the neuron.

Equation (10) has not only a formal similarity with equation (1), but the terms involved in both equations have similar meanings: the terms a_{ik} refer to the substance i, and the terms n_k refer to the neuron dynamics. Thus, equation (10) proves that the "general interactions" sets, found through crisp methods [22], can be found through a fuzzy modeling approach as well.

3.2.2. The synthesis of the fuzzy model

In the following, we outline a method for the synthesis of the fuzzy system introduced above. The method assumes that the fuzzy sets $\tilde{A}_1$ to $\tilde{A}_3$ and $\tilde{B}_1$ to $\tilde{B}_4$ are fixed and known. This implies that the numbers a_{i1} to a_{i3}, as well as the functions μ_1 to μ_4 from equation (6), are known.

The synthesis method finds the matrix **W** of the system, such that the system output y approximates the experimental real time response of the neuron with minimum instantaneous energy error.

If we denote by $d(k)$ the experimental response at the discrete time moment k, and by $y(k,i)$ the system output at the moment k, for the substance i, the instantaneous energy error $E(k)$ is computed with

$$E(k) = [d(k) - y(k,i)]^2 \tag{11}$$

The weights w_{mj}, m = 1,2,3; j = 1,2,3,4 of the fuzzy system are computed with a gradient stochastic algorithm [26], that minimizes the energy $E(k)$:

$$w_{mj}(k+1) = w_{mj}(k) - \eta(k)\frac{\partial E(k)}{\partial w_{mj}(k)} \tag{12}$$

where $\eta(k)$ is an adaptation factor.

From equations (6), (11), and (12) one derives that the weights w_{mj} are computed with

$$w_{mj}(k+1) = w_{mj}(k) + \eta(k)[d(k) - y(k,i)]a_{im}\mu_j(k) \tag{13}$$

Another possibility for the system synthesis appears after regrouping the terms of equation (6) into the form:

$$y(u,i) = \sum_{j=1}^{4} b_{ij}\mu_j(u) \tag{14}$$

where

$$b_{ij} = \sum_{k=1}^{3} a_{ik}w_{kj}; \quad j = 1,2,3,4 \tag{15}$$

If we consider the known functions μ_1 to μ_4, we can directly compute the numbers b_1 to b_4 with the LMS algorithm [26], since the output y has the form of a linear combination

$$b_j(k+1) = b_j(k) + \beta[d(k) - y(k,i)]\mu_j(k); \quad j = 1,2,3,4 \tag{16}$$

where $d(k)$, $y(k,i)$ have the same meaning as in equation (11), and β is an adaptation constant.

Equation (14) shows that the response of a specific neuron to a specific substance i is, in fact, a weighted combination of the fixed functions μ_1 to μ_4. One could use the set $\{b_{i1}, b_{i2}, b_{i3}, b_{i4}\}$ to characterize a neuron's temporal behavior for the substance i. This functional approximation does not reveal the qualitative logical rules (see equation 2) of the response dynamics, as the fuzzy modeling does. Still, this approach has the virtue of allowing a simple and global description of the response of a taste neuron. This description could be useful for unveiling the embedded functional role of the taste neurons within the taste sense system.

3.2.3. Simulating the dynamics of taste neurons

The data used in our simulations were published and analyzed in a previous study of the activity of gustatory neurons, in the nucleus of the solitary tract of the rat [17]. The neural impulses were counted for the first three seconds of the response. The counts were taken during consecutive 100 ms intervals: this gives us 30 integer numbers that represent the impulse counts for the first three seconds.

Twelve stimuli (substances) were used (see Table 1).

Table 1. The stimuli used to test the taste neurons

Code	Stimulus
1	1.0M Sucrose
2	0.01M QHCl
3	0.03M KCl
4	0.1M NaCl
5	0.03M HCl
6	0.1M $NaNO_3$
7	0.1M Na_2SO_4
8	0.1M Li2SO4
9	0.3M $CaCl_2$
10	0.1M LiCl
11	0.1M NH_4Cl
12	0.1M $MgCl_2$

For a specific neuron, and for every substance i, i = 1,2,...,12, we compute three numbers $\{S_{i1}, S_{i2}, S_{i3}\}$ representing the mean of the neuron response over three consecutive time periods [0, 0.5], [0.5, 1.5], and [1.5, 3]:

$$S_{i1} = \sum_{u \in [0, T_1]} d(u); \quad S_{i2} = \sum_{u \in [T_1, T_2]} d(u); \quad S_{i3} = \sum_{u \in [T_2, T_3]} d(u) \tag{17}$$

where $d(u)$ is the experimental time-response of the neuron, $T_1 = 0.5s$, $T_2 = 1.5\ s$, and $T_3 = 3s$.

For the substance i, $i = 1,2,....,12$, we compute the membership function degrees $\{a_{i1}, a_{i2}, a_{i3}\}$ to the fuzzy discrete sets $\tilde{A}_1$ to $\tilde{A}_3$ of the input substance, by normalizing the set $\{S_{i1}, S_{i2}, S_{i3}\}$:

$$a_{i1} = S_{i1} \Big/ \max_i \{S_{i1}\}; a_{i2} = S_{i2} \Big/ \max_i \{S_{i2}\}; a_{i3} = S_{i3} \Big/ \max_i \{S_{i3}\} \tag{18}$$

Finally, every set $\tilde{A}_j$, $(j = 1, 2, 3)$ is formed from all a_{ij}, $(i = 1, 2,..., 12)$, and $\tilde{A}_j = \{a_{1j}, a_{2j}, ..., a_{12j}\}$. The sets $\tilde{A}_1$ to $\tilde{A}_3$ of the neuron #015711 are represented in Figure 6.

The sets $\tilde{A}_1$ to $\tilde{A}_3$ are different for every neuron. They define a "mean" temporal behavior of the neuron over the intervals [0, 0.5], [0.5, 1.5], and [1.5, 3], measured in seconds. From the analysis of these sets, we can directly see the sensitivity of the neuron to different test substances, and one can group the substances giving similar responses.

The membership functions μ_1 to μ_4 are common to all the modeled taste neurons. They are computed as the difference of two sigmoid functions:

$$\mu_j(x) = \frac{G}{1 + \exp(-a_j x + b_j)} - \frac{G}{1 + \exp(-c_j x + d_j)}, \ j = 12,3,4 \tag{19}$$

From Figure 7, one can see that the function μ_1 has a "low pass" form, the functions μ_2 and μ_3 have an asymmetric "band pass" form, and the function μ_4 has a "high pass" form. After some "try and cut" trials, we selected the following coefficients for the functions μ_1 to μ_4:

$$\begin{aligned} &G = 2.75 \text{ (for all the functions);} \\ &a_1 = 32,\ a_2 = 16,\ a_3 = 8,\ a_4 = 4; b_1 = -1.92,\ b_2 = b_3 = b_4 = 1.28; \\ &c_1 = 16,\ c_2 = 8,\ c_3 = 4,\ c_4 = 2; d_1 = 0;\ d_2 = d_3 = d_4 = 1.6. \end{aligned} \tag{20}$$

The coefficients from equation (20) have the virtue that some of them (e.g., a_1 to a_4, c_1 to c_4) respect a simple decreasing law, or are equal to each other (e.g., $b_2 = b_3 = b_4$ and $d_2 = d_3 = d_4$). The coefficients were computed such that the functions μ_1 to μ_4 "cover" the interval [0, 1]. The "time" input x_2 should receive integer values in the interval [1, 30]. We scale these integers by dividing by 30, so that the input x_2 receives real numbers in the interval [0, 1].

We compute the rule matrix **W** with equation (13) where the adaptation factor $\eta(k)$ is given by

$$\eta(k) = 0.5 / E_{in}(k) \tag{21}$$

where

$$E_{in}(k) = 0.5E_{in}(k-1) + 0.5\sum_{m=1}^{3}\sum_{j=1}^{4}[a_{im}\mu_j(k-1)]^2 \tag{22}$$

The sequence $E_{in}(k)$ represents an estimate of the energy of the sequence $a_{im}\mu_j(k)$, $m = 1,2,3$, $j = 1,2,3,4$. This sequence is the "input" sequence in the stochastic gradient algorithm from (13). Thus, computing $\eta(k)$ using equation (21) assures the convergence of the stochastic gradient algorithm from equation (13) [26].

For every neuron, the system is adapted in cycles: during every cycle all the substances are "sent" to the system input ($i = 1,2,...,12$), and the discrete time has the values $k = 1,2,....,30$. Initially, the neuron weights w_{mj} from equation (13) are zero, and $E_{in}(0) = 1$.

Figure 8 presents the coefficients b_{ij} computed, after the model synthesis, with equation (15), for the neuron #015711. Figure 9, panels A to L, depicts the time response of neuron #015711 in the *nucleus of the solitary tract* (NTS) to all the twelve stimuli (the thin curves), and the fuzzy system response computed with the equation (6) (the solid curves). One can make a straight connection between Figures 8 and 9: the approximated neuron response to the substance i is directly related to the "distribution" of the weights b_{ij}. For example, all the weights b_{i3} of the neuron #015711 have low values. Thus, the neuron activity in the second part of the time interval is mainly supported by the function μ_4 (see Figure 7), as can be seen from an inspection of the panels of Figure 9.

The modeling performance was measured, for every neuron, with the number M, which is the mean value of the ratios R_i, computed for every substance i, $i = 1,2,.., 12$:

$$M = \frac{1}{12}\sum_{i=1}^{12} R_i; \quad R_i = \frac{\sum_{k=1}^{30}[d(k,i) - y(k,i)]^2}{\sum_{k=1}^{30} d(k,i)^2} \tag{23}$$

where the sequence $d(k,i)$ is the experimental time-response of the neuron for the substance i, the sequence $y(k,i)$ is the output of the fuzzy system, computed from equation (6), and the numbers R_i represent the ratio between the error energy and experimental data energy. The simulation results are included in Table 2.

In order to measure the neurons' activity, we use their mean sum of impulses, for all the substances, for the entire interval [0, 3.0] seconds:

$$Act = \frac{1}{12}\sum_{i=1}^{12}\sum_{k=1}^{30} d(k,i) \tag{24}$$

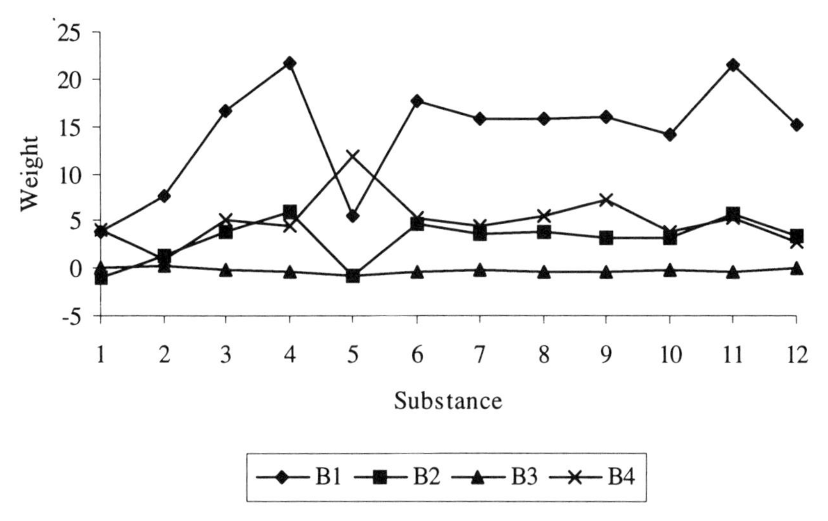

Figure 8. The weights of the membership functions for the temporal variable of the fuzzy model.

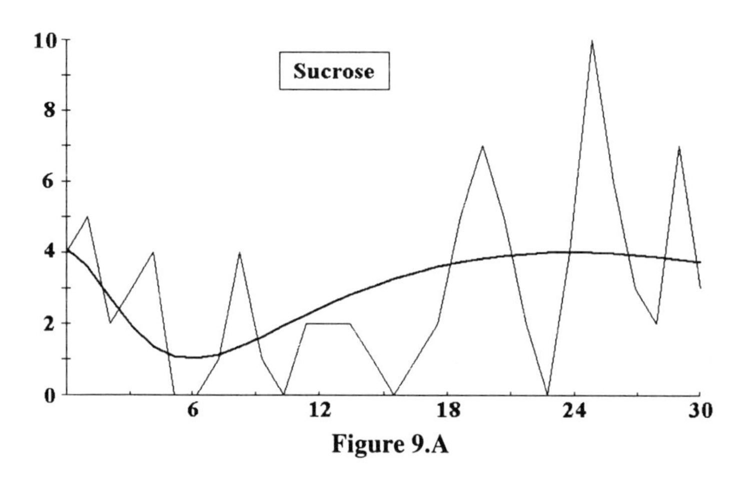

Figure 9.A

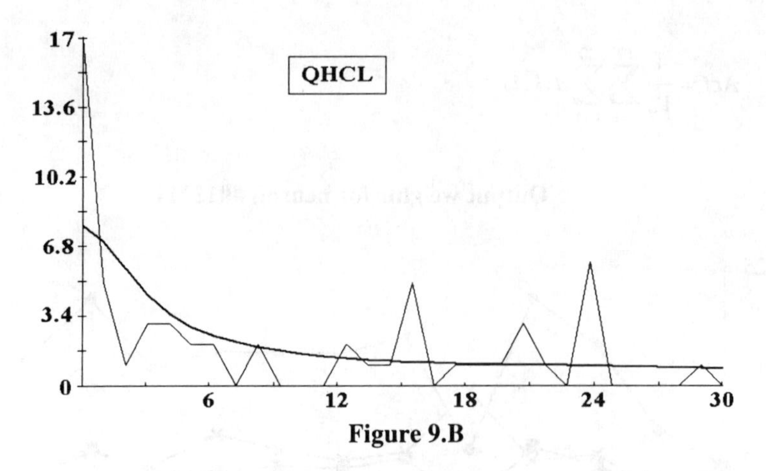

Figure 9.B

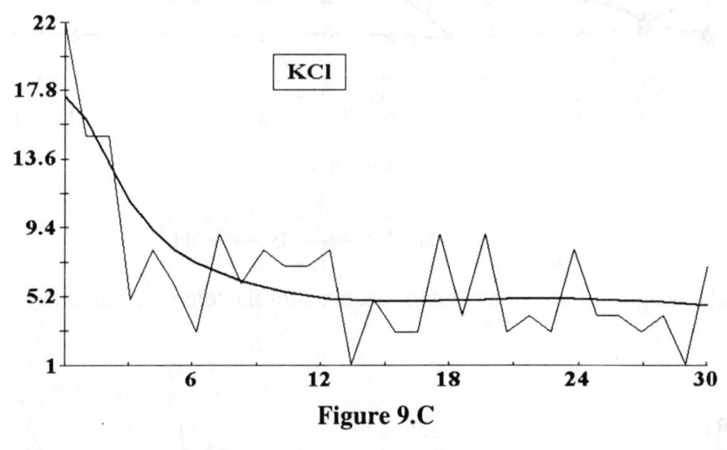

Figure 9.C

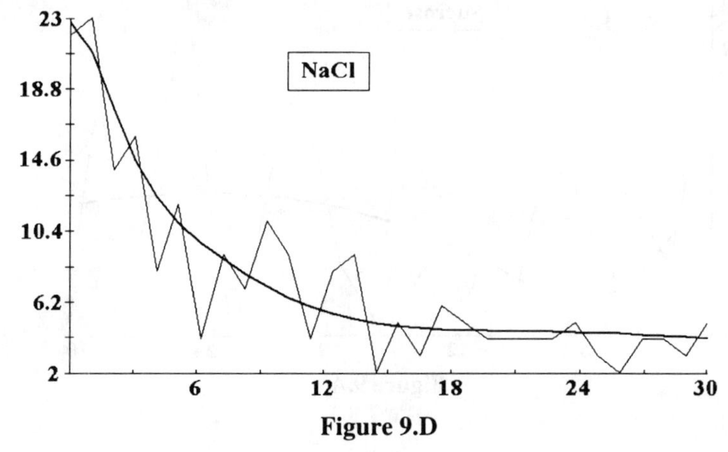

Figure 9.D

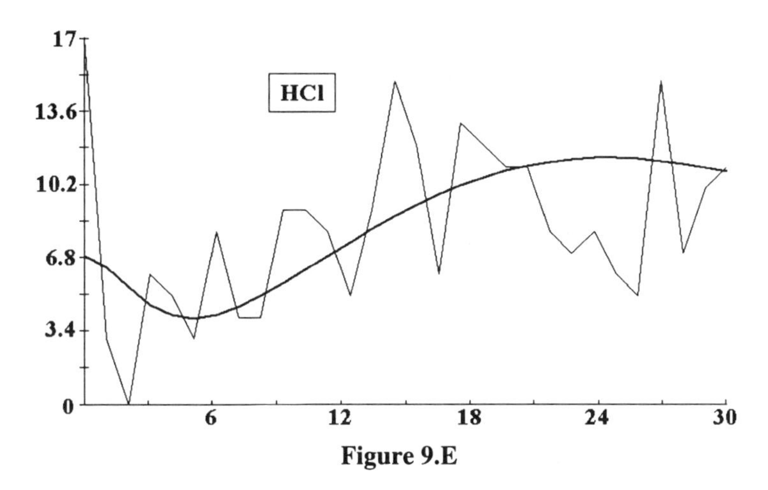

Figure 9.E

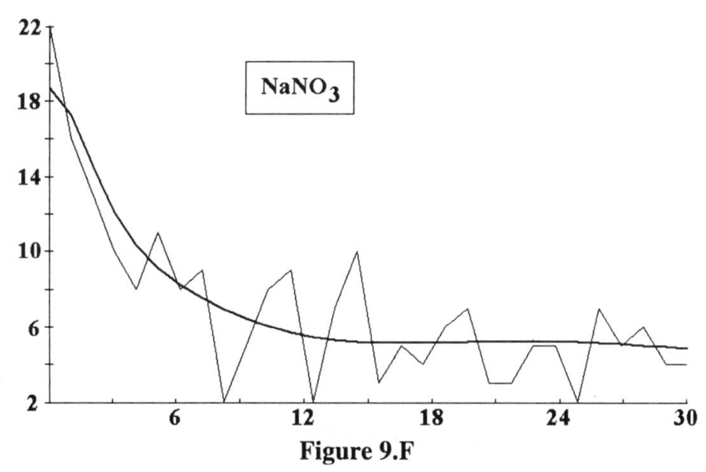

Figure 9.F

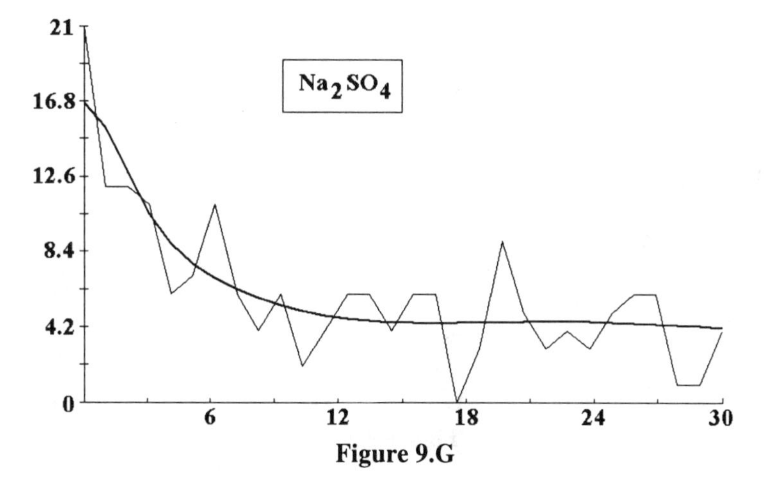

Figure 9.G

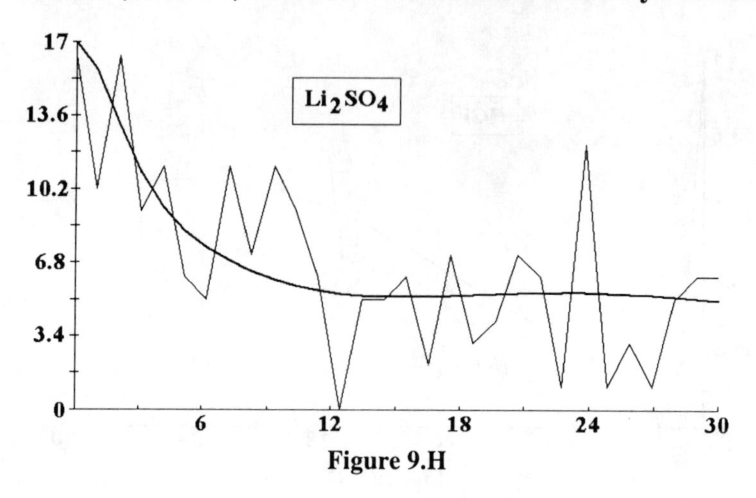

Figure 9.H

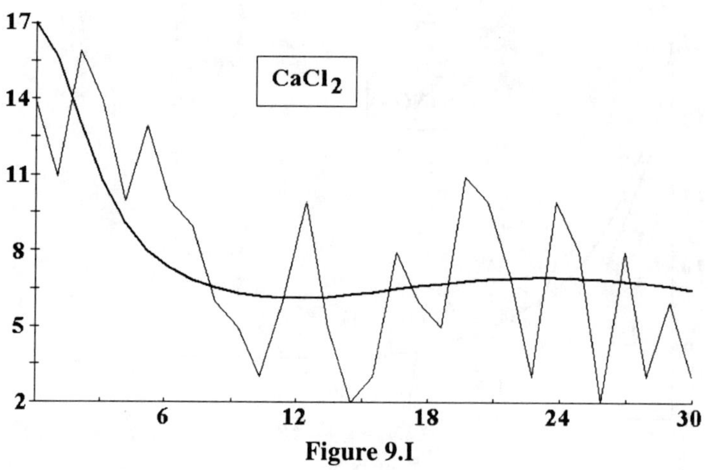

Figure 9.I

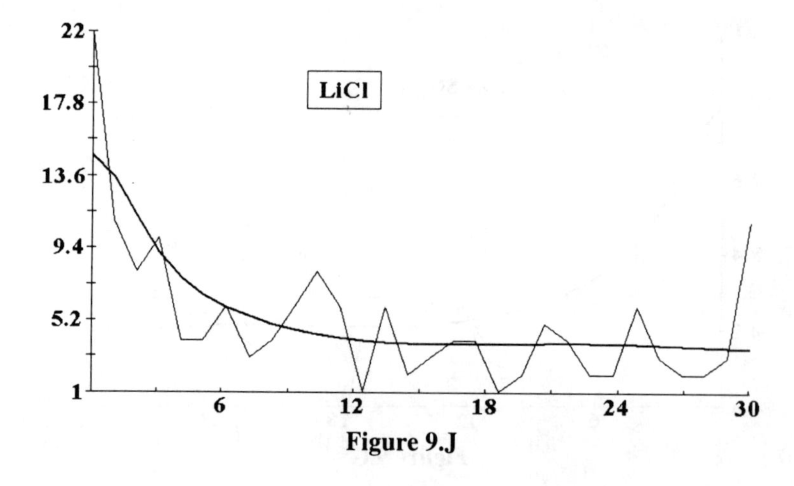

Figure 9.J

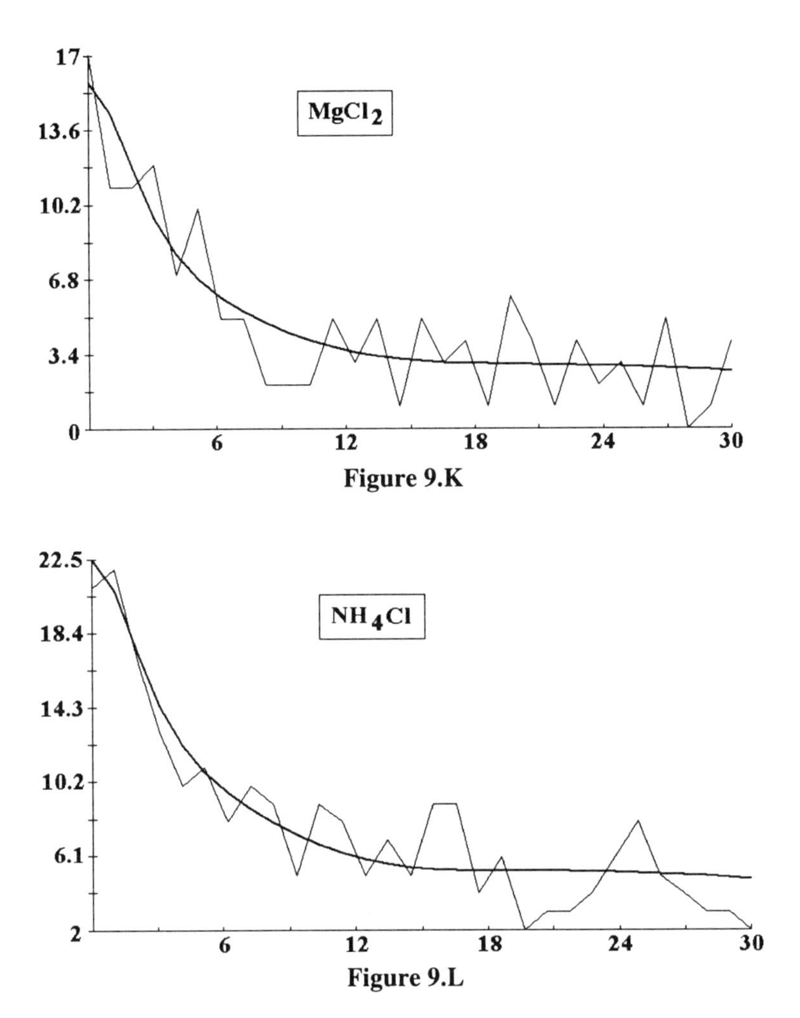

Figure 9.K

Figure 9.L

Figure 9. Simulations with the fuzzy model. Each panel (from A to L) represents the real-time (thin line) and predicted (thick line) response of a neuron to each of the 12 different substances (see text).

For every neuron, we included in Table 2 the sum *Act*, from equation (24). An analysis of Table 2 shows that the level of performance is proportional to the inverse of the amount of the mean activity. Thus, there is a class of neurons (#015911, #016111, #016311, #016612) that have a low level of activity, and their modeling performance is lower than the performance of the neurons (e.g., #016411, #015711) with higher activity. We "tuned" the membership functions μ_1 to μ_4 (see Figure 7) such that the modeling performance was maximized for the neurons with higher activity. Table 2 shows, at least for the most active neurons, that the modeling performance is satisfactory, relative to the low complexity of the model.

Table 2. Modeling performances

Neuron	Performance: M	Activity: Act
#016411	0.09	260.83
#015711	0.15	178.92
#018031	0.18	190.83
#017911	0.20	172.83
#017912	0.21	161.42
#018832	0.28	105.17
#016311	0.32	83.00
#016612	0.33	123.67
#016111	0.49	42.42
#015911	0.51	7.92

The above results suggest that the fuzzy model can be considered a reasonable model for the dynamic of the taste neurons. Since the Sugeno system is a nonlinear function approximator, the taste neuron response can be approximated by it. Still, in the broad sense in which we use fuzzy set theory, the neuron temporal activity could be organized in a fuzzy fashion. Thus, the tastant could start processes whose temporal properties are modeled by the membership functions from Figure 7, and the resulting dynamics could be described with fuzzy rules as defined by the relation in equation (2).

4. FUZZY MODEL FOR BRAIN ACTIVITY

Let us assume for the sake of argument that the brain is conservative in its organizing principles, and that these principles include neurons as fuzzy sets. How might the brain function by using these principles?

It should be noted from the above that both the input and output functions of the brain simulate fuzzy sets. It would seem odd for the nervous system to change principles in mid-stream just where information representation is most complex and, thus, where the value for Young's fuzzy logic is strongest.

While in the above section we demonstrated that the dynamics of taste neurons can indeed be modeled by a fuzzy system, the question of a real fuzzy implementation is still open. Therefore, since the brain is a biological neural network, we address the possible implementation of a fuzzy machine using artificial neural networks. Thus, we suggest that the brain may indeed be a neural implementation of a fuzzy machine.

4.1. A neural network implementing a fuzzy machine?

Since the brain is a *biological neural network*, the question that we are addressing in the following is: Can one conceive (in a broad sense) the activity of any neural network (biological and/or artificial) in fuzzy terms?

The reason we are examining this question here is fourfold. First, the brain can easily be conceived as a neural network whose internal activity is very complicated. A fuzzy description of its activity would greatly improve our understanding of its functioning (as shown above for taste neurons) because it would allow theorists to manipulate linguistic variables instead of specific neural activities.

Second, artificial neural networks were shown to correctly describe various aspects of the behavior of normal animals (see e.g., [27,28,29,30]) and the effect of specific physiological and pharmacological manipulations (see e.g., [31, 32, 33]). Therefore, the interest of psychologists in artificial neural networks has increased over time. Unfortunately, as in biological neural networks, in order to explain a great deal of behavioral data, the artificial neural models have become more and more complicated. In many practical situations, the complexity of the problem is paralleled by the complexity of the solution found by an adaptive neural network. The view that the "internal code" discovered by an adaptive neural network relates (or even implements, as suggested below) a fuzzy rule, or a fuzzy membership function, would allow a simpler analysis of resulting solutions.

Third, given the fact that learning algorithms were devised for artificial neural networks (see e.g., [29], [30], [34], [35], [36]), these algorithms could be used to find the rules of a fuzzy system or, more generally, to finely tune the behavior of a fuzzy system (see e.g., [37], [38], [39]).

Fourth, it is suggested that the robustness of artificial neural networks is related to the robustness of fuzzy sets, in that the parallel distribution of the information in a network is equivalent to the very broad membership functions in fuzzy sets.

4.2. An artificial neuron implements a fuzzy membership function

These findings may be placed in a more formal setting. Figure 10 depicts an artificial neuron. Each input line is excited by input stimulus x_i, and contributes to the activation of the neuron in proportion to the weight of the line, w_i. The activity of the neuron, a, is given by a linear combination of the input stimuli by the weights of the neuron:

$$a = \sum_i x_i w_i = x_1 w_1 + x_2 w_2 + \ldots + x_n w_n \tag{25}$$

One notes the similarity with Equation (1). For example, in the (abstract) space of tastes, a taste neuron responds to a specific chemical combination (the numbers a_1, a_2, and a_3 in equation (1)), that can be imagined as a preferential direction $\mathbf{w}$. When excited with a chemical input, $\mathbf{x}$, the neuron can be imagined to respond in relation to the similarity between the real input and the preferred direction $\mathbf{w}$.

If one represents the inputs and the weights as vectors, the activity of the neuron is given by the inner product of the input vector and the weight vector, which gives a measure of similarity of the two vectors. The similarity between $\mathbf{x} = [x_1 \quad x_2 \quad x_3]^T$ and direction $\mathbf{w} = [w_1 \quad w_2 \quad w_3]^T$ is given by the inner product $\mathbf{x}^\mathbf{T}\mathbf{w}$, i.e., $x_1 w_1 + x_2 w_2 + x_3 w_3$. One notes that the inner product $\mathbf{x}^\mathbf{T}\mathbf{w}$ can be also written as $\mathbf{x}^\mathbf{T}\mathbf{w} =$

$\|x\| \cdot \|w\| \cos\alpha$, where α is the angle between the direction defined by the two vectors, and $\|x\|$ and $\|w\|$ are the norms of the two vectors.

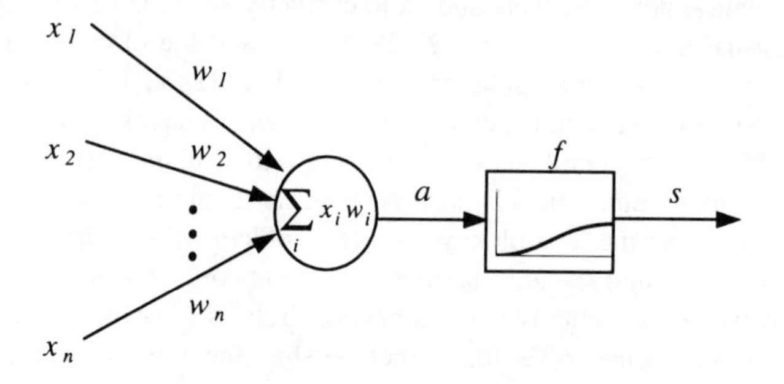

Figure 10. An artificial neuron. Input vector $\mathbf{x} = [x_1, x_2, \ldots, x_n]^T$ excites the n input lines. The activity of the neuron, a, is given by the weighted sum of the input by the weight vector $\mathbf{w} = [w_1, w_2, \ldots, w_n]^T$. The output of the neuron, s, is controlled by the activity function f, usually a sigmoid.

Let us assume that **w** is fixed while **x** sweeps out a sphere in the input space (i.e., **x** changes its direction α, but not its norm, $\|x\|$). The similarity between **x** and **w** will be proportional to $\cos\alpha$, i.e., it will peak when **x** points in the same direction as **w,** will gradually decrease to zero for directions orthogonal to **w**, and to −1 when **x** is the opposite of **w**. The norms $\|x\|$ and $\|w\|$ control the amplitude, but not the shape, of the similarity function. In fact, the above similarity function resembles the output of one of the color neurons to input frequencies.

The output of the artificial neuron (here denoted with s for "similarity") is given by the activation function f, usually a sigmoid. The activation function f changes the shape of the membership function by modifying its width. Thus, the output of the neuron is

$$s = f(a) = f(x^T w) \tag{26}$$

and is a measure of the *similarity* between input vector, **x**, and weight vector, **w**. The left panel of Figure 11 shows the activity a (Equation 25) of the neuron when input **x** sweeps a sphere in the input space. The right panel of Figure 11 shows the output, s (Equation 26) of the neuron when input **x** sweeps a sphere in the input space. Note the resemblance to Young's color membership functions (Figs. 1-2).

4.3. A layer of neurons implements a fuzzifier

In neural network models, the various inputs to hidden layers have different weights. This is analogous to neurons responsive to faces being excited to differing degrees by various shaped lines, aspects of faces, or a variety of whole faces. Thus, one can imagine that faces are coded, memorized, and recognized by a set of "references" or "preferred directions" in the "face space." Indeed, Gaal [40], using a parallel between biological neural networks and artificial neural networks, suggested that population coding might be a way of representing, transforming, and reconstructing external (e.g., visual) stimuli in an intrinsic coordinate system implemented by receptive field weighting functions.

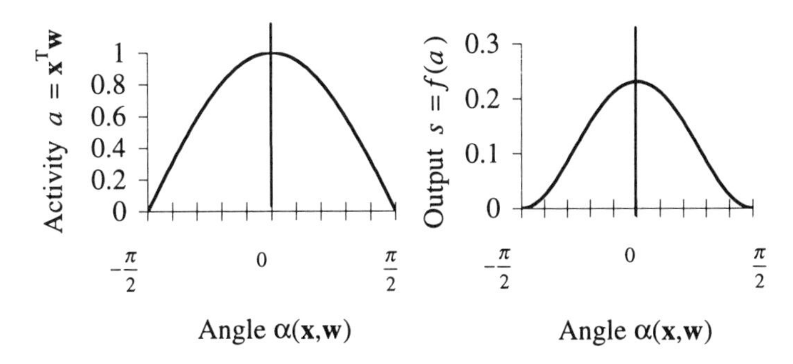

Figure 11. An artificial neuron implements a membership function. The left panel shows activity $a = \mathbf{x}^T\mathbf{w}$ when the angle α between input **x** and weight vector **w** varies. The right panel shows output $s = f(a) = f(\mathbf{x}^T\mathbf{w})$ of the neuron when angle α between input **x** and weight vector **w** varies. Here, $\|x\| = \|w\| = 1$, $f(a) = 1/(1+\exp(-a^2)) - 0.5$ for $a \geq 0$, and 0 elsewhere.

For example, one can imagine that each of the three color-type neurons implements a "preferential direction" in a multidimensional space (a three-dimensional space in the case of colors). These directions form a base in the input space, in that any given input can be represented in relation to these directions. In order to understand this proposal (which is essential for the relation between fuzzy systems and neural networks), let us imagine that each preferential direction i is characterized by three numbers $\{w_{i1}, w_{i2}, w_{i3}\}$ which form column vectors, $\mathbf{w}_1$, $\mathbf{w}_2$, $\mathbf{w}_3$:

$$\mathbf{W} = [\mathbf{w}_1\ \mathbf{w}_2\ \mathbf{w}_3] = \begin{bmatrix} w_{11} & w_{21} & w_{31} \\ w_{12} & w_{22} & w_{32} \\ w_{13} & w_{23} & w_{33} \end{bmatrix} \tag{27}$$

These numbers form a base in the input space if matrix **W** is nonsingular, i.e., if the three directions are independent. Indeed in the case of color vision, the three base colors are unique. In other words, two base colors can be more or less different, but not dependent. Indeed, the loss of one color base (as in "color weak" persons) makes the above matrix singular, and prevents it from uniquely representing all the directions in the input space.

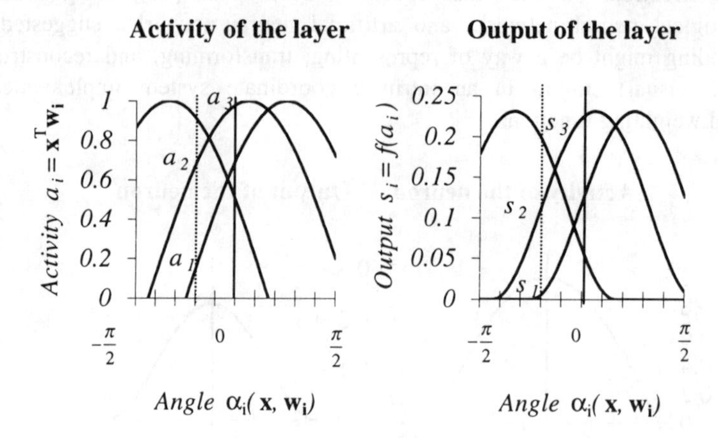

Figure 12. The output of a layer of (e.g., three) neurons implements a fuzzifier. Each neuron responds to the same input vector $\mathbf{x}$ in relation to its preferential direction $\mathbf{w_i}$. The left panel shows the activity $a_i = \mathbf{x}^T\mathbf{w_i}$ of neuron i, $i = 1,2,3$, when the angle α_i between input $\mathbf{x}$ and weight vector $\mathbf{w_i}$ varies. The right panel shows the output $s_i = f(a_i) = f(\mathbf{x}^T\mathbf{w_i})$ of neuron i, $i = 1,2,3$ when α_i between input $\mathbf{x}$ and weight vector $\mathbf{w_i}$ varies. For a given direction (e.g., marked by the dotted line) s_1, s_2, s_3 represent the coordinates of vector x in the vector base $[\mathbf{w_1}, \mathbf{w_2}, \mathbf{w_3}]$. Here, $\|x\| = \|w_i\| = 1$, $f(a) = 1/(1+\exp(-a^2)) - 0.5$ for $a \geq 0$, and 0 elsewhere.

A given input vector $\mathbf{x}$ can be represented in the above space by its coordinates $\{s_1, s_2, s_3\}$. The coordinates represent the "similarity" between the input vector and the three base "directions" in the input space. The coordinates can be determined by applying the inner product operation and the activation function, f

$$s_i = f(\mathbf{x}^T\mathbf{w_i}) \tag{28}$$

Figure 12 shows that three neurons implement a fuzzifier (in terms of directions $\mathbf{w_i}$) when input vector $\boldsymbol{x}$ sweeps a sphere in the input space. Although in this example membership functions have the same height, note that this is not the case in general, since $\mathbf{w_i}$ can have different norms.

Most important, while a standard fuzzifier codes for one dimension only, this is not the case here. For example, it is a fact that color neurons do not code only for the type of color but also for the intensity of the color. Similarly, an artificial neuron is also simultaneously coding not only for the direction of the input vector $\mathbf{x}$, but also for its norm, $\|x\|$. If weight vectors are considered fixed, then the effect of changing the norm of the input vector $\mathbf{x}$ is to amplify or attenuate the shape of the membership functions, without changing the peak positions, at direction corresponding to $\mathbf{w_1}$, $\mathbf{w_2}$, and $\mathbf{w_3}$. Figure 13 shows that when $\mathbf{x}$ sweeps spheres of different radii, although the membership functions peak at the same preferred directions, $\mathbf{w}_1$, $\mathbf{w}_2$, $\mathbf{w}_3$, the three neurons respond more vigorously for $\|x\| = 1$ (thick curves) than for $\|x\| = 0.5$ (thin curves).

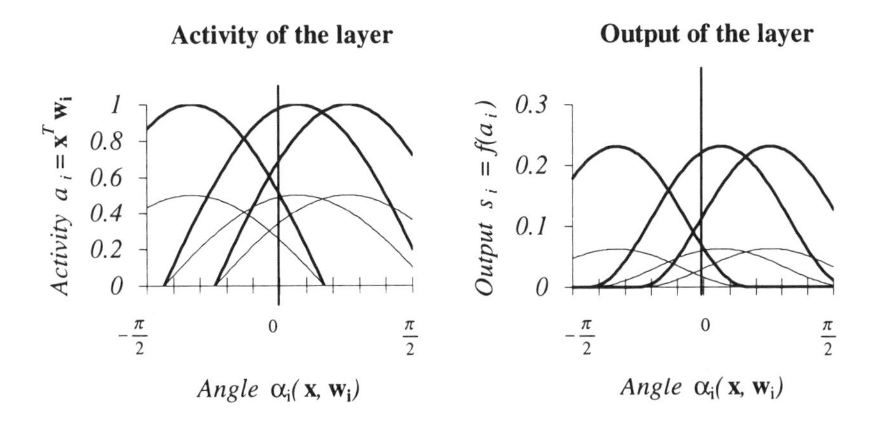

Figure 13. A layer of neurons codes simultaneously for direction and intensity. When input vector $\mathbf{x}$ sweeps spheres of different radii, the shape of the activity (left panel) and output functions (right panel) of a layer of neurons peak at the same preferential directions $\mathbf{w}_1$, $\mathbf{w}_2$, $\mathbf{w}_3$, but the amplitudes are controlled by the input vector $\mathbf{x}$. Thick lines denote activities a_i and outputs s_i for $\|x\| = 1$, while thin lines denote activities a_i and outputs s_i for $\|x\| = 0.5$.

4.4. A "hidden" neuron implements a fuzzy rule

The connections between the layers of a neural network, forward and backward, are not as elusive as they seem. Many things, however, can distract us from a clear view. For example, much is known about the wiring diagram of the brain, but a wiring diagram does not explain what is happening. The difficulty could easily lie in our lack of understanding what is supposed to happen at the various connection points, the synaptic relays between neurons that cluster together into *ganglia* or *nuclei*. For example, in the visual relays it seems reasonable that as successive relays are passed, various elements of vision are dissected out for closer inspection – such as color, and then more complexities are emphasized - such as the perception of faces. Research has suggested that this apparently happens [41], and thus this view of increasing complexity of categorization across synaptic relays is very widely accepted in neurobiology. However, as discussed in Section 2.2.3, these target categories (the words "color," "faces," etc.) which guide neural research may be misguiding us in their crispness. This was noted by some researchers (e.g., [15]) who are aware that these terms are problematic. A potentially more powerful fuzzy approach will be taken here.

As shown above, an artificial neuron implements a preferential direction (vector) in the input space, and its output s is a measure of the similarity between the input and the neuron's weight vector. Thus, it was suggested that a layer of neurons implements a vector base. The output of the layer gives the similarity of the input with each of the neurons' weight vectors, and thus gives the GoM of the input in each fuzzy set. Therefore, in the following we examine a multilayer network to abstract the fuzzy operations implemented by a "hidden" neuron. While Yager [39] showed that fuzzy rules can be implemented by suitably designed neural networks, our approach here is to examine a "standard" network, in order to find out if its normal function can be interpreted in fuzzy terms.

We suggest that a "hidden-layer" neuron implements a fuzzy rule and, in general, the "hidden layer" implements a fuzzy relation. Figure 14 presents two layers of an artificial neural network. Input vectors $\mathbf{x_1}$ and $\mathbf{x_2}$ activate the neurons of the first layer of the network, whose outputs are, respectively, s_1, s_2, s_3, and s_4, s_5, and s_6. Input $\mathbf{x_1}$ activates neurons *A*, *B*, and *C*. Their output, s_1, s_2, and s_3, show how similar $\mathbf{x_1}$ is to the weight vectors of *A*, *B*, and *C*. Similarly, input $\mathbf{x_2}$ activates neurons *D*, *E*, and *F*. Their output, s_4, s_5, and s_6, show how similar $\mathbf{x_2}$ is to the weight vectors of *D*, *E*, and *F*. As suggested above, input vectors $\mathbf{x_1}$ and $\mathbf{x_2}$ are fuzzified in the first layer of the network.

In the following we concentrate on neuron *G*. The activity of this neuron is given by

$$a = \sum_i s_i w_i = s_1 w_1 + s_2 w_2 + \ldots + s_n w_n \tag{29}$$

which is similar to Equation (26). Its input vector, **s**, is the output of some neural layers implementing some fuzzifiers, so that **s** is a vector of similarities. The rule implemented by the neuron is given by its weight vector, **w**. Namely, the neuron fires more at the specific direction **w** in the input space (of similarities), i.e., if $[s_1, s_2, s_3]$ is similar to $[w_1, w_2, w_3]$ and $[s_4, s_5, s_6]$ is similar to $[w_4, w_5, w_6]$. In other words, the "hidden" neuron fires if input vectors $\mathbf{x_1}$ and $\mathbf{x_2}$ satisfy rule **w**.

Assume the output of neurons *ABC* is $[w_1, w_2, w_3]$ when $\mathbf{x_1} = \mathbf{x_1}^*$ while output of neurons *DEF is* $[w_4, w_5, w_6]$ is obtained when $\mathbf{x_2} = \mathbf{x_2}^*$. Then, the "hidden" neuron implements (i.e., its activity peaks for) the following logical premise:

$$\mathbf{x_1} \text{ is } \mathbf{x_1}^* \text{ AND } \mathbf{x_2} \text{ is } \mathbf{x_2}^* \tag{30}$$

Figure 15 shows the output of "hidden" neuron *G* when inputs $\mathbf{x_1}$ and $\mathbf{x_2}$ sweep their input space. The upper panels of Figure 15 show the outputs of the first layer of neurons, i.e., of the fuzzifiers. The left upper panel shows outputs $[s_1, s_2, s_3]$ for each direction α of input $\mathbf{x_1}$, while the right upper panel shows outputs $[s_4, s_5, s_6]$ for each direction β of input $\mathbf{x_2}$. When α and β vary, neuron *G* computes the similarity between $[s_1, s_2, s_3]$ and $[w_1, w_2, w_3]$, and between $[s_4, s_5, s_6]$ and $[w_4, w_5, w_6]$. The lower panel shows the output of the "hidden" neuron *G,* which peaks for certain angles α^* and β^* corresponding to $\mathbf{x}^*$ and $\mathbf{y}^*$.

Although this result is not a demonstration that *any* (e.g., biological) neural network can be immediately interpreted as implementing a fuzzy system, this view can be used toward understanding how neural hardware would compute with fuzzy software. This result suggests that certain fuzzy relations can be immediately implemented in a neural manner, and that neural learning algorithms can be used to discover fuzzy rules. This idea was successfully used (see e.g., [38, 39] etc.) in applications related to image processing [42, 43], pattern recognition [37], and adaptive control [44]. Some problems concerning the process of learning, memory, and retrieval are briefly touched on in Section 5.3.

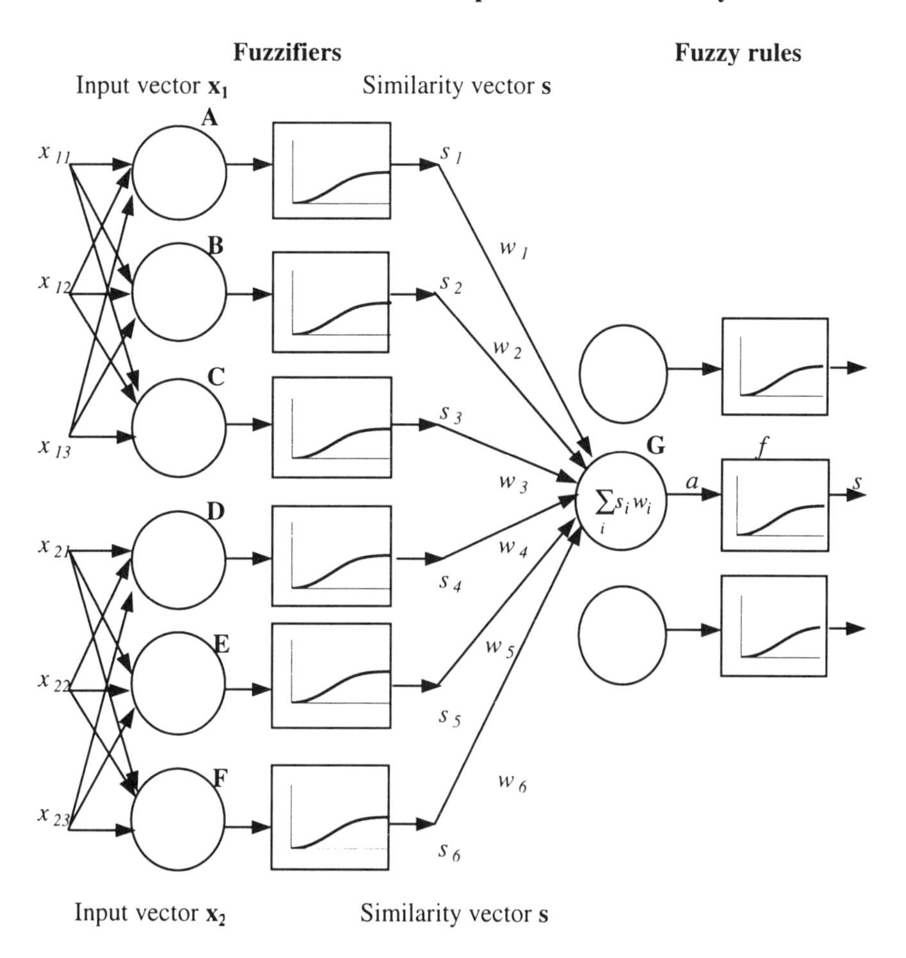

Figure 14. Neural network implementations of relations. The first (leftmost) layer of neurons act as fuzzifiers, while the second layer (rightmost) layer of neurons implement fuzzy relations. Neuron *G* implementing the rule fires preferentially when input similarity vector **s** is oriented similar to **w**, thus selecting a relation between $\mathbf{x}_1$ and $\mathbf{x}_2$. *ABC*: three neurons acting as fuzzifier for input vector $\mathbf{x}_1$; *DEF*: three neurons acting as fuzzifier for input vector $\mathbf{x}_2$; G: "hidden" neuron acting as a fuzzy rule; $[s_1\ s_2\ s_3]$ output of $\mathbf{x}_1$ fuzzifier; $[s_4\ s_5\ s_6]$ output of $\mathbf{x}_2$ fuzzifier; **w**: the fuzzy rule given as a vector in the similarity space.

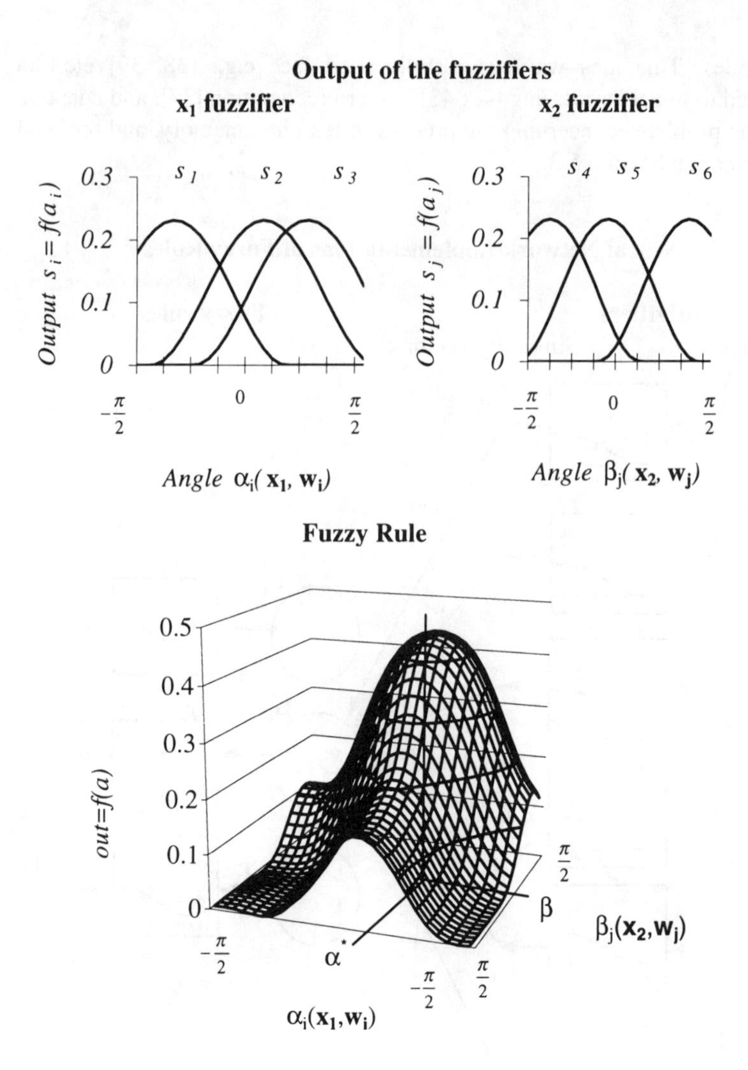

Figure 15. An example of a neural network implementation of relations. Upper panels show the outputs of the first layer of neurons, i.e., of the fuzzifiers. Left upper panel shows outputs $[s_1, s_2, s_3]$ of neurons *ABC* (Figure 14) for each direction α_i, $i = 1,2,3$ of $\mathbf{x_1}$ input vector. The right upper panel shows outputs $[s_4, s_5, s_6]$ of neurons *DEF* (Figure 14) for each direction β_j, $j = 4,5,6$ of $\mathbf{x_2}$ input vector. Neuron *G* (Figure 14) computes the similarity between $[s_1, s_2, s_3]$ and $[w_1, w_2, w_3]$ and between $[s_4, s_5, s_6]$ and $[w_4, w_5, w_6]$. The lower panel shows the output of the "hidden" neuron *G* when α_i and β_j vary. The output peaks for $\alpha = \alpha^*$ and $\beta = \beta^*$ corresponding to $\mathbf{x}^*$ and $\mathbf{y}^*$ (see text). Here we assumed $\mathbf{w}$ fixed, $\mathbf{w} = [-0.1, 4, 0.3, -0.7, 0.2, 3]$.

There is a further interesting aspect of this approach. Neurons sensitive to color are only *primarily* sensitive to color – they also respond to other aspects of the stimulus such as location etc. [15], [41]. This suggests a fuzzy organization on a much broader scale. In this case, location and color occupy the same sets. The same holds for "movement" neurons, neurons for "place" in the visual field, neurons for "fine detail," and neurons for "complexities" (faces). The neurons' responses, even to their named tasks, are fuzzy, in that they also respond to categorically different (e.g., color and form) tasks.

Therefore, AI models might benefit from modifying their terms into a fuzzy status. Instead of crisp words, one would use fuzzy membership functions derived from multidimensional scaling (MDS) solutions [2] or fuzzy statistical analysis [45], from fuzzy synthesis methods as the one described in Section 3, or using neural methods. The fuzzy approach might provide AI models with greater power.

5. APPLICATIONS OF FUZZY LOGIC TO NEURAL SYSTEMS

5.1. Quantitative aspects of the fuzzy neural sets

5.1.1. Neural mass

What might we expect of the breadth of an NRF, i.e., the range of inputs to a fuzzy set neuron? In Young's example, he realized that the color receptors had to be as widely tuned as possible so that the few available neurons at each retinal point could cover the entire wavelength continuum for color. As an interesting parallel, people can identify a wide variety of different tastes with only one taste papilla. Analogous to a point on the retina, a taste papilla is not served by many neurons. And it turns out that typically each of these taste neurons is activated by a wide variety of tastants – i.e., each neuron is broadly tuned. It is as if the nervous system needs each input to activate as many neurons as possible – perhaps to provide a stable message.

However we would be wrong to carry this too far. If we consider space across the retina, or across the skin, each NRF shows sensitivity to only a very limited part of these continua. Looked at this way, it would somehow seem unrealistic that each neuron of the skin responds to very wide areas of the body. And the reason seems obvious: Young needed this breadth of sensitivity to wavelengths to get even a few neurons to respond. On the other hand, there is such a great quantity of neurons available across the skin surface that if the NRFs were very broad – in terms of the percentage of total skin surface – a touch at one point would cause a very large, perhaps overwhelming amount of neural activity.

Such comparisons suggest that the brain computes with certain moderate-sized amounts of neural activity. "Moderate sized" would mean enough to get a clear message given the sensitivity of the neurons (amount of response), their problematic reliability, and the complexity (number of inherent discriminations) of the message (e.g., color vs. face). As an example, we must forgive Young for his lack of appreciation of the limited capability of a neuron to give many usable levels of activity, and the less than perfect reliability of all neural elements. Even the fact that the brain was composed of cells (neurons and others) would not be known for another half-century after his statement. So, although we may accept the fact that there are only three *types* of color receptors, we should not believe that only one group of three will give the brain a clear message. A point source of light (e.g., a star) does not give clear impressions of color. However, we can discern some color in Venus and Mars (white and red). Presumably the brain needs a minimal neural mass to provide a clear message; it is interesting to entertain the idea that, whatever the system, the brain needs the same quantum of neural mass to carry information [20], [21]. This is certainly relevant for the brain, and may also be relevant to adequate models of the brain.

5.1.2. Sensitivity to fine gradations in input

It seems that narrower NRFs would provide finer discriminations. It is interesting to note that with a given number of neurons arranged along a parameter, varying the width of the NRFs does not change the total amount of difference between the patterns of neural activity caused by two different inputs [6]. However, the *total* mass of activity produced by an input increases in parallel with the width of the NRFs, simply because with wider NRFs, more neurons would be activated. But, since broader NRFs have shallower slopes, the differences in response between any two inputs would be less in each neuron. Interestingly, the total differences between two inputs would remain unchanged. The decrease in neural differences in each neuron caused by broader NRFs is offset by the increase in the number of neurons carrying differences. So it may be that NRF breadth is adjusted to be just sufficient to provide enough "neural mass" to provide a reliable and accurate message, and no more.

The quantitatively detailed answer awaits a clear view of the coding process, and of the limits of reliability and sensitivity of the nervous system; we can offer only a definition of the problem, which may provide a starting point toward answers.

5.1.3. Intelligence

The fuzzy width of an NRF may have another very interesting and perhaps profound meaning. One useful meaning of the term "intelligence" is the ability for an organism to perceive uncommon relationships. The more that a person can "think laterally," making comparisons among seemingly orthogonal issues, the more "intelligent" that person can be considered to be. When we were considering the perception of tastes, we noted that information comes in terms of comparisons of (differences between) the across-fiber patterns of activity evoked by the different stimuli. Conversely, the comparisons that we can make require that the inputs to be compared activate the same population of neurons. If we simply follow this logic, intelligence would be a measure of how diverse the input was to a given population of fuzzy neurons. The width of the NRFs would indicate the degree of this diversity. Then, for some convergent part of the brain, "intelligence" is given in the range of inputs to the fuzzy sets and the width of the NRFs [20].

5.2. Defuzzification and responses

An important unresolved point relates to "defuzzification". What does the "output" of the nervous system look like if there are no defuzzified "button-pressing," etc.

neurons as discussed above (Section 2.1)? An arm movement is composed of the relative amounts of activity in many fuzzy motoneuron sets. In the arm and hand, there are 41 muscles all of which essentially partake in any arm response. Actually one must consider almost all the muscles of the body in accounting for a movement since no button pressing could occur without the maintenance of the body in an appropriate posture. And as in color vision, all these motoneurons are broadly tuned across movements, bell-shaped [14] fuzzy sets, and thus all contribute to all arm movements – which is a very large amount of information with few neural elements.

To emphasize this point, it is necessary, in this model, that these neurons do not converge to provide "crisp" representations of stimuli: an example could be a neuron specific for a button-press, or for your grandmother. Young's point was that there are not enough neurons for this kind of coding. It would destroy the economic strength of his idea if there were central neurons that specifically respond to only one wavelength. Then there would be the nightmare of having very large numbers of "color-crisp" neurons at each point in the internal representation of visual space – just what his theory was designed to avoid. And thus, all neural information, where there is any demand for large amounts of information, must remain embedded across populations of fuzzy neurons. It follows from the above discussion of the motor system that there can be no crisp "button-pressing" or "door knob turning" neurons.

Some contrary organization from simple animals may clarify this view of "defuzzification". In the crayfish, there is one "giant" axon driving the tail-flipping escape response known to anyone who startles one of these animals. This response sends the animal backwards, violently, and without aim. But there is only this one response, and the giant axon is dedicated to it. This is not a fuzzy response with gradations. There is a sensory equivalent to the giant axon response; certain receptors in the male moth respond to only one stimulus, the odorous pheromone produced by the female. This is the only function of these receptors: to receive this signal, and direct the male moth to the female. These behaviors represent an extreme in simplicity, and thus may be considered not fuzzy. These examples are useful to us in the instructive difference from other behaviors they present.

From this we conclude that as far as the vertebrate nervous system is concerned, responses are at least largely in fuzzy form. Here, we consider not only motor actions, but also all other "productions" of the nervous system, such as perceptions, memories, etc. Motor acts are special only in that they are overt.

5.3. Memory: input and retrieval

Much attention has been given to the issues of how information is placed into memory (learned), stored, and then used in behavior [e.g., 45, 46]. Something can be added from the fuzzy viewpoint to such assessments of neural function. It may be that these several crisp terms, devised for our use for "memory" (e.g., attention, stimulus, input, configuration, storage, reinforcement, response, etc.) are not exactly the functions used by the brain. We must recognize that they are *our words*, not necessarily the brain's functions. But proper classifications (whether or not they conform to our words) might be *derived* from observation of neural activity (see Sections 2.2 and 2.3). What we know of memory is simply that, after exposure to certain events, organisms give different responses than before: the same input gives a different output. Before adding further rubrics, decomposing this memory process into elements pleasing to us, we can only say that at a minimum the shapes of the NRFs in some, probably many, areas of the brain, must change to provide a changed output.

6. CONCLUSIONS

In this chapter, we argued that, considering current physiological data, neurons at all levels of the brain may be described as fuzzy sets. These sets, usually bell-shaped curves, overlap with their neighbors along their continua.

Neurons give the brain a much greater capacity to handle large amounts of information if they are considered as fuzzy sets rather than as crisp encoders of non-overlapping events. This includes the representation of continua such as the wavelength parameter for color vision, continuous variations in movement, and perhaps such complex events as the representation of faces, memories, "attention," etc. with the relatively few neurons required by fuzzy sets.

The gustatory system is taken here as a case study. In the context of a Sugeno fuzzy model we proposed that sensory inputs (tastes) and temporal dynamics of taste neurons are coded by fuzzy sets. The rules of a taste neuron thus have straightforward interpretations. Using experimental data obtained by measuring the response of some neurons to different substances, we have presented a method to synthesize the above fuzzy model. Computer simulations for 10 neurons and 12 substances demonstrated that the model is successful in modeling the time responses of taste neurons.

The apparent fuzzy nature of neural activity has implications for neural network models of the brain. For example, in the context of artificial neural networks, we showed that the units at each level would represent fuzzy sets that overlap each other along their continua. While this approach has been used to show that fuzzy rules can be implemented by suitably designed neural networks, our approach here is to examine a "standard" network to find out if its normal function can be interpreted in fuzzy terms. In agreement with others, we have shown that input layers can be seen as fuzzifiers, that fuzzifiers implement a vector base in a similarity space, and that "hidden" neurons implement fuzzy rules. In addition to this, we suggest that the activity of the brain can be interpreted in fuzzy terms.

We have attempted to provide a fuzzy set approach to understanding the brain and neural network models of its organization. This approach is consonant with, and based on, the known properties of the neural elements of the brain.

Authors' Note: Each author contributed equally to this chapter.

APPENDIX 1. Abbreviations

GoMs: Grades of Membership.
AFP: Across-Fiber Pattern Theory.
NRF: Neuron Response Function.
CNS: Central Nervous System.
NTS: Nucleus of the Solitary Tract.
MDS: Multidimensional Scaling.

APPENDIX 2. Terminology

Across-fiber Pattern (AFP). The theory that neural codes reside in the relative amount of activity across a population of neurons. This is a concrete statement of "parallel distributed patterns" or "ensemble" coding. This theory depends on fuzzy set logic for individual neurons: "fuzzy neural sets" and AFP theory are two sides of the same coin.

General Interactions. The sets in which the neurons and stimuli have GoMs. These might be reified herein as three receptor mechanisms residing on neurons. Each neuron has these to idiosyncratic degrees, and each stimulus activates these to idiosyncratic degrees. The response of each neuron to each stimulus is a simple function of their GoMs in these General Interactions.

GoMs. Grades of Membership of elements (here, neurons and stimuli) in fuzzy sets.

Neural Mass and Neural Mass Differences. Amount of neural activity summed across neurons and some chosen time interval. Neural Mass Differences computed between the activity evoked by two neural events are important in discrimination between the events, discrimination being the basis of neural codes, and thus of knowledge.

Neural Response Functions (NRFs). Identical to fuzzy membership functions. Describe the responsiveness of a neuron along its parameter. For example, visual color receptors have graded sensitivity across the wavelength parameter. NRFs are approximately Gaussian in shape, with some simple variants. They overlap with other NRFs along the same parameter. This term is included herein since it appears in the biological literature before it was recognized that the activities of individual neurons were the neural equivalent of fuzzy sets.

References

[1] Manton, K.G., Woodbury, M.A., and Tolley, H.D. Statistical Applications using Fuzzy Sets. Wiley: New York, 1994.

[2] Erickson, R. P., Rodgers, J. L., and Sarle, W. S. Statistical Analyses of Neural Organization. *J. Neurophysiol., 70*, 2289, 1993.

[3] Young, T. On the theory of light and colours. *Phil. Tans. Soc. of London*, 92, 12, 1802.

[4] Helmholtz, H.V. Helmholtz's Treatise on Physiological Optics, 3rd ed., Trans. and ed. J.P.C. Southall, Rochester, NY: Optical Soc. Of America, 1860, vol, 2, 1924.

[5] Erickson, R. P. (1963). Sensory Neural Patterns and Gustation. In Y. Zotterman (Ed.), *Olfaction and Taste*, Oxford, England: Pergamon Press, 1963, vol. 1, 205.

[6] Erickson, R. P. Stimulus Coding in Topographic and Nontopographic Afferent Modalities. *Psych. Rev.*, 75, 447, 1968.

[7] Erickson, R. P., Doetsch, G. S., and Marshall, D. A. The Gustatory Neural Response Function. *J. Gen. Physiol., 49*, 247, 1965.

[8] Fernandez, C., Goldberg, J. M., and Abend, W. K. Response to static tilts of peripheral neurons innervating otolith organs of the squirrel monkey. *J. Neurophysiol.*, 35, 978, 1972.

[9] Henry, G. H., Dreher, B., and Bishop, P. O. Orientation specificity of cells in cat striate cortex. *J. Neurophysiol.*, 37, 1394, 1974.

[10] Rose, J.E., Galambos, R. and Hughes, J. R. Microelectrode studies of the cochlear nuclei of the cat. *Bull. Johns Hopkins Hosp. 104,* 211, 1959.

[11] Marks, W. B., Dobelle, W. H., and MacNichol, E.F. Visual pigments in single primate cones. *Science, 143*, 1181, 1964.

[12] DeValois, R. L., Abramov, I., and Jacobs, G. H. Analysis of response patterns of LGN cells, *J. Opt. Soc. Am., 56*, 966, 1966.

[13] Matthews, P. C. B. Where does Sherrington's "Muscle Sense" originate? Muscles, joints, corollary discharges? *Ann. Rev. Neuroscience, 5*, 189, 1982.

[14] Georgopoulos, A.P., Kalaska, J.F., Caminiti, R., and Massey, J.T. On the relations between the direction of two-dimensional arm movements and cell discharge in primate motor cortex. *J. Neurosci.*, 2, 1527, 1982.

[15] DeSimone, R. Face-Selective Cells in the Temporal Cortex of Monkeys. *J. Cog. Neuro., 3,* 1, 1991.

[16] Erickson, R. P. Parallel "Population " Neural Coding in Feature Extraction. In Schmitt, F. O. and Worden, F. G. (Eds.), *The Neurosciences: Third Study Program.* New York: Academic Press, 393, 1974.

[17] Doetsch, G. S. and Erickson, R. P. Synaptic Processing of Taste-Quality Information in the Nucleus Tractus Solitaius of the Rat. *J. Neurophysiol., 33*, 490, 1970.

[18] Erickson, R. P. Neural coding of taste quality. *The Chemical Senses and Nutrition*, Kare, M. R. and Maller, O. (Eds.). The Johns Hopkins Press, Baltimore, 1967, Chap. 22.

[19] Smith, C. V. and Vogt, M. B. The neural code and integrative processes of taste. In *Tasting and Smelling*, Beauchamp, G. K. and Bartochuk, L. M., Academic Press, 1997, Chap. 2.

[20] Erickson, R. P. The "across-fiber pattern" theory: An organizing principle for molar neural function. In W. D. Neff (Ed.), *Contributions to Sensory Physiology,* New York: Academic Press, 1982, vol. 1, 79.

[21] Erickson, R. P. A Neural Metric. *Neuroscience and Biobehavioral Reviews,* 10, 377, 1986.

[22] Erickson, R.P., DiLorenzo, P.M., and Woodbury, M.A. Classification of Taste Responses in Brain Stem: Membership in Fuzzy Sets, *Journal of Neurophysiology, 71*, 2139, 1994.

[23] Erickson, R.P., Woodbury,M.A., and. Doetsch, G.S. Distributed Neural Coding Based on Fuzzy Logic, *Information Sciences, 95*, 103, 1996.

[24] Erickson, R.P., Schiffman, S.S., Doetsch, G.S., DiLorenzo, P.M., and Woodbury, M.A. A fuzzy set approach to the organization of the gustatory system. *Primary Sensory Neuro.*, *1*, 65, 1995.

[25] Sugeno, M. and Yasukava, T. A fuzzy-logic-based approach to qualitative modeling, *IEEE Transactions on Fuzzy Systems, 1*, 7, 1993.

[26] Widrow, B. and Stearn, S. *Theory and Applications of Adaptive Signal Processing*, Prentice-Hall, 1985.

[27] Hebb, D. O. *The Organization of Behavior: A Neuropsychological Theory.* New York: Wiley, 1949.

[28] Grossberg, S. (Ed.). *The Adaptive Brain. vol I and II.* Amsterdam: Elsevier/North-Holland, 1987.

[29] Kohonen, T. *Self-Organization and Associative Memory.* New York: Springer-Verlag, 1977.

[30] Kohonen, T. (1993). Physiological interpretation of the self-organizing map algorithm. *Neural Networks, 6*, 895, 1993.

[31] Buhusi, C.V. and Schmajuk, N.A. Attention, configuration, and hippocampal function. *Hippocampus*, 6, 221, 1996.

[32] Buhusi, C.V., Gray, J.A., and Schmajuk, N.A. The perplexing effects of hippocampal lesions on latent inhibition: A neural network solution. *Behavioral Neuroscience*, in press.

[33] Gray, J.A., Buhusi, C.V., Schmajuk, N.A. The transition from automatic to controlled processing. *Neural Networks*, 10, 1257, 1997.

[34] Rescorla, R. A. and Wagner, A. R. (1972). A theory of Pavlovian conditioning: Variations in the effectiveness of reinforcement and nonreinforcement. In A. H. Black and W. F. Prokasy (Eds.), *Classical conditioning II.* New York: Appleton-Century-Crofts, 1972, 64.

[35] Rumelhart, D.E., Hinton, G.E., and Williams, G.E. Learning internal representations by error propagation. In D.E. Rumelhart and J.L. McClelland (Eds.), *Parallel Distributed Processing: Explorations in the Microstructure of Cognition*, Vol. 1: Foundations. Cambridge, MA: Bradford Books, MIT Press, 1986.

[36] Widrow, B and Hoff, M.E. Adaptive switching circuits. *IRE WESCON Convention Record*, 96, 1960.

[37] Buhusi, C.V. and Evans, D.J. On building a Kohonen net parallel simulator. *Journal of Parallel Algorithms and Applications*, 2, 229, 1994.

[38] Sushmita, M. and Sankar, K.P. Logical operation based fuzzy MPL for classification and rule generation. *Neural Networks*, 7, 353, 1994.

[39] Yager, R.R. Modeling and formulating fuzzy knowledge bases using neural networks. *Neural Networks, 7,*1273, 1994.

[40] Gaal, G. Population coding by simultaneous activities of neurons in intrinsic coordinate systems defined by their receptive field weighting functions. *Neural Networks*, 6, 499, 1993.

[41] Zeki, S. The Visual Image in Mind and Brain. *Scientific American*, 28, 1992.

[42] Buhusi, C.V. On fuzzy image processing. *Fuzzy Systems & A.I.*, 2(2), 48. Romanian Academy Publishing House, Bucharest, 1993.

[43] Buhusi, C.V. Dynamic self-organizing fuzzy system for image processing. *Advances in Modeling & Analysis, B: Signals, Information, Data, Patterns*, 30B(1-2), 55, 1994.

[44] Buhusi, C.V. Dynamic fuzzy control. *The 2nd European Congress on Intelligent Technologies and Soft Computing EUFIT'94*, Aachen, 1408, 1994.

[45] Barnes, C. A. Involvement of LTP in memory: Are we "Searching under the streetlight?" *Neuron, 15,* 751, 1995.

[46] Larkman, A. U. and Jack, J. J. B. Synaptic plasticity. *Current Opinion in Neurobiology, 5*, 324, 1995.

Chapter 3

Brain State Identification and Forecasting of Acute Pathology Using Unsupervised Fuzzy Clustering of EEG Temporal Patterns

Amir B. Geva and Dan H. Kerem

1. INTRODUCTION

The electroencephalogram (EEG) signal, being the superficially recorded gross electrical activity of the brain, is a non-stationary, continuously fluctuating signal, characterized both by the frequency distribution of its ongoing background pattern and by the existence and form of single waves or complexes of physiological or pathological origin. Both characteristics are, on the one hand, state specific and, as such, amenable to classification by brain state but, on the other hand, possess enough variability, overlap, and vague transition to require fuzzy classification. In addition, the number of underlying semi-stationary states or processes in the continuously sampled signal is both unknown and time-varying, a fact that requires an adaptive selection of the number of classes.

Following a brief description of the basics of the EEG, as related to physiological and pathological brain states, we argue for the particular suitability of time-frequency analysis methods to process and characterize it. We then introduce an algorithm, which, in an unsupervised manner, partitions the features extracted from the signal into fuzzy clusters by the above methods. The algorithm re-projects the classification results on the sampled time-series. It overcomes the aforementioned limitations of the EEG signal by assigning successive time epochs to one or more concurrent classes (brain-states).

We finally show examples of how such classification allows brain state identification, the dynamic tracking of state transitions, the forecasting of a

0-8493-9806-1/98/$0.00+$.50

pathological state from premonitory signal patterns, and discrimination between transient, stimulus-induced pattern changes, obtained under different experimental regimes.

2. BACKGROUND

2.1 The Electroencephalogram (EEG) Signal [1], [2]

Ever since its introduction by Berger in 1929 [3], this non-invasively acquired bio-potential has remained a major indicator of the (patho)physiological state of the brain. It is obtained by recording the time series of spontaneous potential differences between any point on the scalp and a reference point removed from it (monopolar recording) or between any two points on the scalp (bipolar recording). It is conceded that the EEG is a field potential resulting from the summated activity of neuronal assemblies in the upper few millimeters of the brain's cortex, an area of high macroscopic isotropy. The major contributors are the pyramidal cells, which form a highly ordered layer with an orientation perpendicular to the surface. The potential fluctuations of the EEG signal have amplitudes of the order of 100 microvolts and durations of 0.01–2 seconds.

The unit field-potential generator is the neuron which produces both "resting," stationary and rather slow potential oscillations of the order of tens of millivolts (post-synaptic potentials) and "active," propagating brief potentials of over 100 millivolts (action potentials), across the cell's membrane. These rather high electric potentials are recorded only with intracellular electrodes. Extracellular potentials produced by the same trans-membrane ionic currents will be much smaller because the resistance of the extracellular fluid is so much smaller than that of the cell membrane. Since, like the electrocardiogram, the EEG is a record of extracellular potentials spreading in a volume conductor, EEG amplitude is correspondingly low.

The fact that the ongoing EEG signal has features at all, albeit time-variant, speaks of large populations of cortical cells either firing or changing their resting membrane polarization in the same direction, in synchrony. The signal, as recorded from the scalp, is smeared both temporally by the less than perfect synchrony, causing the EEG wavelengths to be much larger than those of the single-cell potentials, and spatially by the high resistivity of the skull, causing spatial averaging of underlying sources. The fact that the features are not just noisy but often, in a state-dependent manner, display spectral peaks of dominant frequencies or "rhythms" provides essential information. Namely, it speaks either of deeper cerebral pacers which radiate into the cortex and entrain its electrical activity [4], or else, of several intrinsic resonances within which this activity naturally reverberates, depending on the underlying state of the brain.

The basic rhythms usually do not have uniform frequency peaks, not within the same individual at different states and certainly not between subjects. Rather, their peak-by-peak frequency fluctuates between limits, which define a frequency band. These bands, listed in Table 1, actually fill out the entire continuum of the observed

EEG frequencies. Slow rhythms usually involve a higher degree of synchrony and therefore contain waves, which are larger in amplitude, as indicated in the table. In addition to the rhythms, certain universally recurring waveforms or double-wave complexes are described (Table 2). These may appear as single events embedded in the ongoing signal or in a train, which takes over the entire time series. The origin of these waveforms is believed to be sub-cortical. The exact mechanism of their formation is still unclear.

2.2. Brain States and the EEG

The major physiological brain states that have distinct EEG and behavioral correlates are concerned with the degree of vigilance. They range from extreme alertness to very deep sleep or unconsciousness (the latter may be considered as an insult-induced defensive state). The EEG states may be defined by means of a predominant frequency band, a specific composition of bands, the presence of distinct waveforms, or combinations thereof. Some totally different states could present apparently similar EEG patterns. They may be told apart by combining the EEG information with simultaneously occurring muscle-generated potentials (electromyograms or EMG), mainly of postural muscles in the neck and the muscles moving the eye balls (electrooculograms or EOG). The different states or "sleep stages," based on the Rechtschaffen Kales or R-K formulation [6], are listed in Table 3. A portion of a night's sleep hypnogram, which is the graphic partitioning of sleeping time into sleep stages, is shown in Figure 1.

The most remarkable of the pathological brain states are the epilepsies. Human epilepsy is for the most part an intrinsic brain pathology. Its major manifestation is the epileptic seizure, which may involve a discrete part of the brain (partial) or the whole cerebral mass (generalized). In the latter instance, seizures are recurrent, with between-seizure periods ranging from several minutes to many days. The behavioral facet of the seizure affects the body as a whole, typically with a strong motor involvement in the form of convulsions. Seizure EEG is characterized by repetitive high amplitude activity (paroxysmal), either fast (spikes), slow (waves), or spike-and-wave (SPW) complexes. This activity varies, depending on the type of epilepsy. It may take the form of two-60- second periods of very regular and symmetric 3 Hz SPW discharges in absence or *petit mal* epilepsy. The tonic-clonic or *grand mal* epilepsy, has 40-60 second periods with fast polyspike activity, gradually decreasing in frequency and increasing in amplitude (tonic or rigid phase) interrupted by slow waves (clonic or jerky phase) and followed by post-seizure EEG depression to the point of an almost flat signal (flaccid state). Other less regular patterns occur in myoclonic, clonic, tonic, atonic, and atypical absence epilepsies [2], [7].

Table 1
Frequency Bands of the EEG

BAND	FREQUENCY RANGE (Hz)	PROPERTIES
Delta	0.5 - 3.5	High Amplitude
Theta	4.0 - 7.0	High Amplitude
Alpha	7.5 - 12.0	Peak in Frequency Range
Beta	15.5 - 30.0	Low amplitude

Table 2
Single Waves and Complexes in the EEG

EVENT	AMPLITUDE	DURATION	CONTEXT	OTHER PROPERTIES
Spindle	up to 150 μV	50-100 msec	Sleep	monophasic, >0.5 sec waxing and waning bursts
K-complex	up to 250 μV	> 500 msec	Sleep	bi- or polyphasic, spontaneous or evoked
Spike	up to 500 μV	20-80 msec	Epilepsy	sharp, mono- or biphasic, single or burst (polyspike)
Wave	up to 250 μV	150-300 msec	Epilepsy	monophasic
Spike and Wave (SPW)	as above	170-380 msec	Epilepsy	complex of adjacent spike and wave, in that order

Table 3
Conventional Scoring Of Sleep Stages

STAGE	EEG			EOG	EMG
	DOMINANT BAND	SPINDLES	K-Com.[b]		
Awake	Beta (alert); Alpha (relaxed, eyes closed)	–	–	High, sharp, and irregular	High, often in bursts
1	Low amplitude Theta, some Alpha	–	–	Low, smooth	Intermediary
2	Intermediate amplitude, Theta, rare Alpha	+	+	Low	Low
3-4	High amplitude Delta waves	(+)	–	Very low	Very low
REM[a]	Desynchronized, Theta, some lower frequency Alpha	–	–	High, rhythmic, rapid bursts	Very low

[a] Rapid eye movement (“dream sleep”). [b] K-Complex.

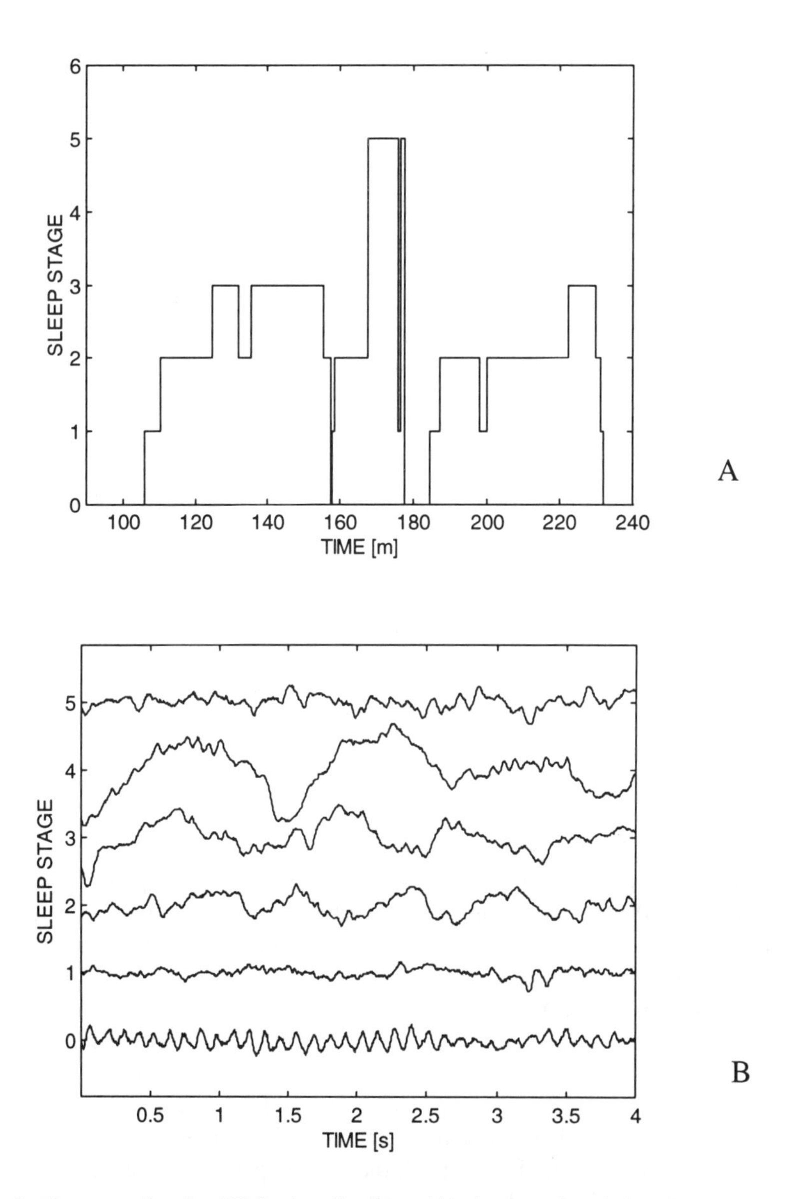

Figure 1. Sleep staging by EEG visualization. (**A**) A part of a night's hypnogram of a 45-year- old male patient in a sleep laboratory, as visually staged by a human expert. Shown are two sleep cycles with evolving stages. 0 - Awake state, 1- Sleep stage 1, 2 - Sleep stage 2, 3 - Sleep stages 3 and 4 (often lumped together), 5 - Rapid eye movement (REM) Sleep stage. Shown are two sleep cycles which, unlike normal sleep, are bracketed by periods of wakefulness. The first cycle is also interrupted by a brief awakening. (**B**). Examples of 4-second EEG segments picked from each of the stages. Note close similarity of stage 1 and REM sleep EEG, the latter distinguishable by the accompanying rhythmic eye rolling (see also Table 3).

The EEG in the between-seizure periods ranges from normal, through a normal background with riding isolated epileptic activity in the form of single events or brief bursts, to an abnormal, usually slow, background with or without riding isolated epileptic activity. Other than frank paroxysmal activity such as spikes or SPW complexes, the riding short transients may include abnormal rhythms, sub-clinical discharge patterns, and non-paroxysmal complexes [8]. Allusion to a pre-seizure period is sometimes made, but usually as an afterthought. While abnormal EEG patterns occasionally abound just before the seizure, more often the latter strikes unheralded with a dramatic sharp transition of the EEG signal from its ongoing background pattern into the characteristic impulse-train of spikes or SPW complexes, coincidental with the motor involvement (Figure 2).

2.3 Stimulus-Evoked EEG Patterns

So far, we have described the continuous ongoing electrical activity generated by the cortical neuronal mass as it is being bombarded by the grand sum of all internal and external inputs. Presenting the brain with a single short stimulus of one of the sensory modalities will produce, after a given delay, a related response which will stand out from the spontaneous background activity. The response may resemble a spontaneous single event, but, averaging out, a train of stimulus-triggered EEG response segments will flatten the quasi-random background and convince the observer of the presence of an "evoked potential" (EP) of constant and distinct features (Figure 3).

It is customary to subdivide evoked potentials into early, endogenic components, and organ and late or exogenic components. The early ones have latencies of < 200 ms from the stimulus trigger, which subconsciously arise within and along the sensory pathway of the stimulated sense organ. The late or exogenic components have latencies > 200 ms, which are purely cortical in origin and reflect higher cognitive functions, directly or indirectly related to the stimulus. The early components are the only ones observed during a perceptual reaction to a neutral stimulus.

In situations or actual tasks involving stimulus evaluation (as, for example, image projections with occasional, random inclusion of subject-meaningful pictures or making decisions about items stored in the short-term memory, in response to a test stimulus or "probe"), the responses, often termed Event Related Potentials (ERP), will contain both early and late components as exemplified in Figure 3.

Unlike most of the EEG signal, the earliest part of the EP reflects the electrical activity of rather deep, sub-cortical cell assemblies. The reasons that a surface scalp electrode often succeeds in faithfully tracing the instantaneous electrical activity traveling in deep structures *en route* to the cortex are twofold. First, the course includes structures having a very packed and oriented geometry as well as a high temporal synchronization (nerves, nuclei, and tracts), which generate substantial electromagnetic fields. Second, the spread of the electromagnetic waves through the volume conductor is much faster than the propagation of the neural activity. As the stimulus activates more than one-cell assembly, i.e., paired sense organs, the surface EP is a superposition of several co-active sources.

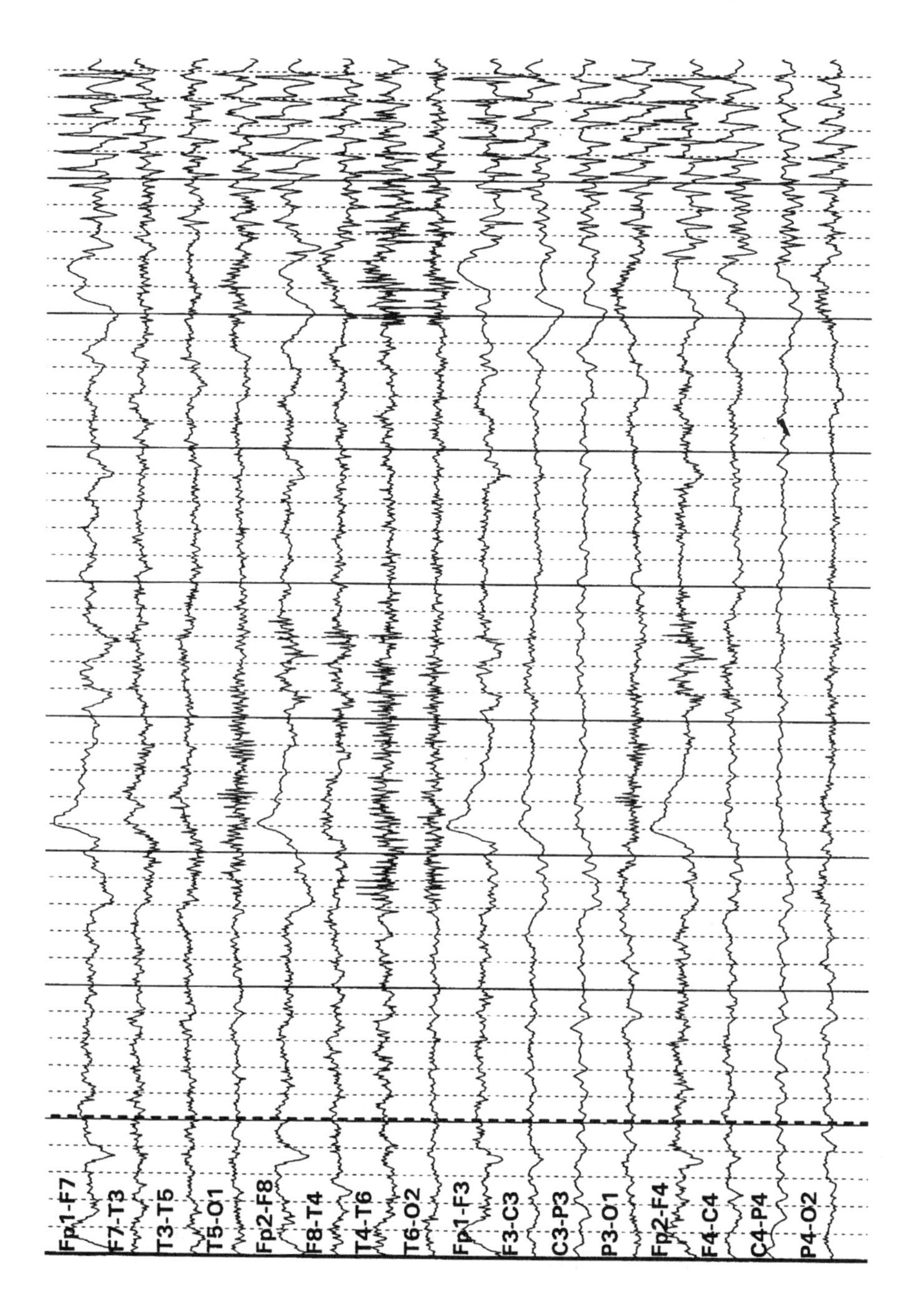

Figure 2. EEG in epilepsy. A 9 second, 16 bipolar channel EEG recording of an epileptic patient. Pre-epileptic, sharp polyspike activity is evident in most but not all scalp site combinations, during seconds 3-5. Generalized epileptic activity emerges at second 8 and develops into a full-blown, 7/sec SPW seizure. Font size of lettering at left is equivalent to 40 μV.

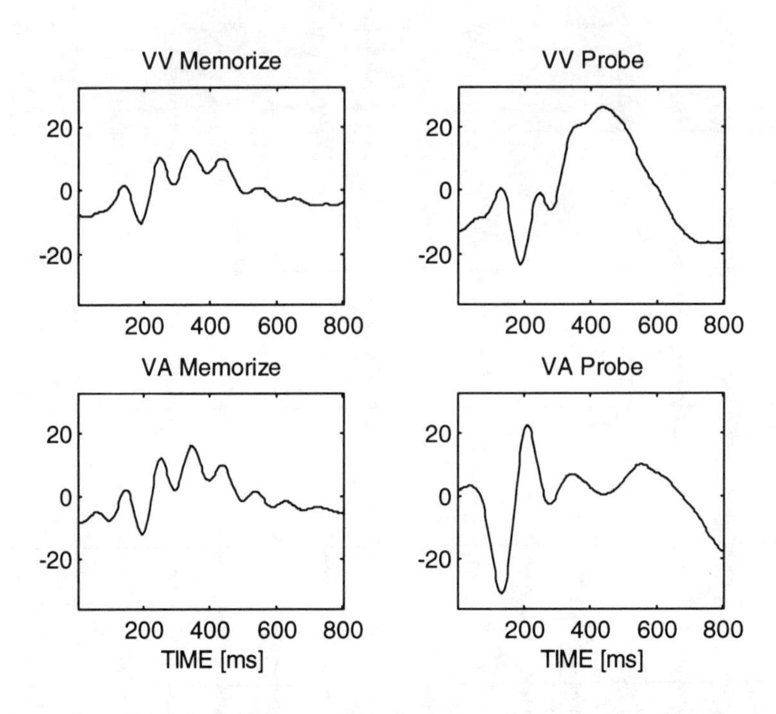

Figure 3. The event related potential waveform. This is a composite graph, which shows results from a short-term memory testing experiment, detailed in Section 3.3. Each panel depicts a grand-average ERP, obtained from the same scalp recording site, in 24 subjects. The Y axes are amplitude in microvolts and the X axes show milliseconds from the stimulus trigger. The two left-hand panels are responses to visual stimuli (V), which the subject is asked to memorize. The two right-hand panels are either visual (upper) or auditory (A) "probe" stimuli, which the subject has to match to the memorized ones. Upper deflections are termed positive (P) and lower, negative (N) and are numbered consecutively. The early (up to 250 ms) part of the waveform is essentially an event-unrelated EP, comprised of a P_1 - N_1 - P_2 complex in the visual responses and a N_1 - P_1 in the auditory. The latter, event related, part is much more prominent in the probe responses and different in the two modalities.

Figure 3 also shows the EP complex for a given stimulus, as recorded from a given site, to be rather uniform. The exact shape will be subject and state-specific and will vary with any pathology along the course of its spread. Until recently [9], no attempt was made to quantify and categorize evoked potentials by their detailed pattern. Rather, each complex was dissected according to its individual peaks and troughs which, numbered consecutively, are characterized by their latencies and by their relative amplitude.

2.4 Underlying Processes

The main theme of electroencephalographic research has been the effort to extract information from the EEG signal, pertinent to the dynamics of psycho-physiological as well as pathological brain functions. Modern imaging methods, such as functional Magnetic Resonance Imaging (fMRI) or Positron Emission Tomography (PET), can respectively portray the metabolic or the hemodynamic correlates of neural activity in the brain as a whole. The reason that EEG recording still resists competition with these methods is that it is a direct measurement of the electrical activity and, as such, provides 4-5 orders of magnitude higher temporal resolution.

Classical neurophysiology describes the activity of single neurons and predicts the response of simple cell interactions with rigorous mathematics and stochastic statistics, respectively. Ascending into larger neuronal assemblies, the basic notion of the brain being a master controller promoted their early description in terms of machine control theory as an aggregation of feedback loops, set to minimize perturbations and maintain stability. More recently, it was recognized that the necessary abilities to anticipate, adapt to, and incorporate the consequences of totally new external conditions, are inherent features of the central nervous system not readily explained by deterministic feedback-control systems. Physical models of turbulence, the mathematical approximations of nonlinear dynamics and low-dimensional chaos, as well as the flexible behavior of artificial neuronal networks, have all since found ever-growing footholds in descriptions of the gross pathophysiological activity of the brain, as evident from the EEG [10].

Extreme examples of seemingly chaotic behavior of physiological neuronal nets are the sudden unpredictable onsets of a generalized epileptic seizure or of a full-blown paranoid delusion. Such rapid state changes or "bifurcations" are characteristic occurrences in chaotic systems in which widespread changes, in this case an instantaneous lowering of the excitatory thresholds of vast neuronal populations, may be caused by subtle changes in initial conditions. Evidence in support of the contention that brain networks behave as nonlinear systems with chaotic dynamics during epileptic seizures is steadily accumulating [11-13]. Non-linear dynamics theory may help to determine the likelihood of bifurcations under a given set of conditions, but fails to forecast every outburst [12].

2.5 FUZZY SYSTEMS AND THE EEG

The last few decades have witnessed a massive invasion of signal-processing expertise into the field of neurology. While the standard reference of EEG interpretation and categorization is still the eye of the human expert, a lot of effort has been invested in automated EEG-interpreting systems. Most are not meant to better the expert human eye, but rather to match and replace it, especially when dealing with large accumulated data bases, or with the need to process on-line the output of up to 128 electrodes spread over the scalp. These endeavors have three main interrelated aspects: the EEG as a state-related time series, the EEG as a spatio-temporal signal which contains information on the locale of deep generators, and the EEG as a mirror of global mechanisms of large neuronal networks.

In this chapter, we will concentrate on the first aspect of time series analysis. Emphasis may be placed either on the ongoing signal or on transients such as epileptic events and evoked responses. In the former case, the interest lies in assigning quasi-stationary sections of the signal to specific brain states and, in the latter, in obtaining information on existing or impending pathology from the mere appearance of, or from, the change in form of the single events.

Categorizing brain states from the EEG is not always a straightforward procedure. Taking sleep stages as an example, the typical sequence as outlined in Figure 1 may never be expressed in (a) given subject(s) during (a) given episode(s). Physiologically or pathologically aberrant sequences, lacking in stages and/or differing in stage-signal characteristics, are very common. They may be subject-specific or state-specific in a given subject. Transitions between stages may be very gradual, making it difficult to pinpoint their occurrence. A fuzzy environment which includes stages, the number and exact properties of which are *a priori* unknown, with parts of the signal unequivocally belonging to one distinct stage and other parts to two or more, is a very realistic representation of an EEG sleep episode. Fuzzy rules, such as the ones inherent in the initial definitions (Table 3) may still serve to accommodate sleep episodes within the conventional templates. Categorizing short single events, such as EPs, seems easier, but is not so in practice. In EP studies, the waveforms are affected by a variety of experimental factors. Consequently, while specific changes in waveform within a given experimental setting may be followed and quantified, the considerable variability which exists in a large pool of normal waveforms (see Figure 7) will confound the identification of abnormal forms.

When the professed aim of EEG interpretation is brain state identification, what is needed is an automated categorization device, and the natural choice in a fuzzy environment is a procedure termed fuzzy clustering [14]. This procedure is applied on an assembly of points scattered in an N-dimensional space, each dimension representing a common feature, which are believed to be classifiable into groups. Attempting to classify the points into discrete, non-overlapping categories will leave much of the content unclassified or classified by default. The method of *fuzzy* clustering overcomes this by allowing items to share membership in more than one category. By some optimization criteria, one or more group or cluster centers are located and each point is assigned a degree of membership in each of these clusters. The procedure may be supervised in the sense of forcing the number of clusters, and/or the size, form, or even location of any of them, or else it could be made to run without preliminary assumptions.

In the case of spontaneous EEG, the ongoing time series is broken up into segments which are long enough to convey even the slowest rhythms and short enough so as not to dilute the contribution of single events which are deemed important. Next, one of several possible feature-extracting methods is applied to each of the segments and the values of a selected list of resulting parameters are assigned to each segment. If the values of a properly chosen combination of parameters, so produced over time, are pooled and fed to the fuzzy clustering procedure, they would be expected to be naturally classifiable into fuzzy clusters representing EEG states. Then, if the results of the procedure (degree of membership in each cluster/state of each segment) are reproduced as a time-series, EEG states will stand out as strings of segments sharing a common cluster list and/or a similar partitioning of membership in various clusters. The evolution of states, the gradual (or sharp) transitions between states and the emergence of abnormal states, may then all be followed.

Another obvious use for EEG automated processing systems is as warning devices. An automated early warning of an impending epileptic seizure may be important to some in-patients, and especially to out-patients, who may take action (medication) to avert it or lower its potential risks (halt driving). In some patients, the obvious transient pathological activity heralding the seizure, could, in principle, be taught to automated pattern-recognition devices (neuronal nets), but, in practice, a high patient-specificity of such patterns precludes a universal system. In many other instances, even the expert eye fails to notice specific changes in the minutes preceding the seizure, the pre-seizure state or PSS. Yet, if we believe that the cortex is somehow "primed" toward the outburst of the full-blown seizure, the PSS may be defined by some variation in the apparently normal EEG pattern. Allowing an unsupervised process to perform a fuzzy classification of selected features of the signal may succeed in such a definition.

In this chapter, we will present three applications of unsupervised fuzzy clustering, in conjunction with various feature-extraction methods, to the EEG signal. The first is aimed at an automated scoring of sleep stages which will match that of a human expert, while the second prepares the background for an automated forecaster of generalized epileptic seizures which will match or improve on the human expert. The last one is a means of categorizing evoked potentials. It may have a time-variant aspect as an additional input to the epilepsy forecaster or it may be used on an existing signal pool, gathered from one or more subjects.

Feature-extracting methods may be tailored to the case in question. The rather stereotyped, single-event evoked potential is described using Eigenvector Principle Component Analysis (PCA); sleep EEG and pre-seizure EEG are either segmented and described by Linear Prediction, or dealt with using the Fast Wavelet Transform. The latter two methods are preferred over spectrum estimation methods such as the Fast Fourier Transform because of the inherent non-stationarity of the EEG signal. The resultant set of feature vectors derived for each consecutive segment or event, are then fed to an Unsupervised Optimal Fuzzy Clustering algorithm, to be described next.

3. TOOLS

The principal tool to be presented and described is the weighted version of the Unsupervised Optimal Fuzzy Clustering (UOFC) [15], [16] and its application to temporal pattern classification and identification. A general block diagram of the procedure is presented in Figure 4. Candidates for its application are EEG signal sections, which either originated from a single segmented time series or, in the case of evoked responses, are an ensemble obtained from multiple recordings from the same subject under different conditions or from many subjects. Prior to the UOFC application, section features are extracted, and some of them are selected for the classification. The output of the process allows, in the first instance, state identification that could lead to the forecasting of pathology and, in the second instance, to normal or to diagnostic labeling (see Figure 4). Although strict mathematical rules may govern the feature extraction and selection stages, there are stages that could best benefit from expert experience and are the ones to be modified by trial and error. We shall first outline the general procedure of EEG and EP temporal pattern recognition and classification by fuzzy clustering, and then go on to the three examples of its use.

3.1. Data Acquisition

3.1.1 Spontaneous Ongoing Signal

In general, one or more tape-recorded channels of suitably filtered and amplified EEG signal, extending over the period of interest, are digitized at a sampling frequency f to form the database. The T second long and L $(= T\times f)$ sample long digitized signal $s(n)$, $n = 1,\ldots,L$, , is split into Δt second long overlapping segments or epochs which are arranged as the columns or pattern vectors of an $N \times M$ matrix **S**:

$$\mathbf{S} = \begin{bmatrix} s(1) & s(D+1) & \cdot & \cdot & s((M-1)\cdot D+1) \\ s(2) & s(D+2) & \cdot & \cdot & s((M-1)\cdot D+2) \\ \cdot & \cdot & \cdot & & \cdot \\ \cdot & \cdot & \cdot & & \cdot \\ s(N) & s(D+N) & \cdot & \cdot & s((M-1)\cdot D+N) \end{bmatrix}$$

where N $(= \Delta t \cdot f)$ is the dimension of each pattern vector, M $(= \lfloor (L+D-N)/D \rfloor)$ is the number of patterns, and D is the displacement (in sample points) between patterns.

3.1.2 Evoked Responses

When dealing with EPs, the digitized data base will be an ensemble of isolated, equal sized and synchronized signal sections containing the EP, of size Δt long enough to bracket the latest component of interest. The included EPs can be responses from the same subject recorded under different conditions, from different scalp sites, or from different subjects under similar conditions, recorded from the same site, or any combination of the above. The $N \times M$ data matrix **S** will now have M columns or pattern vectors according to the number of events in the ensemble, each column having the dimension $(\Delta t \cdot f =)$ N.

3.2 Feature Extraction

The feature extraction procedures outlined in the next sections are applied to the columns of **S** (which contain either consecutive ongoing EEG sections or isolated EPs) to produce the column vectors of an $F \times M$ feature matrix **X**. Here, F denotes the number of extracted features, to which the UOFC algorithm is applied.

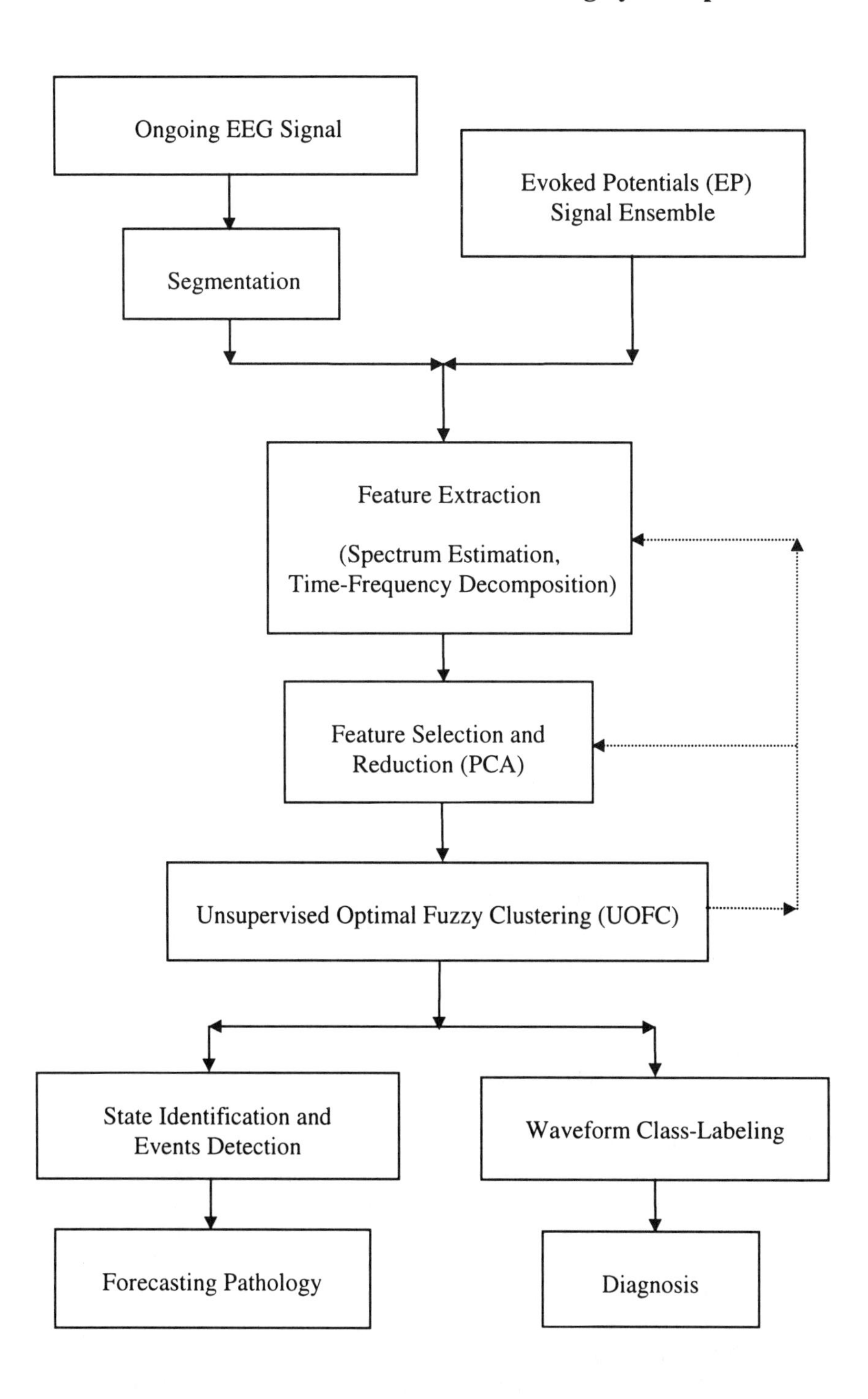

Figure 4. UOFC as the heart of EEG state and event classification.

An exception to the above would be an EEG data base which is presented to a feature extracting procedure such as adaptive segmentation, which, by its nature, segments the signal into M epochs of unequal length. In this case, the **S** matrix step is skipped and the feature extraction procedure will directly produce **X**, with an equal number of features F for the unequally long segments.

3.2.1 Spectrum Estimation

The power spectrum of each of the columns can be estimated by any of the spectrum estimation methods (such as short-time FFT, AR, Eigenvectors, or high-order spectrum analysis) [17], which should be chosen by the signal characteristics and the specific categorization task. Spectrum-derived parameters, such as the relative power content of the different EEG frequency bands, could then be used to construct the columns of the $F \times M$ features matrix **X** for feeding the UOFC algorithm. On the whole, while the power spectrum is a direct, robust, and phase-invariant means to describe the structure of the ongoing signal, the requirement for signal stationarity makes it insensitive to transients and single events.

3.2.2 Time-Frequency Analysis

Two rather new methods, which rapidly became established as important tools in signal analysis and which are better suited to dealing with non-stationary signals and with signals having identifying features that include isolated single events, are the wavelet transform and spatio-temporal model-based decomposition by matching pursuit.

3.2.2.1. Multiscale Decomposition By The Fast Wavelet Transform

The wavelet transform produces a good local representation of the signal in both the time domain and the frequency domain. Unlike the Fourier transform, which is global and provides a description of the overall regularity of signals, the wavelet transform looks for the spatial distribution of singularities [18-20]. This is essentially a band-pass filtering process, which passes the signal simultaneously through several filters whose characteristics overlap on the frequency axis.

This transform particularly suits the EEG signal which, at any instant, is a mixture of (usually up to 5) discrete "rhythms" (Table 1 and Figure 1), with occasional embedded single events (Table 2) which can be viewed as being composed of elements of these same rhythms. The simultaneous multi-convolution of the wavelet transform ensures that in any given epoch, single events will be captured near the central frequency of one or more band-pass filters. Moreover, it ensures that any temporal spectral variation (or switch of dominant rhythm) will differently influence the product of the several filters on consecutive time segments and will activate changes in the feature space [21]. For the EEG signal, with a suitable mother-wavelet, the first five to seven scales (depending on the sampling frequency f) of the wavelet coefficients obtained by the fast wavelet transform are sufficient to capture information within the frequency range of interest [21].

In practice, each column of the time series matrix, **S**, is decomposed into waveforms that are the dilations, translations, and modulations of the single window wavelet. The wavelet transform is computed by convolving the signal, s, with these dilated wavelets (see Appendix 1). Information about the signal is then contained in the wavelet coefficients, which are computed for each time point. Feature reduction is achieved by computing, for each scale, the statistical values of the moments of the variance (energy), skewness, and kurtosis of the wavelet coefficients. Also computed are the correlations between the wavelet coefficients of all the scales. In addition to the above statistics, the number of extrema (zero crossings) per unit time in the transform, which contain other, nonlinear, information about the signal, may be obtained and added to the list of possible inputs to the clustering.

3.2.2.2. Multichannel Model-Based Decomposition by Matching Pursuit

Mallat and Zhang [22] have introduced an algorithm, called matching pursuit, that decomposes any signal into a linear expansion of waveforms that are selected from a redundant dictionary of functions. These waveforms are chosen to match the signal structures. The matching pursuit algorithm can sometimes better isolate specific signal structures, which are not necessarily coherent with the wavelet form. At each iteration of the algorithm, a waveform that is best adapted to an approximate part of the signal is chosen. If a signal structure does not correlate well with any particular dictionary element, namely, a noise component, it is sub-decomposed into several elements and its information is diluted. Although matching pursuit is a nonlinear procedure, it does maintain an energy conservation, which guarantees its convergence [22].

This algorithm has been generalized into spatio-temporal matching pursuit (SToMP), and adapted to multiple source estimation of EEG complexes such as EPs, which are known to be summations of simultaneous electrical activities of deeper generators [23]. By using a physiologically motivated time-frequency dictionary of waveforms, the number and the temporal activity pattern of the signal generators may be estimated. An adaptation of the SToMP algorithm can be directly applied to the matrix **S**, and the F waveform coefficients of each column may then be arranged in an $F \times M$ matrix **X**, which is the subject of classification, with or without secondary feature extraction (see Appendix 2).

3.3 The Unsupervised Optimal Fuzzy Clustering (UOFC) Algorithm.

The **X** features matrix is now subjected to a clustering procedure, which is utilized to find matching groups within. The matrix can either be directly used as the input or undergo feature reduction. A basic method for feature reduction is the Karhunen-Lo'eve transform (KLT) which is also commonly referred to as the eigenvector, principal component (PCA), or Hotteling transform. The reconstruction error is minimized by selecting the eigenvectors associated with the largest eigenvalues (see Appendix 3). Thus, the KLT is optimal in the least-square-error sense, or, from the point of view of information theory, the KLT achieves the lowest overall distortion of any orthogonal linear transform for a fixed number of coefficients [24]. The KLT can follow any specific appropriate feature extraction method (e.g., spectrum or wavelet analysis) or it can be applied directly to the **S** matrix as a feature extraction procedure [25].

The clustering procedure of choice utilizes the unsupervised optimal fuzzy clustering (UOFC) hybrid algorithm introduced by Gath and Geva [15], [16], which partitions the data by a combination of the modified fuzzy k-means (FKM) algorithm [26] and the fuzzy maximum-likelihood estimation (fMLE) algorithm [15], [16],[26]. The superiority of the fuzzy k-means algorithm with respect to speed of convergence is realized by having *it* compute the cluster centers. These are then fed as initial values for the MLE algorithm, which better deals with unequal cluster features. The advantage of the UOFC algorithm is the unsupervised initialization of cluster prototypes, and the criteria for cluster validity using fuzzy hypervolume and density functions. It performs well in a situation of large variability of cluster shapes, densities, and number of data points in each cluster.

In its weighted version, the two inputs to the UOFC algorithm are a data matrix, **X**, and a column vector, **w,** comprised of the weights of each column-pattern in the matrix **X.** No other parameters are given to the procedure. The final outputs are K fuzzy clusters (where K is estimated by the algorithm). The general scheme of the weighted version of the UOFC algorithm (WUOFC) is iterated for an increasing number of clusters in the data set, calculating a new partition of the data set, and computing performance measures in each run, until the optimal number of clusters is obtained:

U.1. Choose a single initial centroid at the weighted (by **w**) mean location of all data patterns.
U.2. Calculate a new partition of the data set by two phases (see Section 3.4):
U.2.1. Cluster with weighted fuzzy K-means with Euclidean distance function.
U.2.2. Cluster with weighted fuzzy K-means with an exponential distance function; (this is a fuzzy modification of the maximum likelihood estimation, MLE).
U.3. Calculate performance measures for cluster validity.
U.4. Add another centroid equally distant (with a large number of standard deviations) from all data points (see step **F.2** in the modified fuzzy K-means algorithm presented in Section 3.4).
U.5. If the number of clusters is smaller than the maximal feasible number of clusters, go to **U.2**.
U.6. Choose the optimal partition by the performance measure criteria.

The number of clusters may or may not be forced on the clustering procedure but other than that, no *a priori* assumptions about the characteristic features of the clusters, including the approximate locations of the group means, are made. Instead, included in the algorithm is a procedure for unsupervised tracking of the initial cluster centers. The basic principle, given a k-partition, is to position the $(k+1)$-th cluster-center in a region where many data points exhibit a low degree of membership in the present k clusters. This is done by placing the new center equally far away (by a given large number of standard deviations) from all data points for the first iterative step. This ensures that during the following steps, the center will indeed converge toward the desired data points rather than to points that happen to reside near its initial placement.

3.4 The Weighted Fuzzy K-Mean (WFKM) Algorithm

The weighted version of the fuzzy *K*-mean algorithm, which is used in stages **U.2.1** and **U.2.2** of the WUOFC algorithm, is derived from the minimization with respect to **P**, a set of cluster centers, and **U**, a membership matrix, of a weighted fuzzy version of the least-squares function:

$$J_q(\mathbf{U},\mathbf{P})=\sum_{i=1}^{M}\sum_{k=1}^{K} w_i \cdot u_{k,i}^{q} \cdot \mathrm{d}^2(\mathbf{p}_k,\mathbf{x}_i) \tag{1}$$

where $\mathbf{x}_i$ is the *i*-th pattern, the *i*-th column in the **X** data matrix, $\mathbf{p}_k$ is the center of the *k*-th cluster, $u_{k,i}$ is the degree of membership of the data pattern $\mathbf{x}_i$ in the *k*-th cluster, w_i is the weight of the *i*-th pattern (as if w_i patterns which are equal to $\mathbf{x}_i$ were included in the data matrix **X**), $d^2(\mathbf{p}_k,\mathbf{x}_i)$ is the square of the distance between $\mathbf{x}_i$ and $\mathbf{p}_k$, M is the number of data patterns, and K is the number of clusters in the partition. The parameter q (commonly set to 2) is the weighting exponent for $u_{k,i}$ and q controls the "fuzziness" of the resulting clusters [26]. The weighted fuzzy *K*-mean clustering algorithm with the modified centroids initialization [15], [16] includes the following steps:

F.1. Use the final centroids (prototypes) of the previous partition as the initial centroids for the current partition: in stage **U.2.1** of the WUOFC algorithm use the *K*-1 (*) final centroids of its previous stage and for phase **U.2.2** use all the *K* final centroids $\mathbf{p}_k$, $k = 1,..,K$, of stage **U.2.1.**

F.2. Calculate the degree of membership $u_{k,i}$ of all data patterns in all clusters by:

$$u_{k,i}=\mathrm{d}^2(\mathbf{x}_i,\mathbf{p}_k)^{1/(1-q)} \Big/ \sum_{j=1}^{K}\left[\mathrm{d}^2(\mathbf{x}_i,\mathbf{p}_j)\right]^{1/(1-q)} \quad , k = 1,..,K,\ i = 1,...,M(*) \tag{2}$$

(*) Only for $k = K$ and in the first iteration of the stage **U.2.1** of the UOFC algorithm use the following distance:

$$d^2(\mathbf{x}_i,\mathbf{P}_k) = 10\cdot\mathrm{Sum}(\ \mathrm{Diagonal}\ (\ \mathrm{Covariance}\ (\mathbf{X})\ )\ ),\ i = 1,...,M$$

Otherwise, use the Euclidean distance (Eq. 4) in stage **U.2.1** or the exponential distance (Eq. 5) in stage **U.2.2** of the WUOFC algorithm.

F.3. Calculate the new set of cluster centers from

$$\mathbf{p}_k=\sum_{i=1}^{M} u_{k,i}^{q}\cdot w_i\cdot \mathbf{x}_i \Big/ \sum_{i=1}^{M} u_{k,i}^{q}\cdot w_i \quad , k = 1,...,K. \tag{3}$$

F.4. If $\max_{k,i}\left[\left|u_{k,i}-\left(\textit{previous } u_{k,i}\right)\right|\right] > \varepsilon$, go to step **F.2.**

In the first phase **U.2.1** of the WUOFC algorithm, the fuzzy weighted *K*-mean algorithm is performed with the Euclidean distance function

$$d^2(\mathbf{p}_k, \mathbf{x}_i) = \left[(\mathbf{p}_k - \mathbf{x}_i)^T \cdot (\mathbf{p}_k - \mathbf{x}_i)\right] \tag{4}$$

The final cluster centers of the first phase **U.2.1** are used as the initial centroids for the second phase. In the second phase **U.2.2**, a fuzzy modification of the maximum likelihood estimation is utilized by using the following exponential distance function in the weighted fuzzy *K*-mean algorithm:

$$d^2(\mathbf{p}_k, \mathbf{x}_i) = \frac{\left[\det(\mathbf{F}_k)\right]^{1/2}}{a_k} \cdot \exp\left[(\mathbf{p}_k - \mathbf{x}_i)^T \cdot \mathbf{F}_k^{-1} \cdot (\mathbf{p}_k - \mathbf{x}_i)\right] \tag{5}$$

where

$$a_k = \sum_{i=1}^{M} u_{k,i} \Big/ \sum_{i=1}^{M} w_i$$

is the sum of membership within the *k*-th cluster, which consists of the *a priori* probability of selecting the *k*-th cluster and

$$\mathbf{F}_k = \sum_{i=1}^{M} u_{k,i} \cdot w_i \cdot (\mathbf{p}_k - \mathbf{x}_i) \cdot (\mathbf{p}_k - \mathbf{x}_i)^T \Big/ \sum_{i=1}^{M} u_{k,i} \cdot w_i \tag{6}$$

is the fuzzy covariance matrix of the *k*-th cluster.

By applying these two phases, the robustness of the fuzzy *K*-mean algorithm with the Euclidean distance function is used to find a feasible initial partition, and the adaptiveness of the fuzzy modification of the maximum likelihood estimation is utilized to refine the partition for normally distributed clusters with large variability of the covariance matrix (shape, size, and density) and the number of patterns in each cluster. Note that other distance functions can be used according to the intrinsic characteristics of the data.

3.5 The Clustering Validity Criteria

In the next step **U.3** of the WUOFC algorithm the following criteria for cluster validity are calculated [15], [16]:

C.1. The fuzzy hypervolume criterion (HPV):

$$V_{HV}(K) = \sum_{k=1}^{K} h_k \tag{7}$$

where the hypervolume of the k-th cluster is defined by $h_k = \left[\det(\mathbf{F}_k)\right]^{1/2}$.

C.2. The partition density (PD):

$$\mathrm{v}_{\mathrm{PD}}(K)=\sum_{k=1}^{K} b_k \Big/ \sum_{k=1}^{K} h_k \cdot \tag{8}$$

C.3. The average partition density (APD):

$$\mathrm{v}_{\mathrm{AD}}(K)=\frac{1}{k}\sum_{k=1}^{K}\left[b_k/h_k\right], \tag{9}$$

where

$$b_k=\sum_{i\in \mathbf{I}_k} u_{k,j}\cdot w_j$$

and I_k is a set of indexes of the "central members" in the k-th set:

$$\mathbf{I}_k=\left\{i\left(\mathbf{p}_k-\mathbf{x}_i\right)^{\mathrm{T}}\cdot\mathbf{F}_k^{-1}\cdot\left(\mathbf{p}_k-\mathbf{x}_i\right)<1,\quad i=1,\ldots,M\right\}\cdot$$

The choice of the criterion or combination of criteria to be the measure of performance is driven by the specific distribution of the signal.

4. Examples Of Uses

4.1 Sleep-Stage Scoring

Data for this example is obtained from a database of digitized EEG-monitored sleep sessions of subjects in a sleep laboratory (Polysomnographic Database, Biomedical Engineering Center of the Harvard-MIT Division of Health Sciences and Technology, Cambridge, Ma, U.S.A.). While the polysomnograms, or multi-signal sleep session records, usually include EOG and chin muscle EMG, in addition to an EEG channel, only the latter is used for classification. All signals in the database are digitized at a rate of 250 Hz. The polysomnographic record is annotated with respect to sleep stages in sections of 30 s and was used to construct Figure 1.

Features of the EEG signal may be extracted by any of the several methods mentioned above. In this example, we present results obtained by using wavelet transform decomposition on 4s non-overlapping epochs. Thus, for the **S** matrix, N = 1000, $D = N$ = 1000 and M = 900 for each hour analyzed. The fast wavelet transform as proposed by Mallat and Zhong [20] and detailed in Appendix 1 is applied. Six wavelet scales, obtained by dilating the mother wavelet by $a = 2^j$, j = (1,2,3,4,5,6), are used. The variance (energy), skewness, kurtosis and number of maxima were computed for each scale of wavelet coefficient to produce an initial list of 24-feature vectors. Features in the vectors were reduced to 3 by the KLT, before being fed the UOFC algorithm.

Global optima in the curves of the fuzzy hypervolume and average partition density versus cluster number will actually define the unsupervised optimal partitioning. In the example that follows, the optimal cluster number came out to be 5, quite close to the R-K convention of six states, or five, if stages 3 and 4 are combined as is often done and as is the case in the example. This match, however, may be somewhat coincidental, as will soon be evident. The output of the unsupervised procedure should be regarded as a subject-night-specific sleep print.

Figure 5 shows the results of the 5-cluster fuzzy classification of the same 2.5 h sleep section shown in Figure 1. The hypnogram, construed from the staging included in the data, is projected on the reconstructed time series. It is immediately evident that there is no one-to-one correspondence between clusters and sleep stages. For instance, cluster 1 spans the sleeping state regardless of stage and cluster 3 predominates in both stages 2 and 3. In fact, there is no stage that is represented by time points having 100% membership in a single cluster. However, a reasonably good stage scoring may be obtained by noting the membership combinations of each of the sleep stages as listed in Table 4. Closer scrutiny will reveal hints of gradual stage transition, which is not evident from the hypnogram.

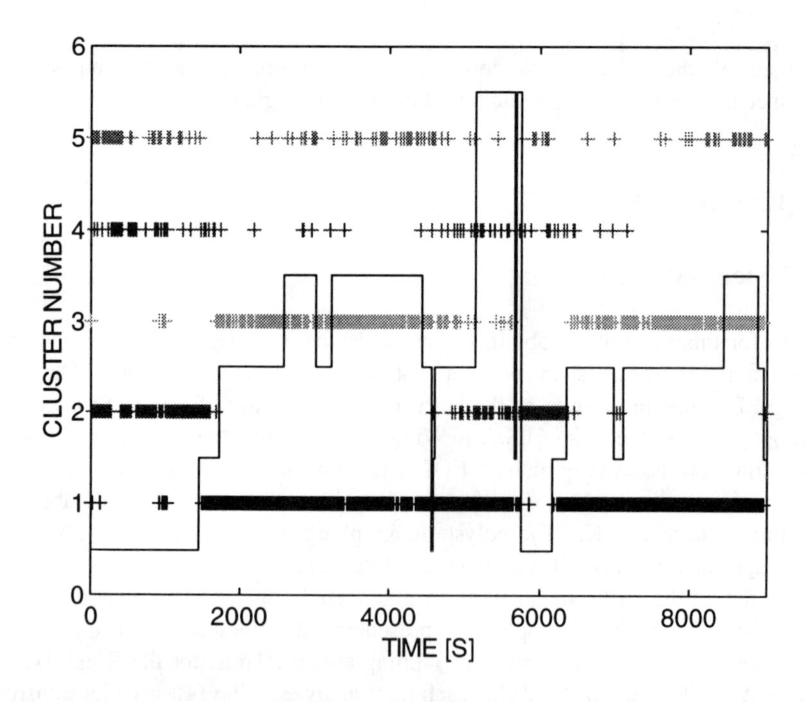

Figure 5. Sleep standing by UOFC. A single EEG channel from the same section of a night's sleep shown in Figure 1, as partitioned to six fuzzy clusters. The expert-produced hypnogram is included for reference. Sleep stages are not identified by distinct clusters but by different combinations of membership in several clusters, as shown in Table 4.

The example presents a promising approach to a goal not yet fully achieved. Optimizing the clustering input is bound to refine the partition, yet undeniable advantages of unsupervised fuzzy partitioning of a sleep episode over template-matched automated systems should already be evident [27]:

1. It accommodates completely new (pathological as well as physiological, such as in non-human subjects) sleep states.
2. It allows fine-structuring of the sleep episode, in at least two aspects:
   a. Temporally, by not being confined to the conventional 20 or 30 second long "page" units.
   b. In content, by avoiding hard partitioning according to a "maximal membership" principle and taking account of degrees of membership, state transitions, state variations, etc. can be defined.

Table 4
Sleep-stage scoring by cluster-membership combinations

SLEEP STAGE	CLUSTER 1	CLUSTER 2	CLUSTER 3	CLUSTER 4	CLUSTER 5
Wake	– [a]	+	–	±	±
1	+	+	–	+	–
2	+	–	+	–	±
3-4	+	–	+	–	+
REM	+	+	±	±	–

[a]) – denotes 0 or very low degree, ± denotes a low degree, and +, a high degree of membership.

4.2 Forecasting Epilepsy

The data for the following example was obtained from an animal model. Recordings of generalized seizures were made by exposing laboratory rats, implanted with chronic cortical electrodes, to pure oxygen in a pressure chamber. By mechanisms not completely elucidated, all mammals studied so far suffer *grand mal*-type motor and electroencephalographic seizures when exposed to high-pressure (hyperbaric) oxygen. The model is convenient, since the oxygen pressure can be controlled to induce the seizures after a predictable mean time lag [28].

Two bipolar channels of EEG are amplified, filtered to pass between 1 and 30 Hz (-6 dB/octave), notch-filtered at 50 Hz (-20 dB/octave), and recorded on tape. The replayed tape recordings are visually screened and six ~2-minute segments free of gross motion artifacts are picked for analysis: two from the control normobaric state (C_1 and C_2), two from the early stay at pressure (P_1 and P_2), and two (S_1 and S_2) from the 5 minutes preceding the first electric seizure, the latter (lasting 10-20s) included in S_2. The segments are digitized at a sampling rate of 128 Hz. Each channel of a segment is divided into 1s epochs with a 50% overlap. Thus, for the data matrix **S**, where N = 128, $D = N/2$ = 64, the overlapping ensures that all transient events are completely represented and dominant in at least one of the epochs.

In this example, we again extract features by the fast wavelet transform, using five wavelet scales. Wavelet analysis is applied to baseline EEG segments of many animals; then we tracked clustering-optimization criteria as cluster number progressively increases. Finally, partitioning segments which include the seizure with different numbers of clusters showed that the optimal partitioning for accommodating the normal as well as the epileptic EEG states is obtained by the number of clusters which maximizes the APD criterion, plus 1 or 2. Accordingly, either one or two is added to the global maximum of this optimization criterion to obtain the optimal number of clusters for partitioning the EEG-data of each exposure.

Rats are nocturnal animals and, when undisturbed, tend to dose off during the day. During the chamber exposure, they alternate between states of light sleep, restful wakefulness, grooming, locomotion, and other alert states. Behaviorally, it seems that most animals sense the approaching seizure. They become very awake and attentive and groom their faces. As in human patients, single or groups of spikes may be seen in the EEG record during the PSS. These are usually in the half minute preceding the seizure. In the rest, no visual clues herald the first electric seizure.

Two works used adaptive segmentation by linear prediction as the feature-extractor, with [29] and without [30] subsequent fuzzy clustering, succeeded in identifying EEG segments containing single epileptic events in patients and in forecasting, by several seconds, an impending drug-induced seizure in rats, respectively. Yet, what is expected from a categorization, which professes to form a basis for a *universal* automated epilepsy forecaster, is to accurately identify the normal states, the seizure and, most important, the PSS, especially in those instances where no obvious clues exist for its identification. Accordingly, the number and nature of the coefficients chosen for the clustering is determined by trial and error, searching for the minimum that would succeed in categorizing the observed states, including the seizure and the PSS. Figure 6 is an example of an exposure where the identification of a 4-minute- long PSS, devoid of obvious visual epileptic activity, was achieved with eight coefficients, comprising the variances (energies) of the first four wavelet scales of the two channels.

Panel (b) shows the 3 (out of 8) dimensional projections of a 7-cluster partitioning of the six EEG segments from an exposure terminating with a seizure. Shown are the cluster centers surrounded by the variances of their respective point-clouds.

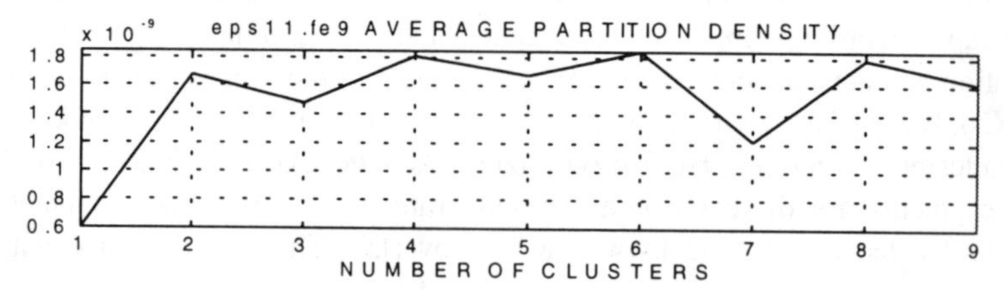

Figure 6a. See figure legend on page 80.

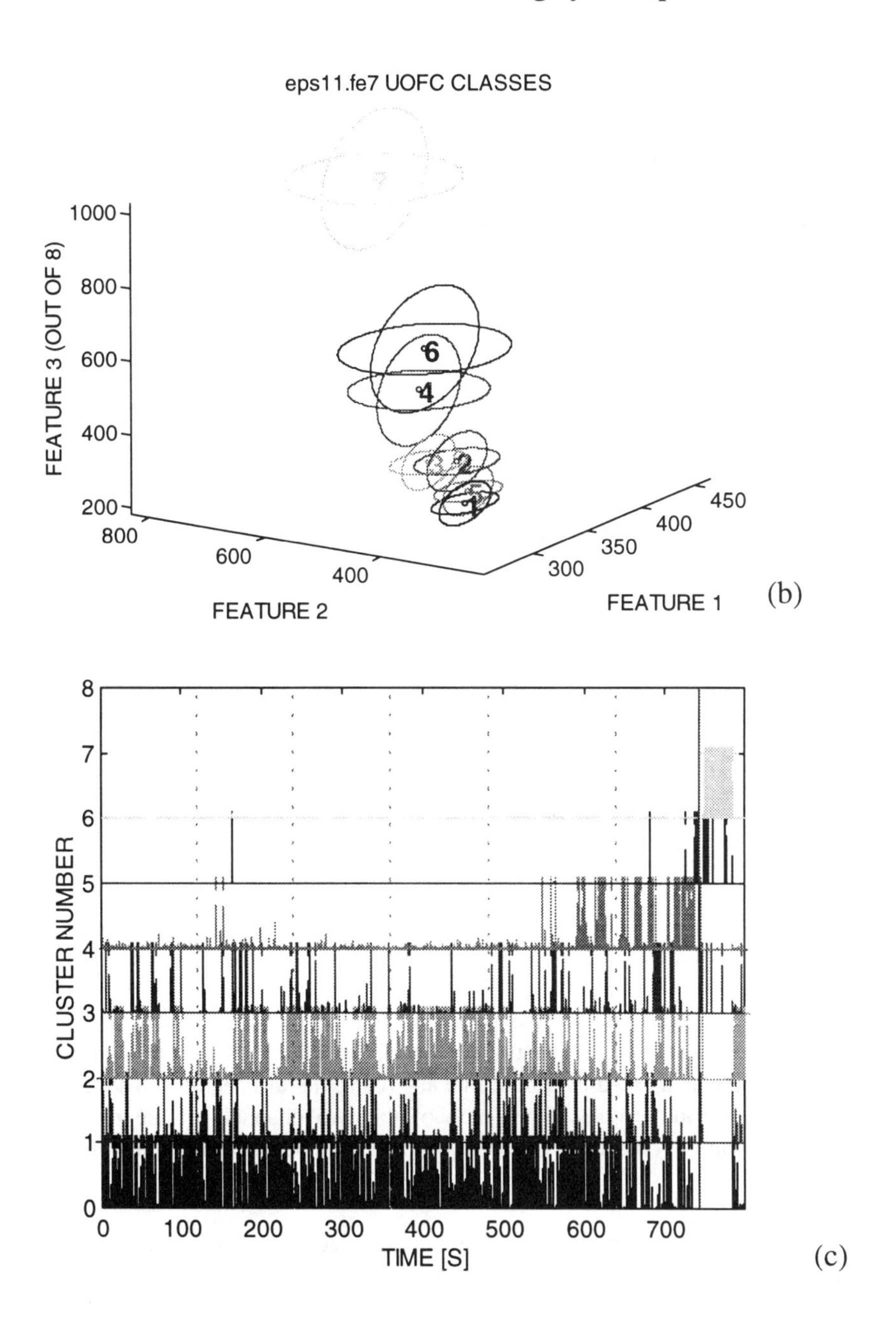

Figure 6b and c. See figure legend on page 80.

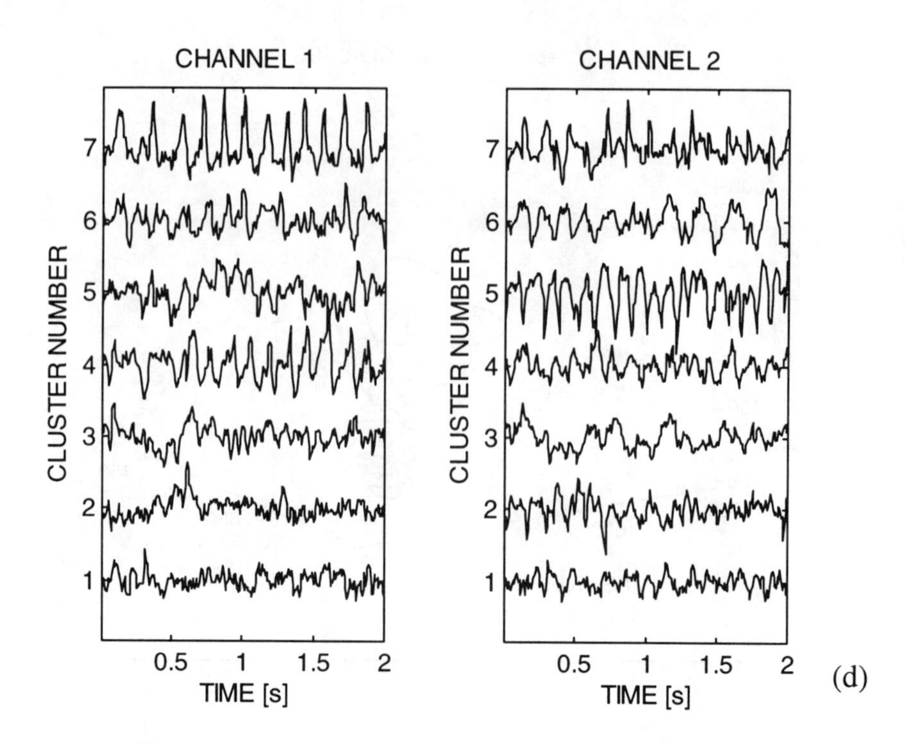

Figure 6. EEG state categorization by UOFC. Results of application on a concatenated EEG stretch from an HBO-exposed rat. Partitioning is based on the variances (energies) of the first four wavelet scales of two EEG channels. Panel (a) shows the Average Partition Density Criterion optimizing at 6 clusters. Panel (b) is a 3- (out of 8) dimensional projection of the 7-cluster partitioning. Panel (c) is the time-series reconstruction by cluster membership, the space between integers on the Y axis being equal to 100% membership. Dashed vertical lines mark the borders of (from left to right) two control segments, two segments at early and mid part of pressure exposure and last, two pre-seizure segments. Beginning of seizure (as located by a human expert) is marked by a bold line at 710s in the last segment. The last panel (d) shows two-second epochs, from each of the two channels of the original signal, corresponding to points in time with 100% membership in each of the 7 clusters.

Panel (c) shows the membership of each successive point in the reconstituted time series, in each of the clusters (the space between two y-axis integers being 100% membership). Examples of 2-second signal epochs from periods with a clear predominance of a given cluster are shown, for each channel, in the lower panel. This animal was alert most of the time and the dominant rhythms are β, θ, and low-frequency α, the latter apparently specific to the PSS, mainly defined by cluster 5. Just prior to the seizure, a 5 Hz θ rhythm predominates (cluster 6), which is also the frequency assumed by the ensuing seizure (cluster 7).

4.3 Classifying Evoked and Event-Related Potentials by Waveform

Evoked Potential (EP) and event-related potential (ERP) waveforms are typically characterized by latency and amplitude measures of peaks and troughs along the curve, ignoring all other features of the record. Features that are thus ignored include latencies and amplitudes of interposed data points, which comprise the waveform of the recording. Consequently, much of the recorded information is lost. The UOFC procedure in this instance is applied to a pool of digitally sampled EP records. The pool may be comprised of evoked potentials to the same stimulus, recorded from the same electrodes in different subjects, or evoked potentials to differing stimulus parameters, recorded from the same electrodes in the same subject, or evoked potentials from the same subject in response to the same stimulus, recorded from different electrodes, or any combination of the above.

In this instance, when dealing with an ensemble of EEG sections all time locked to the stimulus and all having a common basic shape, we may use the KLT (PCA) as a feature-extracting method (Appendix 3). This method has the ability to calculate the basis waveforms from which all members of the ensemble can be derived. A set of basis waveforms (i.e., principal components) common to all the records is computed and arranged in decreasing order of their contribution to reconstruction of all the records in the pool. In more formal terms, the basis waveforms are the eigenvectors of the record set's covariance matrix, which represents the correlation between all records. The eigenvectors, which constitute an orthogonal basis of the set of records, are arranged in decreasing order of their eigenvalues. Thereafter, each record can be exclusively reconstructed by a linear combination of the basis waveforms, each multiplied by an appropriate weight or reconstruction coefficient (Appendix 3). As all EPs in the pool have many common features, they may be reduced to a small number of coefficients. Finally, each EP record in the pool is assigned a feature column vector, characterized by F coefficients of the basis waveforms by which it is reconstructed, and all M pool members are arranged in the $F \times M$ feature matrix for classification [25].

In the following example, an ensemble of ERP waveforms was obtained from 24 normal subjects in a study, which tested the interaction of auditory and visual stimuli during short-term memory scanning [31]. The experimental paradigm involved the presentation in short succession of three lexical stimuli to be memorized, followed by a probe stimulus. This experiment generated the grand-average waveforms shown in Figure 3. The subject is passive throughout the presentation of the set to be memorized and then is required to react in one of two ways (press different buttons), depending on whether or not the probe matches one of the memorized items. Memorized-sets and probes were pseudo-randomly chosen from the pool of digits 1-9. All the memorized stimuli were visual, i.e., digits projected on a screen. Probes were either visual (VV) or auditory (VA), with the subject hearing the verbal digit through earphones. The two kinds of probes were administered on separate sessions, such that subjects knew to expect one or the other.

To capture the entire ERP response, a time span of 0.8 second post-stimulus is sampled at 256 Hz and 40 repetitions are averaged. The averaged ERP has an early part, which reflects the path of the visual stimulus from the retina to the visual cortex and a late, event-related part which reflects its memorization or matching decision (Figure 3). Each EP is recorded from several scalp electrodes; any number of them may be chosen for analysis. In the case presented, each of the 24 subjects contributed 8 averaged EP waveforms (three memorized and one probe for each of the VV and VA combinations, recorded from a single scalp site, Pz). The different modality combinations of sets and probes as well as the in-set and out-of-set probing (50%) are ground for potential grouping of the pooled EPs. A feel for intersubject variability, evident from even one (and the same) scalp site recording, may be gleaned from Figure 7.

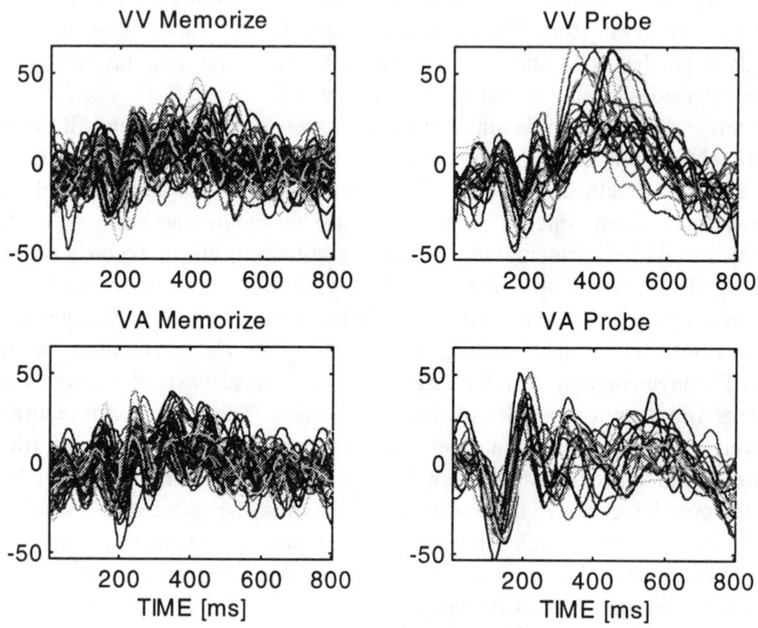

Figure 7. Intersubject and intersite variability in ERP waveform. Shown are the same 4 panels as in Figure 3, with the individual curves which yielded the grand average. Each curve is an average ERP of one subject, recorded from one mutual scalp site. The greater number of curves in the two left-hand panels reflect the fact that there are 3 memorized stimuli for each probe in each experiment. Note high variability, mainly in the late components.

Each waveform was split into the early and late parts, to form two pools of waveforms for feature extraction and clustering. In this case, the data matrix **S** is composed of M = 192 (24x8) signal columns, each having a dimension N = 200 (0.8x250) samples. The KLT (PCA) was then performed on **S** and coefficients of the first 5 eigenvectors (with an 83.6% cumulative reconstruction contribution) were taken.

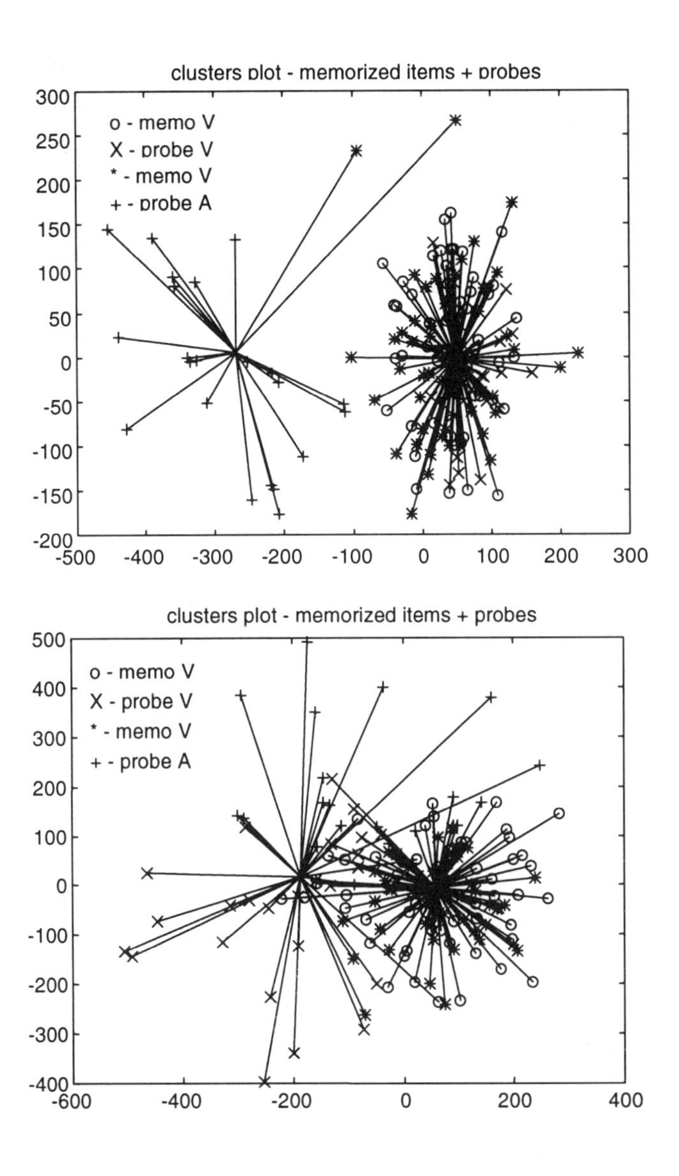

Figure 8. ERP classification by UOFC. A two- (out of six) dimensional projection of a 2-cluster partitioning by eigenvector coefficients of an ensemble of 156 ERP waveforms from 24 subjects. The upper panel shows classification of the early part of the ERP, which follows the different waveforms of auditory- and visual-evoked responses. The second panel shows classification of the later part of the ERP which less perfectly separates memorized and probe-evoked responses. Legend insets in both panels from top down: memorized visual stimuli with anticipated visual probes, visual probes, memorized visual stimuli with anticipated auditory probes, auditory probes.

The UOFC is then performed on the ensemble in a 5-dimensional space. In Figure 8, the upper panel shows the result of the clustering on the early part of the EP. It is evident that the visually evoked and the auditory evoked early potentials are sufficiently different that a crisp clustering separates them with only 1.3% error. Results of the clustering of the late part are shown in the lower panel. While considerable overlap appears, two fuzzy clusters separate with 13.5% error the memorized and the probe-evoked responses, irrespective of modality. It was found that by using other scalp recording sites, grouping of the late part data by modality or by probe-match could be obtained.

A different use for fuzzy clustering of evoked potentials has recently been suggested [32]. In this approach, clustering of the single responses to the stimulus is performed prior to averaging. Results show that the single responses fall into several distinct but overlapping groups. Global averaging causes smearing and loss of features that can be conserved by separately averaging responses showing a high degree of membership in a certain group.

In both cases, we may take advantage of the adaptive and machine-learning nature of the UOFC procedure, to maintain an updated classification of an evolving data base of patient EPs. The procedure is first applied on a large enough pool of EP records. Then, by adding (a) new record(s) to the existing pool and re-applying the procedure, the clustering is updated and the classification is improved and refined with the growing experience of the machine-based learning system.

5. Concluding Remarks And Future Applications

As demonstrated by the examples, the general method outlined in this chapter may have wide applications in the field of EEG signal processing. The combination of time-frequency analysis and fuzzy clustering, both of which are efficient in unsupervised learning, seems particularly useful for dealing with this signal. In the domain of brain state identification, numerous, so far ill-defined, EEG states, which may be expressed during different mental activities, could possibly be picked up. Warnings about physiological as well as pathological state transitions with potential dire consequences (drowsiness, psychotic attacks, etc.) may become available. In the area of single event categorization, the method could transcend the boundaries of the EEG. Other bioelectric signals that could be subjected to beneficial classification for diagnostic or research purposes are action potentials, various electrical events of the cardiac cycle, other EMGs, and responses of the sense organs themselves to their respective stimuli.

In addition to time-distribution, EEG signals are also spatially distributed on the scalp. All of the above presentation relied on one or two derivations of the EEG signal, foreseeing the impracticality of multi-channel monitoring for most ambulatory purposes. However, clinical and research applications routinely use many channels and the results of feature extraction and fuzzy clustering of multi-channel signals could just as well be projected on a spatial map of the scalp, gaining information in quite another dimension.

Two specific developments that deserve more detailed mention are presented in the following sections.

5.1 Dynamic Version of State Identification by UOFC

UOFC may be used in two ways (or their combination) for automated definition of EEG states. The first, the one we have described above, is its application *a posteriori* on a gathered data pool. The second means, which better serves an actual automated forecasting device, is to have the procedure operate from the beginning of an exposure expected to lead to pathology and to observe the dynamics of the classification as consecutive data are being introduced. A combination of adaptive, on-line, classification and the use of past information can be achieved by applying a weighting function to the time series, to place greater weight on incoming information.

Classification is initialized on the first few pattern vectors and then proceeds at consecutive steps of one or more vectors (the computational consequences, although not being prohibitive, are, for the moment, disregarded). The temporal change in the position of cluster centers and in the magnitude of the average density criterion are then followed with time. An abrupt and considerable change in the validity criterion for n classes, a sudden indication of a higher optimal number of classes, and a distinct spatial shift of one or more cluster centers - could all signal the advent of a new state. Use of constraining rules, for instance, that a direct transition from the waking state to sleep stages 3-4 or 5 is not possible, is vital in interpretation and building the reliability of the automated warning system.

5.2 Data Fusion

The UOFC could naturally be used as a data fusion tool, by adding different features to the basic feature vector. For example, the insensitivity of spectrum estimation methods to short transients and single events may be partly overcome by also processing the signal with a pattern recognition tool and adding its output to the UOFC procedure. Or, expanding the classification input to include extra-cerebral signals may aid in EEG state identification. Natural candidates in the case of sleep-stage scoring would be EMG and EOG. Electrocardiographic features and/or parameters relating to the fractal irregularities in heart rate, or heart-rate variability, and even evoked responses induced at regular intervals, could be expected to vary during state transitions and to contribute to a better grouping [33], [34]. The new input could first be used to split clusters that are suspected to lump physiologically unrelated but similarly-featured sections and later, if successful, could be fused with the initial list of classifiers.

APPENDIX 1: THE FAST WAVELET TRANSFORM

Any function $\theta(t)$ whose integral is equal to 1 and that converges to 0 at infinity is called a smoothing function. In our case, $\theta(t)$ is chosen to be a Gaussian function. Let $\psi(t)$ denote the first derivative of $\theta(t)$ (which results in a biphasic waveform). By definition, the function $\psi(t)$ can be considered to be a wavelet because its integral is equal to 0. Let $\psi_a(t)$ be the dilation by scaling factor a of the function $\psi(t)$:

$$\psi_a(t) = \frac{1}{a}\psi(\frac{t}{a}) \qquad \text{(A1)}$$

The wavelet transform is computed by convolving the signal, s, with these dilated wavelets. To allow fast numerical implementations, the scaling factor a is varied along the dyadic sequence (2^j, j = 1,2,...). The wavelet transform, $T_{2^j}S(t)$, is defined by

$$T_{2^j}S(t) \equiv S(t) * (2^j \psi_{2^j})(t) = S(t) * (2^j \frac{d\theta_{2^j}}{dt})(t) = 2^j \frac{d}{dt}(S * \theta_{2^j})(t) \qquad \text{(A2)}$$

The wavelet function $\psi(t)$, can be characterized by the discrete filters H (low pass) and G (high pass) [20], where the low pass filter H is a product of the Gaussian $\theta(t)$, and the high pass filter G is a product of its derivative. The discrete filters H_{2^j}, G_{2^j} can be obtained by placing $(2^j - 1)$ zeros between each of the coefficients of the filters. The transfer function of these filters are $H(2^j w)$ and $G(2^j w)$. The wavelets corresponding to each scale can be obtained by convoluting the appropriate filters.

The discrete wavelet transform of a signal can be estimated using the fast wavelet transform proposed by Mallat and Zhong [20]. At each scale, j, the wavelets' coefficients, $T_{2^{j+1}}f$, and the smoothed signal, $O_{2^{j+1}}S$, for the next scale, $j+1$, are calculated. For scale zero, j = 0, the signal, S, is used as the smoothed signal. The algorithm proposed is

$$
\begin{aligned}
& j = 0 \\
& O_{2^j}S = S \\
& While\ \ (j < \mathrm{J}) \qquad \text{(A3)} \\
& \quad T_{2^{j+1}}S = \frac{1}{\lambda_j} O_{2^j}S * G_j \\
& \quad O_{2^{j+1}}S = O_{2^j}S * H_j \\
& end
\end{aligned}
$$

where J is the total number of scales and the constant λ_j compensates for the discretization errors [20].

APPENDIX 2: MULTICHANNEL MODEL-BASED DECOMPOSITION BY MATCHING PURSUIT

Let $y^m(t)$, $m = 1,2,...$ be a *family* of "mother" (prototype, model) wavelets. The waveform corresponding to the m-type wavelet with scale (dilation) a and translation b is, in the continuous case,

$$y^m_{a,b}(t) = \frac{1}{\sqrt{a}} \cdot y^m\left[\frac{t-b}{a}\right] \tag{B1}$$

For our physiological realization we can use the following family of "Hermite-type mother wavelets" [5]:

$$y^m(t) = \left[\frac{d}{dt}\right]^m \cdot e^{-t^2}, m = 0,1,2,... \tag{B2}$$

We thus obtain the following **non**-orthogonal dictionary of wavelets:

$$\mathbf{Y} = \{ y^m_{a,b}(n) = \frac{c^m_{a,b}}{\sqrt{a}} \cdot \left[\frac{d}{dt}\right]^m \cdot e^{-\left[\frac{t-b}{a}\right]^2} \quad \textit{such that:} \tag{B3}$$

$$t = 1,...,N, \quad m = 0,1,2,... \quad (a,b) \in Z \quad \textit{and} \quad 0 < a \le A, \quad 0 < b < N\},$$

where A is determined by the spectrum (bandwidth) of the signals. To make sure that we get the best match with the best fitting signal, we normalize the wavelets by $C^m_{a,b}$ such that: $\left\langle y^m_{a,b}, y^m_{a,b} \right\rangle = 1$ where $< , >$ denotes the inner product.

Our algorithm [5] is similar to the matching pursuit algorithm of Mallat and Zhang [22], but in our case we conduct multichannel analysis and choose more then one waveform at any iteration. In our physiological realization, we try to find the minimum number of localized waveforms that can represent our $N \times M$ data matrix, **S**, according to some predetermined criteria:

0. Set $p = 1$ and $\mathbf{S}^p = \mathbf{S}$.

1. For each translation b, find the best matching wavelet $\mathbf{w}_b \in \mathbf{Y}$ and a column temporal pattern $\mathbf{S}_j^p$ over all patterns and waveforms set:

$$\mu^p(b) \equiv \left|\left\langle \mathbf{w}_b, \mathbf{s}_j^p \right\rangle\right| = \max_{m,a,i} \left\{ \left|\left\langle \mathbf{y}_{a,b}^m, \mathbf{s}_i^p \right\rangle\right| \right\}; \quad 0 < b \le N \tag{B4}$$

Thus, we obtain a reduced subset of N wavelets best matched to one of the temporal patterns at translation b:

$$\mathbf{Y}_1^p \equiv \left\{ \mathbf{w}_b \in \mathbf{Y} : \ \mu^p(b) = \left|\left\langle \mathbf{w}_b, \mathbf{s}_j^p \right\rangle\right|; \quad 0 < b \le N \right\} \tag{B5}$$

2. Select from the subset $\mathbf{Y}_1^p$ all the wavelets that correspond to local maxima of the function $\mu^p(b)$:

$$\mathbf{Y}_2^p \equiv \left\{ \mathbf{w}_l \in \mathbf{Y}_1^p : \ \mu^p(l-1) < \mu^p(l) > \mu^p(l+1); \quad 0 < l \le N \right\} \tag{B6}$$

Thus, we obtain a reduced subset $\mathbf{Y}_2^p$ of F (*the number of local maxima* in $\mu^p(b)$) non-orthogonal localized waveforms, which we select for decomposition of our signals.

We can restrict the number of local minima in $\mu^p(b)$ and the gap between them. For example, if we choose one local minimum, we get one localized waveform in $\mathbf{Y}_2^p$ by:

$$\mathbf{Y}_2^p \equiv \left\{ \mathbf{w}_l \in \mathbf{Y}_1^p : \ \left|\left\langle \mathbf{w}_l, \mathbf{s}_j^p \right\rangle\right| = \max_{m,a,b,i} \left\{ \left|\left\langle \mathbf{y}_{a,b}^m, \mathbf{s}_i^p \right\rangle\right| \right\} \right\} \tag{B7}$$

namely, the best matching between all waveforms and all temporal patterns.

3. Let **B** denote an $N \times F$ matrix which consists of the set of all the column vectors $\mathbf{w}_l \in \mathbf{Y}_2^p$. We can use the Optimal Bi-Orthonormal Reconstruction theorem [35] to reconstruct all the signals $\mathbf{S}^p$ by the reduced subset of wavelets $\mathbf{B}^p$

$$\hat{\mathbf{S}}^p = \mathbf{B}^p \cdot pinv(\mathbf{B}^p) \cdot \mathbf{S}_p \tag{B8}$$

where

$$pinv(\mathbf{B}^p) \equiv \left((\mathbf{B}^p)^T \cdot \mathbf{B}^p \right)^{-1} \cdot (\mathbf{B}^p)^T.$$

We can define a coefficient matrix

$$\mathbf{X}^{p} = pinv\,(\mathbf{B}^{p})\cdot\mathbf{S}_{p} \tag{B9}$$

so we can write

$$\hat{\mathbf{S}}^{p} = \mathbf{B}^{p}\cdot\mathbf{X}^{p} \tag{B10}$$

4. The residual signals are

$$\mathbf{R}_{v}^{p} = \hat{\mathbf{S}}^{p} - \mathbf{S}^{p}. \tag{B11}$$

If $(\mathbf{R}_{v}^{p})^{2} > \varepsilon$, we set $\mathbf{S}^{p+1} = \mathbf{R}_{v}^{p}$, $p = p+1$, and start from step (1) again.

5. The resultant reconstruction (model-based approximation) of the signals matrix is given by

$$\hat{\mathbf{S}} = \sum_{j=1}^{P} \hat{\mathbf{S}}^{j} = \sum_{j=1}^{P} \mathbf{B}^{j}\cdot\mathbf{X}^{j} = \mathbf{B}\cdot\mathbf{X} \tag{B12}$$

where

$$\mathbf{B} = [\mathbf{B}^{1},\ldots,\mathbf{B}^{P}]$$

is a set of waveforms and

$$\mathbf{X} = \begin{bmatrix} \mathbf{X}^{1} \\ \cdot \\ \cdot \\ \cdot \\ \mathbf{X}^{P} \end{bmatrix}$$

is a reduced coefficients matrix which can be used as a feature matrix for the UOFC algorithm.

APPENDIX 3: FEATURE EXTRACTION AND REDUCTION BY PRINCIPAL COMPONENT ANALYSIS

For any $N\times M$ matrix **S**, the $N\times N$ covariance matrix of **S** is defined as

$$\mathbf{C}^{s} = \mathrm{E}\left\{(\mathbf{S} - \bar{\mathbf{s}})\cdot(\mathbf{S} - \bar{\mathbf{s}})^{T}\right\}, \tag{C1}$$

where E is the expected value, and the N dimensional column vector $\bar{\mathbf{S}}$

$$\bar{s}_{n} = \mathrm{E}\left\{\mathbf{s}_{n}\right\} = \frac{1}{M}\cdot\sum_{i=1}^{M} s_{ni}\ , n = 1,2,\ldots,N \tag{C2}$$

is the mean of all data patterns.

Let the N dimensional column vector $\mathbf{g}_j$ and λ_j, $j = 1,2,\ldots, Q$, be the eigenvectors and the corresponding eigenvalues of $\mathbf{C}^S$, where Q ($\leq N$) is the number of non-zero eigenvalues. The eigenvalues are arranged in decreasing order so that

$$\lambda_1 \geq \lambda_2 \geq \ldots \geq \lambda_Q$$

An $N\times Q$ transformation matrix whose columns are the eigenvectors of $\mathbf{C}^S$ can be given by:

$$\mathbf{G} = \begin{bmatrix} g_{11} & g_{12} & \cdot & \cdot & g_{1Q} \\ g_{21} & g_{22} & \cdot & \cdot & g_{2Q} \\ \cdot & \cdot & \cdot & \cdot & \cdot \\ \cdot & \cdot & \cdot & \cdot & \cdot \\ g_{N1} & g_{N2} & \cdot & \cdot & g_{NQ} \end{bmatrix}$$

A QxM coefficients matrix $\mathbf{X}$ is then computed as follows:

$$\mathbf{X} = \mathbf{G}^T \cdot (\mathbf{S} - \bar{\mathbf{s}}) \tag{C3}$$

As it can be shown that the eigenvectors of a covariance matrix are orthogonal [36],

$$\mathbf{G} \cdot \mathbf{G}^T = \mathbf{I} \quad \text{or} \quad \text{inverse}(\mathbf{G}) = \mathbf{G}^T,$$

S can be reconstructed by

$$\mathbf{S} = \mathbf{G} \cdot \mathbf{X} + \bar{\mathbf{s}} \tag{C4}$$

If we now form an $N\times F$ matrix $\mathbf{G}_F$ from the first F eigenvectors that correspond to the largest eigenvalues, and the respective $F\times M$ matrix $\mathbf{X}_F$ from the first F rows of **X**, we can approximate **S** by

$$\hat{\mathbf{S}} = \mathbf{G}_F \cdot \mathbf{X}_F + \bar{\mathbf{s}} \cdot \tag{C5}$$

It can be shown that the mean square error between **S** and $\hat{\mathbf{S}}$ is given by

$$e = \sum_{j=F+1}^{L} \lambda_j \tag{C6}$$

The M reduced columns of the $\mathbf{X}_F$ coefficient matrix are then used as the features vectors for the clustering procedure.

LIST OF ACRONYMS

APD	average partition density (clustering validity criterion)
EEG	electroencephalogram (brain electrical activity)
EMG	electromyogram (muscle electrical activity)
EP	evoked potential (stimulus-induced change in EEG)
ERP	event-related potentials (EP induced by a conditional stimulus)
(W)FKM	(weighted) fuzzy k-means (algorithm)
FHPV	fuzzy hypervolume (clustering validity criterion)
FMLE	fuzzy maximum likelihood estimation
KLT	Karhunen Lo'eve Transform
PCA	principal component analysis
PD	partition density (clustering validation criterion)
PSS	pre-seizure (brain) state
SPW	spike-and-wave (epileptic EEG pattern)
SToMP	spatio temporal matching pursuit
REM	rapid eye movement (sleep stage)
R-K	Rechtschaffen Kales (sleep stage formulation)
(W)UOFC	(weighted) unsupervised optimal fuzzy clustering (algorithm)
VA	visual-audio (stimulus presentation sequence)
VV	visual-visual (stimulus presentation sequence)

Acknowledgments
The authors would like to thank Pnina Nahamoni of Professor Hillel Prat's Evoked Potentials Laboratory, Faculty of Medicine, The Technion, Haifa, for her contribution to the analysis of the Evoked Potential data and Mor Reich of Professor Arnon Cohen's Biological Signal Processing Laboratory, Electrical Engineering Department, Ben Gurion University, Be'er Sheva, for help with the processing of the sleep EEG data.

REFERENCES

[1] Berger, H., Uber das Elektroenzephalogram des Menschen, *Arch. Psychiat. Nervenkrankheiten,* 87, 527, 1929.

[2] Kandel, E. R., Schwartz, J. H., and Jessel, T. M., Eds., *Principles of Neural Science, 3rd Edition,* Prentice Hall International Inc. London, 1991, Chapters 50 and 51.

[3] Nunez, P. L., Ed., *Neocortical Dynamics and Human EEG Rhythms.* Oxford University Press. Oxford, 1995, Chapters 1 and 2.

[4] Sejnowski, T. J, McCormick, D. A., and Steriade, M., Thalamocortical oscillations in sleep and wakefulness, in *The Handbook of Brain Theory and*

Neural Networks, Arbib, M. A., Ed, The MIT Press, Cambridge, Massachusetts, 1995, 976.

[5] Geva, A. B., Pratt, H., and Zeevi, Y.Y., Spatio-Temporal Source Estimation of Evoked Potentials by Wavelet-Type Decomposition, in *Advances in Processing and Pattern Analysis of Biological Signals*, Gath, I., and Inbar, G. F., Eds., Plenum Press, New York, 1996, 103.

[6] Rechtschaffen, A. and Kales, A., *Manual of Standardized Terminology, Techniques and Scoring System for Sleep Stages of Human Subjects,* N.I.H. publication No. 204, U.S. Government Printing Office, Washington, D.C., 1968.

[7] Loiseau, P., Epilepsies, in *Guide to Clinical Neurology*, Mohr, J. P. and Gautier, J. C., Eds., Churchill Livingstone, New York, 1995, 903.

[8] Dumermuth, G., Possibilities of Electronic EEG Processing in Epileptology, in *Epileptology: Proc. 7th Intl. Symp. on Epilepsy*, Georg Thieme Publishers, Stuttgart, 1976, 365.

[9] McGillem, C. D. and Aunon, J. I., Analysis of Event-Related Potentials, in *EEG Handbook Vol. 1: Methods of Analysis of Brain Electrical and Magnetic Signals,* Gevins, A. S. and Remond, A., Eds., Elsevier Science Publishers, Amsterdam, 1987, 131.

[10] Elbert, T., Ray, W. J., Kowalik, Z. J., Skinner, J. E., Graf, K. E., and Birbaumer, N., Chaos and Physiology: Deterministic Chaos in Excitable Cell Assemblies. *Phsiol. Rev.,* 74, 1, 1994.

[11] Babloyantz, A. and Destexhe, A., Low dimensional chaos in an instance of epilepsy, *Proc. Natl. Acad. Sci. U.S.A.,* 83, 3513, 1986.

[12] Iasemidis, L. D. and Sackellares, J. C., The evolution with time of the spatial distribution of the largest Lyapunov exponent in the human epileptic cortex, in *Measuring Chaos in the Human Brain,* Duke, D. and Pritchard, W., Eds., Singapore World Scientific, 1991, 49.

[13] Pjin, A. P., Van Neerven, J., Noestt, N., and Lopes Da Silva, F. H., Chaos or noise in EEG signals: dependence on state and brain site, *Electroencephalorgr. Clin. Neurophysiol.*, 79, 371, 1991.

[14] Zadeh, L. A., Fuzzy Sets, *Information and Control* 8, 338, 1965.

[15] Gath, I. and Geva, A., Unsupervised optimal fuzzy clustering, *IEEE Trans. vol. PAMI* 7, 773, 1989.

[16] Gath, I. and Geva, A., Fuzzy clustering for estimation of parameter of the components of mixtures of normal distributions, *Pattern Recog. Lett.*, 9, 77, 1989.

[17] Bronzino, J. D., Ed., Handbook of Biomedical Engineering. Section VI: *Biomedical Signal Analysis*. CRC Press. Boca Raton, Florida, 1995, 802.

[18] Daubechies, I., Orthogonal Bases of Compactly Supported Wavelets, *Commun. Pure Appl. Math., 41*, 909, 1988.

[19] Mallat, S. G., A Theory for Multiresolution Signal Decomposition: The Wavelet Representation, *IEEE Trans. PAMI-11,* 7, 674, 1989.

[20] Mallat, S. G. and Zhong, S., Characterization of signal from multiscale edges, *IEEE Trans. PAMI,* 10, 710, 1992.

[21] Sun, M. and Sclabassi, R. J., Wavelet feature extraction from neurophysiological signals, in M. Akay, Ed., *Time-Frequency and Wavelets in Biomedical Signal Processing*. IEEE Press, New York, 1998, 305.

[22] Mallat, S. and Zhang, Z., Matching persuits with time-frequency dictionaries, *IEEE Trans. on Signal Processing* 12, 3397, 1993.

[23] Geva, A. B., Pratt, H., and Zeevi, Y.Y., Multichannel Wavelet-Type Decomposition of Evoked Potentials: Mode-Based Recognition of Generator Activity. *Med. and Biol. Eng. and Comput.* 35, 40, 1997. Gersho, A. and Gray, R. M., Eds., *Vector Quantization and Signal Compression,* Kluwer Academic Publisher, 1992.

[24] Geva, A. B. and Pratt, H., Unsupervised Clustering of Evoked Potentials by Waveform. *Med. and Biol. Eng. and Comput.* 32, 543, 1994.

[25] Bezdek, J. C., *Pattern recognition with Fuzzy Objective Function Algorithms*, Plenum Press, New York, 1981.

[26] Gath, I., Feuerstein, C., and Geva, A., Unsupervised Classification and Adaptive Definition of Sleep Patterns, *Pattern Recognition Letters,* 15, 977, 1994.

[27] Clark, J. M., Oxygen toxicity, in *The Physiology and Medicine of Diving.* Bennett, P. B. and Elliott D. H., Eds. 3rd Edition, Best Publishing Co. San Pedro, California, 1982, 200.

[28] Krajca, V., Petranek, S., Patakova, I., and Varri, A., Automatic identification of significant graphoelements in multichannel EEG recordings by adaptive segmentation and fuzzy clustering, *Int. J. Biomed. Comput.* 28, 71, 1991.

[29] Rogowski, Z., Gath, I., and Bental, E., On the prediction of epileptic seizures, *Biol. Cyber.*, 42, 9, 1981.

[30] Pratt, H., Erez, A., and Geva, A. B., Effects of Auditory/Visual and Lexical/Non-lexical Comparisons on Event-related Potentials in a Memory Scanning Task. *Memory,* 5, 321, 1997.

[31] Zouridakis, G.,. Jansen, B. H., and Boutros, N. N., A Fuzzy Clustering Approach to EP Estimation. *IEEE Trans. Biomed. Engin.* 44, 673, 1997.

[32] Sartene, R., Poupard, L., Bernard, J. L., and Wallet, J. C., Sleep Images Using the Wavelet Transform to Process Polysomnographic Signals, in *Wavelets in Medicine and Biology,* Aldroubi A., and Unser, M., Eds., CRS Press, Roca Baton, Fla., 1966, Chapter 13.

[33] Brandt, M. E., Jansen, B. H., and Carbonari, J. P., Pre-stimulus Spectral EEG Patterns and the Visual Evoked Response, *Electroenceph. Clin. Neurophysiol.*, 80, 16, 1991.

[34] Genossar, T. and Porat, M., Optimal Bi-Orthonormal Approximation of Signals, *IEEE Trans. on Systems, Man, and Cybernetics,* vol. 22, 3, p. 449, 1992.

[35] Noble B., *Applied Linear Algebra*, Prentice-Hall, Englewood Cliffs, NJ, 1969.

Chapter 4

Contouring Blood Pool Myocardial Gated SPECT Images with a Sequence of Three Techniques Based on Wavelets, Neural Networks, and Fuzzy Logic

Luis Patino, André Constantinesco, and Ernest Hirsch

1. INTRODUCTION

Medical imaging currently makes use of a wide range of non-invasive modalities such as Magnetic Resonance Imaging (MRI), Ultrasound, Computerized Tomography, Nuclear Imaging, and Radiography. Although each acquisition modality has its specific application domain, a common point relates them all. From the images acquired, the medical specialists have to recognize a given pathology after adequate presentation or processing of the data. However, due to various reasons (acquisition conditions, distortion, noise, attenuation of the signal, etc.), the images contain only variable amounts of approximated or sometimes incomplete information. In such situations, fuzzy logic [1], which permits us to deal with inaccurate or ill-defined data, is often applied in the field of medical imaging. Over the past years, an increasing number of applications using this approach have been reported in the literature. For example, Boegl et al. [2] have implemented a computer-assisted on-line diagnosis system, based on fuzzy reasoning, to detect rheumatic diseases in radiological images. The assessment of the age of bones based on features automatically extracted from hand radiographs is another example in the field [3]. Other applications of fuzzy logic and reasoning can be found for diagnosing chronic liver diseases in liver scintiscans; see, e.g., Shiomi S. et al. [4]. Further, in the domain of MRI, image segmentation to achieve tissue differentiation is described in several publications [5], [6]. Measuring the volumes of cerebrospinal fluid, and white matter and gray matter in brain images (see, as an example, [7]) is also frequently investigated. Bezdek published an interesting survey

0-8493-9806-1/98/$0.00+$.50

paper showing the fuzzy methods currently employed in medical imaging [8] and, in particular, emphasizing the fuzzy c-means clustering method.

Use of Neural Networks [9] appeared in the late seventies as a complementary method to fuzzy logic based approaches. Since the parameters used to build a fuzzy system usually need to be tuned heuristically, neural networks, by virtue of their self-learning capability, can provide valuable help in the determination of the values of these parameters. Indeed, several contributions combining the two methods for evaluating medical images have been reported. For instance, Lin et al. used fuzzy set theory combined with a convolution neural network for reducing the detection of false-positives in digital chest radiographs [10]. The so-called neuro-fuzzy techniques have been developed for tissue classification in MRI data [11].

This chapter is devoted to the description of the development of an alternative automated method to extract the contours of the left ventricle in blood pool Single Photon Emission Computed Tomography (SPECT) images. The method combines wavelets theory, neural networks, and fuzzy logic for the extraction of contours for use in computing the Left Ventricle Ejection Fraction (LVEF). The outline of the contribution is as follows. In Section 2, we describe how the kind of images we work with are obtained and how the information they encode has to be interpreted. The overall strategy of the method we suggest for processing the image sets to determine the LVEF is introduced in Section 3. This section is composed of five parts. The three parts following the introduction summarize the aim of the techniques combined into a single system. These are a wavelets based pre-processing of the data, a segmentation into regions of interest using neural network approaches and a recognition technique built around fuzzy logic components. The section ends with the description of the parallel learning procedure developed and implemented to train the system and automatically fix its parameters. Representative results obtained using our method are given in Section 4, and Section 5 concludes our contribution and gives a short outlook.

2. ANATOMY OF THE G-SPECT IMAGES

Blood pool myocardial Gated Single Photon Emission Computed Tomography (G-SPECT) images are acquired in our laboratory with an Elscint Helix gamma camera equipped with a low-energy, high-resolution collimator. After radiolabeling the red blood cells with technetium 99m, a set of 30 electrocardiographic (EGC) gated projections of 40 sec with 8 phases of the cardiac cycle is obtained, while rotating the camera 180° from the right anterior oblique position to the left posterior oblique position, the patient being in a supine position. Each projection image is corrected for non-uniformity and the center of rotation is also adjusted. After filtered backprojection using a Butterworth filter with a cut-off frequency of 0.35, three kinds of heart axis reoriented slices are obtained: Horizontal Short Axis (HSA), Horizontal Long Axis (HLA), and Vertical Long Axis (VLA) images, as shown in Figure 1. Each of these sets of slices of 1.2 cm thickness is further divided into a Diastolic Subset and a Systolic Subset.

The most informative of the three available data sets is the HSA image set. The main reason is that this latter set can be meaningfully used to calculate the Left Ventricle Ejection Fraction (LVEF). This ratio is defined as the difference between the diastolic and systolic heart volumes, divided by the diastolic heart volume. It is now well known that the left ventricle ejection fraction is one of the most useful parameters describing cardiac function. A normal value of the left ventricle ejection fraction should lie in the range 59%-65%.

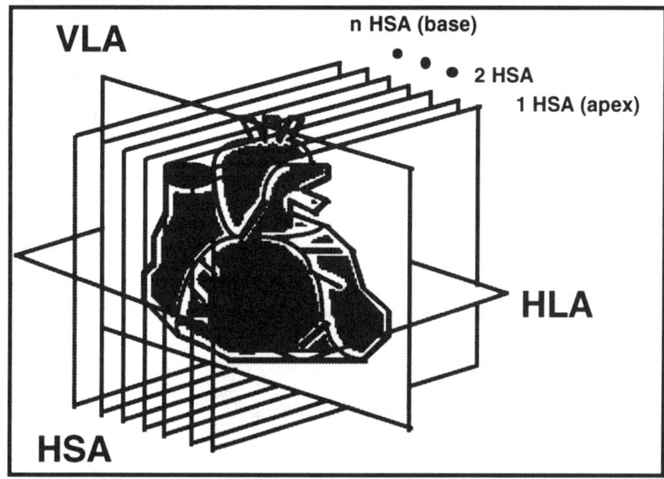

Figure 1. Schematic representation of the different acquired image sets after application of a backprojection algorithm.

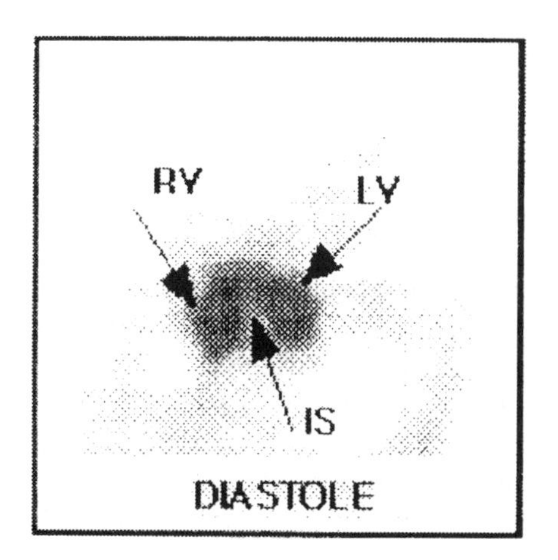

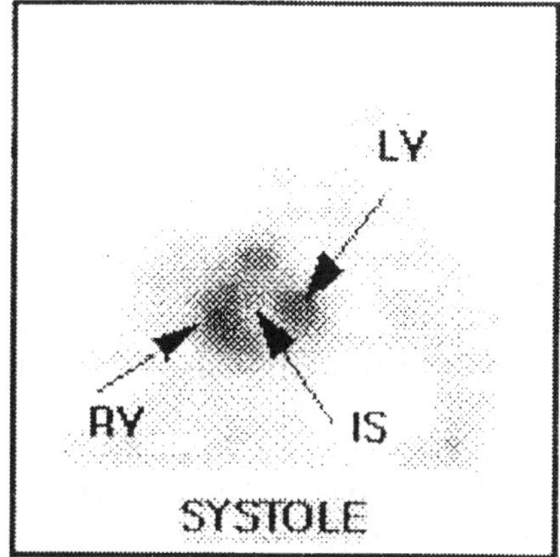

Figure 2. Typical HSA images from the same midventricular slice with Left Ventricle (LV), Right Ventricle (RV), Interventricular Septum (IS), and residual radiation, showing the diastolic and systolic phases of the cardiac cycle.

In our case, the main parts to recognize are the left and right ventricles, the interventricular septum, the atriums, and the valve planes. In addition, due to the specific acquisition and backprojection processes used, a residual radiation contours the whole myocardium (see Figure 2). Further, for the first one or two images in the HSA set, only the heart in its diastolic phase may be seen. This can be understood if one recalls that the ventricular volume is greater in this phase, and that the apical motion of the left ventricle is important. On the other hand, in the last HSA images, usually only the atriums can be seen and not the right and left ventricles (see Figure 3). Thus, for the whole set of images, the actual location of the ventricles in the different images must be established. Conventionally, the last image to be evaluated is the one in which the left ventricle appears and is recognized, because its area is approximately the same in the diastolic and the systolic phases, corresponding to the mitral valve plane.

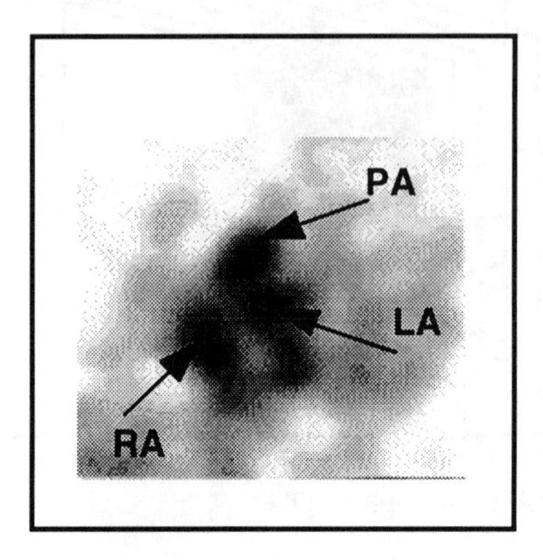

Figure 3. Example of an HSA image at the base of the heart showing the atriums: Right Atrium (RA), Left Atrium (LA), and the Pulmonary Artery (PA).

One of the major problems in the analysis of blood pool myocardial scintigraphy image data is to accurately separate the right and left ventricles. As experimentally evidenced, when the septum is too thin, the gamma diffusion radiation coming from the adjacent ventricles superpose each other, so that the images contain one unique uniform region including the right ventricle, the left ventricle, and the septum. In most of these cases, the expert observer corrects the images manually by erasing the right ventricle and closing the contour of the left ventricle. A second problem encountered when trying to precisely establish the boundaries of the ventricles is the noise introduced into the acquired data during both the acquisition and the reconstruction processes. This leads to blurred boundaries, which complicates separating the two ventricles. A further important problem is related to the automated recognition of the left ventricle. This pattern recognition problem has a high degree of complexity due to the great variety of observed heart forms, which can considerably vary according to the characteristics of a patient and of her/his pathology. Only one fully automated recognition procedure for this task has been reported in the literature [12], using myocardial instead of volume blood pool images. However, some semi-automated methods have also been proposed (see, e.g., [13]).

3. Strategy Of The Proposed Method

3.1. Overview of the method

The proposed algorithm consists of three processing steps, executed sequentially. The first operation is a wavelet-based pre-processing algorithm ([14], [15]), whose task is to remove as much noise as possible from the images, to separate the two ventricles when these organs form a single uniform region, and to determine approximately the border of the left ventricle image region. The second procedure segments the images, and is performed using a neural network implementing the ART (Adaptive Resonance Theory) algorithm [16-18]. Last, a neuro-fuzzy system is used for recognition of the left ventricle and of its contour. For this last procedure, an off-line learning process step is required, to create the rule base for reasoning in the neuro-fuzzy system.

3.2. Wavelets-Based Image Pre-Processing

Wavelets are a mathematical tool for decomposing a signal according to different resolutions. Their great advantage is their ability to analyze a signal with the same accuracy in both the time and frequency domains. This is not the case when applying Fourier analysis, where accuracy is favored either in the frequency domain or in the temporal domain. In other words, increasing accuracy in one domain implies a decrease in precision in the other domain. Further, wavelets are also known for their capacity to identify singularities associated with fine variations of the signal to be evaluated (the image intensity in our case) [19], [20]. Finally, wavelets provide an inherent smoothing property, due to the fact that noise can be filtered out at low resolutions. The difficulty at this stage of image analysis is thus to choose the appropriate resolution at which the images should be analyzed, to achieve the best compromise between noise reduction capabilities and the extraction power of singularities. The multiresolution scheme proposed by Mallat [21] is a possible candidate for solving this difficulty, and enables us to extract signal singularities and to reduce noise at high frequencies. Going from one resolution to the next finer resolution is achieved through doubling the band frequency of the filters associated with the two resolutions, as a result of the underlying sampling operations. Other authors, see for example Nguyen [22], first analyze the frequency response of the family of filters to be used, and then derive a multiresolution analysis scheme by changing the sampling rates of the filters. This permits the analysis of different bands of frequencies. In our method, we propose to analyze the images with two different high pass filters, defining two frequency bands.

In our implementation, we chose to use the first and second derivatives of a Gaussian function as the wavelet functions. The coefficients of the filters are obtained after sampling both functions at a frequency of 0.5 pixel^{-1}. The resulting frequency bands are, respectively, 0.06-0.22 pixel^{-1} for the filter corresponding to the first derivative of the Gaussian function and 0.04-0.16 pixel^{-1} for the one associated with the second derivative of the Gaussian function. The second derivative-based filter is applied first to enhance the gray-level variations between the endocardium and the septum. As a result, the second derivative of the Gaussian function leads to a filter, also known as the Mexican hat filter [15], which has one central positive peak and a negative lobe on each side of this peak. This particular form allows us, while applying the filter by convolution, to correct the filtered values to decrease the values adjacent to the central point of convolution. Consequently, in the images and particularly in the septal regions, this allows us to reduce the image intensities that have been influenced by the nuclear emissions coming from the adjacent heart cavities.

In a second step, the filter based on the first derivative of the Gaussian function is applied to detect the contour points. This filter avoids false detections, as it is well known that detection of the local maxima of the resulting gradient is more optimal than extracting the zero crossings of the second derivative [19], [20]. Further, an averaging low pass filter with constant coefficients of value 0.4 is used to smooth the image after the application of each high pass filter.

The filters are applied as suggested by Mallat. First, all the image lines are processed, and then the columns. After each filtering step, the low pass filter is applied in a direction orthogonal to the high pass filter used.

Choosing the appropriate resolution at which the images should be analyzed depends on both the intrinsic resolution of the acquisition system and on the level of detail to be detected in the images. After evaluation of a rather exhaustive set of images and considering the acquisition conditions, the frequency characteristics of the filters could be defined as indicated above. Accordingly, the wavelets coefficients have been derived enabling optimal processing of the images obtained in our laboratory. Although our algorithm has been specifically tailored to suit our acquisition system, it can also work with other systems, the design of the filters being related primarily to resolution and not to the internal parameters of the camera (in our case, a gamma camera).

3.3. Neural Network Based Image Segmentation

Neural networks, which attempt to emulate biological neural networks [9], are known as non-linear cognitive systems that are able to learn and store new models, which can then be used for subsequent pattern recognition or classification operations. Neural networks are usually implemented as parallel structures, to perform the often large amounts of computations. Further, a wide variety of neural network architectures can be found in the literature. The implementation proposed and developed by Carpenter and Grossberg ([16], [17]) is among the most well known, and is particularly suited for classification tasks. Their algorithm works as follows. Assuming no *a priori* knowledge about the different classes to be defined during the process, for each element in an initial set of data to be classified the neural network performs a first classification operation according to a similarity measure. For that purpose, one element in the set is chosen as the representative of a first class and designated as its leading element or leader. Then, if another element from the initial set can be considered similar to this leader, it is classified in the corresponding class. If this element fails to be similar enough, it is used as the leader of a new class. After all elements are processed by the neural network, the result is a set of leaders, each representing a class defined by the elements fulfilling the similarity criterion associated with the corresponding leader.

In our application, for segmenting the pre-processed images, we chose to implement a neural network similar to the structure proposed by Carpenter and Grossberg. As a result, each class defined after completion of the processing is associated with a specific region in the image, the area corresponding to the first initial class being defined as the region corresponding to the left ventricle. Further, the dimensional characteristics of each region depend strongly on the gray level value of the pixel chosen as the leader of the initial class. Accordingly, if the leader of the first class is chosen randomly, there is no usable criterion available to certify that the dimensions of the corresponding region are indeed those of the left ventricle. To improve the accuracy of these measures, we chose to initialize the classification step with the image intensity value of a given pixel, previously extracted by the wavelet-based filtering step. This pixel is defined, in our case, as the contour point located on the boundary between the region corresponding to the left endocardium and the region associated with the septum. This choice defines the gray level value of the initial class to be used for the neural network based classification and segmentation step.

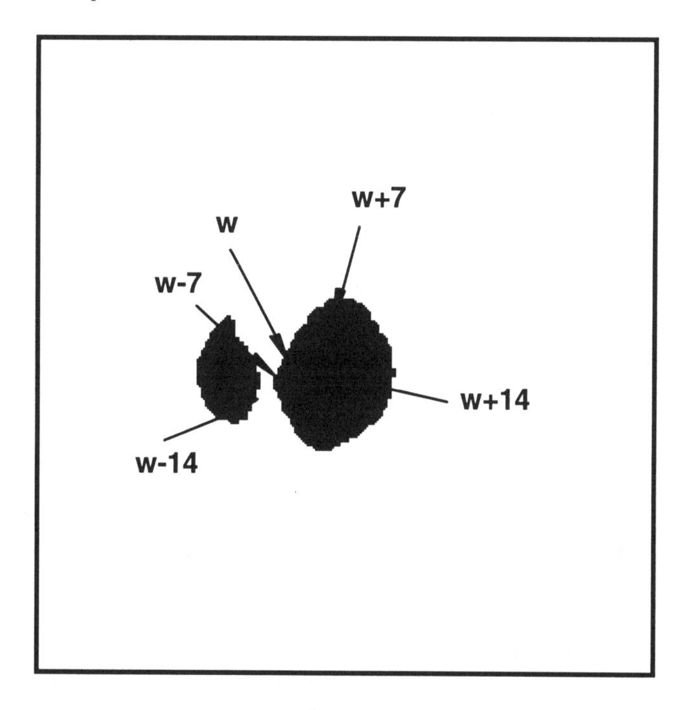

Figure 4. Typical resulting segmented image (zoom: ×2) after neural network based processing (the corresponding input HSA diastolic slice is given in Figure 2).

Our implementation of the neural network uses two layers, the first layer being the so-called input or comparison layer, and the second one being the output or classification layer. Each layer is composed of five neurons, whose transfer functions are defined as Gaussian functions. The third of these five Gaussian functions has its central value located at the gray level value *w*, which is given by the specific contour point output determined in the wavelets-based processing step. The other central peak values are then defined, respectively, through displacements by multiples of seven. The central values of the Gaussian functions are thus placed as follows: at *w*-14, *w*-7, *w*, *w*+7, *w*+14. (The value 7 has been chosen heuristically to achieve a good segmentation resolution around *w*). This allows us to establish guidelines, to segment the image and classify its contents, without establishing the total number of classes and their centers in advance. This approach for classification is referred to as "partially supervised classification" [23]. The neural network-based segmentation method developed has the advantage of not needing the operator to initialize the ART procedure. In addition, the implementation is straightforward. Even though both initial class and gray level resolution are important parameters of the segmentation task and control the quality of the results, it has to be recalled that the initial leaders (acting as some kind of seed points) correspond to meaningful points delivered by the wavelet-based extraction procedure. The implementation of other methods, such as the so-called *k-means approach*, would require more effort without providing results of better quality. The same comment applies to the completely supervised *k* nearest neighbors method. In comparison to more standard region segmentation techniques (such as, e.g., region growing or split and merge approaches), our approach has the advantage of delivering results of comparable quality in much shorter times, as there are, strictly speaking, no iterations to execute. A typical segmentation result, corresponding to the HSA diastole slice shown in Figure 2, appears in Figure 4.

3.4. Fuzzy logic-based recognition of the regions of interest (ventricles)

3.4.1. Definition of the required fuzzy sentences

One of the most difficult steps in medical image evaluation systems that attempt to automatically recognize organs, such as ventricles in G-SPECT images, is to differentiate the organ of interest (in our case, the left ventricle) from the other organs represented in the image, such as vessels, etc. As a matter of fact, for our specific application, even in the cases where just the right and left ventricles are visible in the images, it is hard to establish without ambiguity which of the two is the left ventricle. This is due to changes, from one patient to another or even from one image slice to another for a given patient, in both the position and form of the ventricle.

In order to associate a given region in the images with the left ventricle, we can assume that its cavity should be contained within the limits defined by two concentric segmented regions of respectively minimal and maximal size, which define some kind of dimensional tolerances for the imaged ventricle. The neural network based segmentation process generates regions of nearly the same gray level value, whose borders are poorly defined "level" lines in the original gray value images. Each of these regions is a tentative candidate for being recognized as the left ventricle. If we take each of these level lines separately (as illustrated in Figure 4) for determining the corresponding enclosed region and its association with the myocardium, several hypotheses may be generated for relating the left ventricle to a segmented region. In Figure 5, we show the different associations that can be issued after a typical image segmentation step. We call this series of images "exploded images." The first image at the top left will be immediately discarded because of its shape and dimensions, and the four remaining images remain candidates.

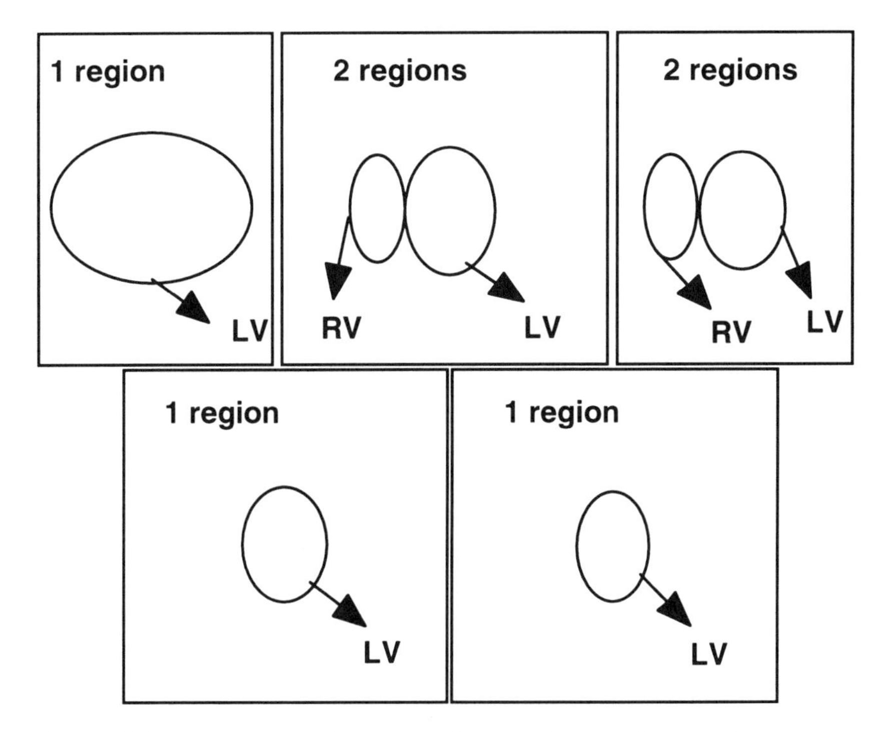

Figure 5. Left-ventricle region association hypotheses after a typical segmentation step (the input image is given in Figure 4).

To permit the automated recognition of the region to be considered as the left ventricle, each region is described with a series of measured geometric parameter values. These measures, considered the most descriptive for the image contents given our final goal, are the following:

• *Number of slices in the image set.* This enables us to determine which region of the myocardium (in the direction of the short axis) is currently being analyzed. This region usually is either the apex, the base of the heart, or the area in between. For the three types of regions, both the shape and the dimensions of the myocardium may vary quite largely.

• *Number of regions in the image.* This value can be helpful for locating the left ventricle in the images and for accurately defining its borders with the atriums. As demonstrated experimentally, when we recognize the apex of the myocardium in the image, then we surely have only one region in the image, the left ventricle, with a high degree of confidence. Continuing with this reasoning, when we are in a region corresponding to the middle of the myocardium, we should encounter two other regions corresponding, respectively, to the left and right ventricles. Also, near a region corresponding to the base of the heart, where the boundaries with the atriums are located, the number of segmented regions should increase to three or even more. Of course, smaller segmented regions produced by arteries and other organs (such as the great vessels) can also be observed and make recognition an even more difficult task.

- *Area of a region.* This parameter becomes very meaningful when associated with the left ventricle region. It is, on one hand, very informative about the growth of the region associated with the left ventricle from the apex to the base, and thus can serve to maintain coherence when assigning sense to the regions in the various slices. On the other hand, its value can also be ultimately used to calculate the ejection volume fraction, as a pixel corresponds, in fact, to a spatial elementary volume of the organ (i.e., a voxel).
- *Circularity of a region.* This value describes the shape of the left ventricle region in each slice of the set of HSA images to be evaluated. At the beginning of the image sequence, the region corresponding to the apex appears more or less like a circle. Further on in the sequence, this region progressively takes on an increasingly elliptical form. In our implementation, the circularity value is determined as follows. First, the area and the perimeter of each region are computed. Then, we calculate the two radii corresponding respectively to this area and this perimeter. Last, the ratio between these two radii is defined as the desired circularity value (the coefficient being equal to 1 if the region is circular). We have implemented an automated procedure to compute the area and perimeter of each region without operator manipulation, using some basic functions from the public domain NIH image processing software [24].
- *x and y Coordinates of the center of gravity of a region.* These values allow us to locate the region under evaluation in the image. It should be noted here that all evaluations are carried out in the short axis plane.

After having evaluated this parameter set for all the segmented regions present in the set of slices under investigation, we obtain a data structure called *heartinfo*, an example of which is given in Table 1. With the exception of the data of column 2, added here to provide a better understanding of the measures shown, this data structure is the basic input for the subsequent recognition process based on fuzzy logic techniques. The data in the fourth column is normalized in the following way. In the first exploded image of the first slice of the set, the area of the region considered as the left ventricle is taken as a reference. The area of the regions in all the other exploded images (and thus also in the other slices) is divided by this reference value. Thus, a relationship can be established between the slices, since the resulting normalized coefficient generally grows from the first to the last slice. In this sense, the coefficient acts as some kind of inter-slice information. Data in the sixth and seventh columns are normalized with respect to the maximum value in the respective column.

Table 1. Example of measures summarized in the data structure *heartinfo*. The parameters correspond to the regions obtained after segmentation of a set of five HSA slices. The fifth slice corresponds to the image given in Figure 4.

Slice number	Exploded image	Number of regions	Area of possible LV	Circularity of possible LV	*X* Position of possible LV (pixels)	*Y* Position of possible LV (pixels)
1.0000	1	1	1.0000	1.8066	0.7143	0
1.0000	2	1	0.3613	1.9333	0.7143	-0.2000
1.0000	3	1	0.2514	1.9450	0.7143	-0.2000
2.0000	1	1	0.7948	1.8963	0.7143	-0.1000
2.0000	2	1	0.4306	1.9669	0.7143	-0.2000
2.0000	3	1	0.3006	1.9541	0.5714	-0.2000
3.0000	1	1	0.9480	1.8883	0.5714	0.1000
3.0000	2	1	0.4249	1.9536	0.5714	-0.1000
3.0000	3	1	0.2659	2.0001	0.5714	-0.1000
3.0000	4	1	0.1474	2.0253	0.5714	-0.2000
4.0000	1	2	2.0318	1.8612	0.5714	0.3000
4.0000	2	2	1.6936	1.8860	0.4286	0.3000
4.0000	3	2	1.4017	1.8588	0.4286	0.3000
4.0000	4	2	1.1705	1.8530	0.4286	0.3000
4.0000	5	2	0.7081	1.8192	0.4286	0.2000
5.0000	1	1	2.3555	1.9094	0.5714	0.2000
5.0000	2	2	2.0058	1.8866	0.4286	0.3000
5.0000	3	2	1.6792	1.9201	0.4286	0.3000
5.0000	4	1	1.3671	1.9036	0.4286	0.3000
5.0000	5	1	0.8873	1.8541	0.2857	0.3000

To analyze and interpret this data structure, sentences defined using fuzzy logic approaches are applied. Since there is no commonly agreed model available for representing the myocardium in every shade of representation that clinicians encounter in medical routine, fuzzy logic is used to circumvent this weakness. As a prerequisite, a knowledge base has to be devised. This knowledge base should apply to all interpretation contexts encountered when evaluating a sequence of slices. It should allow us to identify the right values (fixed or in specified sub-ranges) for the parameters used to recognize the left ventricle among all the segmented regions. This is achieved by taking advantage of the fuzzy logic framework, which allows us to specify knowledge in the form of a set of rules expressed in a natural language. These rules will constitute the inference engine based on the well-known *generalized modus ponens* law summarized below:

Implication : **IF** fuzzy_antecedent **THEN** fuzzy_consequence
Premise : fuzzy_antecedent **has** degree_of_truth
Conclusion : consequence **has** certainty_value

Using this mechanism, we can build rules following the model given in the example below:

IF `Area_of_the_region is Big` **THEN** `Region is The_left_ventricle`

`Area_of_the_region` and `Region` are antecedent variables. `Big` and `The_left_ventricle` are the fuzzy antecedent and consequent sets respectively. When fuzzy sets have only one constant element, they are called singletons. In our application, `The_left_ventricle` would be a singleton.

Combining all the measured parameters encoded in *heartinfo* into one consistent rule enables defining a powerful tool to discriminate image data. In our application, the measured parameters have been combined as follows:

IF
`Number_of_slice_in_the_image_set` is
`Member_of_the_first_acquired_images`
and `Number_of_particles_in_the_image is Small`
and `Area_of_the_region is Big`
and `Circularity_of_the_region is High`
and `x_Coordinate_of_the_region is`
`Well_horizontally_centered`
and `y_Coordinate_of_the_region is`
`Well_vertically_centered`
THEN
`Region is The_left_ventricle`

In the above rule, new fuzzy sets, such as `Small`, `Big`, `Well_vertically_centered`, `Member_of_the_first_acquired_images`, `High`, and `Well_horizontally_centered`, have been introduced. In fact, fuzzy logic makes no assumption, nor defines rules, on how many fuzzy sets must be established to meaningfully determine the value of a fuzzy variable. In the same way, the membership functions describing the fuzzy sets may take on various user-defined forms. The Gaussian, triangular, and trapezoidal membership functions are among the most popular and commonly used.

An experienced user can take advantage of this freedom in defining the membership function shapes and in choosing the number of fuzzy sets, allowing her/him to set or to modify the fuzzy system according to its accumulated experience when applying the system on real data.

In earlier contributions, we heuristically established the number of fuzzy sets and the parameters (coefficients) of the membership functions [25], [26]. Although the corresponding implementations were able to deliver rather good results when evaluating real-life G-SPECT image sequences, sensible application required a long time to manually tune the system parameters. Further, a clear disadvantage is that this parameter value determination procedure must be evaluated again every time the knowledge base is updated, as either new rules are added or new membership functions are devised. To overcome this limitation and to make the implementation more versatile, we recently developed a new module that automatically creates rules and membership functions, based on evaluated image sequences of patients being diagnosed. Using this learning set of data, the knowledge base is computed by this module using an approach that combines the so-called neural network ART algorithm and the fuzzy system FUNNY [27]. The automated determination of the knowledge base and the associated learning phase will be detailed in Section 3.5.

3.4.2. Combining neuronal approaches and fuzzy logic-based inference systems

Neural networks can contribute decisive and important advantages to the efficiency of fuzzy logic based systems. Among other interesting features of neural networks, the ability to perform computations in parallel, the ability to model a nonlinear function, and the invaluable learning capacity can be mentioned. The neuro-fuzzy systems are based on various types of learning processes. These approaches can tentatively be classified as follows:

- The capacity to define (learn) new membership functions for the manipulation of the fuzzy sets. However, the rule base is defined *a priori* and thus cannot be changed.
- The capacity to define (learn) new rules. However, the membership functions are considered specified *a priori* and thus remain constant.
- The capacity to adapt the parameters defining the membership functions in "real time" (that is during execution of the application).
- The capacity to change the weights of the rules in the rule base in "real time."

Several approaches combining the learning capacities descibed above have been proposed. For example, Berenji and Khedkar [28], [29] have suggested a system based on what they call "reinforcement learning algorithm". The approach consists of evaluating the actual and next possible states of the system, based on the history of chosen actions. The evaluation results are then used to determine or modify the membership functions, thus implementing the learning capacity. The same learning behavior can be achieved using the FUNNY (FUzzY or Neural Net) system developed by Bersini and Gorrini [27]. This latter system uses the gradient based learning method, first described by Nomura, Hayashi, and Wakami [30]. A comparable approach is used in the ANFIS (Adaptive Neuro-Fuzzy Inference System) method proposed by Jang [31], [36]. Other contributions go a step further and suggest approaches providing not only a learning capacity for the membership functions, but also the ability to dynamically change the dimension of the rule base. The work of Sulzberger et al. [32] and of Nauck and Kruse ([33], [34]) are representative examples of this method.

Most of the implemented systems that include a learning capacity for the membership functions are designed using the so-called Sugeno fuzzy model [30], [35]. Sugeno fuzzy rules are characterized by the fact that rules are of the form:

IF *x is A and y is B* ***THEN*** $z = f(x,y)$

In the rule template above, x and y are antecedent variables, A and B are fuzzy antecedent sets and $z = f(x,y)$ is a so-called crisp function in the consequent part of the rule. The function f is a polynomial in x and y, and the order of the polynomial crisp function determines the order of the resulting Sugeno fuzzy system. The most widely used are first order Sugeno systems. The Sugeno system of order zero is identical to the system described above, except the consequent part of the rule that is a constant.

Jang ([35], [35b]) built an equivalent to the Sugeno first order inference system using neural networks. In his approach, the following two Sugeno fuzzy rules *R1* and *R2* find their equivalent in Jang's ANFIS system as indicated in Figure 6.

R1 : ***IF*** *x is A1 and y is B1* ***THEN*** $z1 = f1(x,y)$
R2 : ***IF*** *x is A2 and y is B2* ***THEN*** $z2 = f2(x,y)$

Jang has further suggested differentiating the nodes in the neural network, in order to use a part of them as adaptive nodes for which the parameters may be changed, and to use the other part as fixed nodes having no parameters. The former type of node is graphically represented by squares while the latter type is shown as circles in Figure 6.

The first layer or input layer of the neural network structure suggested by Jang evaluates the set of premises of the antecedent parts in the fuzzy implications defined for our system. To achieve this, each input premise is evaluated by the node in the first layer associated with a fuzzy set. Thus, each of these nodes acts as an adaptive node, and has, as a transfer function, the membership function defining the associated fuzzy set. The set of parameters defining the different membership functions in this layer is the set of the corresponding premise parameters. The second layer is constituted by fixed nodes each defined by the fuzzy function "and." The algebraic product and the minimum between two numbers are the most popular operators used for implementing the "and" functions. The value output by each node of the second layer represents the so-called fired strength of the associated rule. The task of the nodes in the third layer is only to normalize the set of fired strengths. This is achieved using a fixed node, which simply computes the ratio between the fired strength of its corresponding rule and the sum of the fired strength of all the rules manipulated by the system. The fourth layer is again an adaptive layer, the nodes of which evaluate functions of the type $z = f(x,y)$ which represent the consequent parts of Sugeno fuzzy rules as previously described. These functions take as input the normalized fired strengths delivered by the third layer. The parameters defining the polynomial functions are called consequent parameters. The last or output layer of the neural network is built using a fixed node which simply computes the sum of all the consequents parts output by layer four.

In our implementation, we followed a similar approach, the inference system being slightly different in comparison to the ANFIS system. In Figure 7, we give the graphical representation of our system, making use of the ANFIS graphical convention.

We also decided to combine the last three layers of the neural structure described above into one layer acting as the output layer. This layer, placed at the end of the network, uses a fixed node whose transfer function is called max operation. The reason for our choice is that, considering our ultimate goal, we want to use only rules enabling us to recognize the left ventricle. In our implementation, on one hand, if a region in a given image does not correspond to the ventricle, then there will be no rule fired.

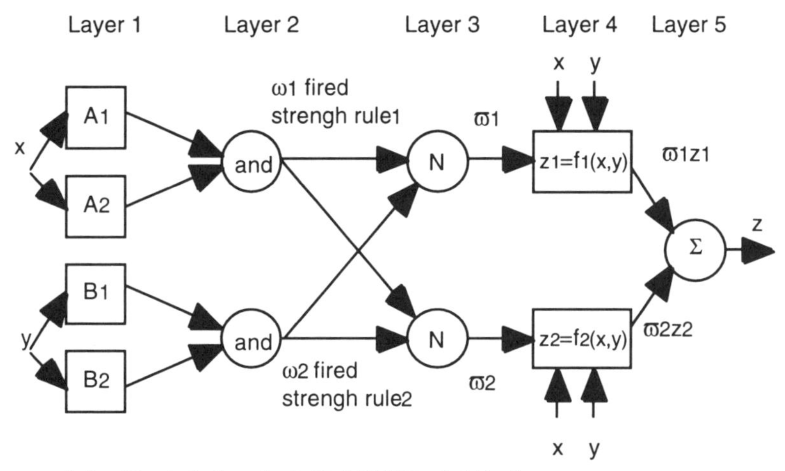

Figure 6. Equivalent ANFIS rule for a two-input first-order Sugeno fuzzy model with two rules (after J.-S. R. Jang in [35b]).

On the other hand, we still need several rules describing the various situations in which the left ventricle has to be recognized. Accordingly, when a region actually corresponding to the left ventricle is input to the system, the neural network will fire one or several rules of the rule base, corresponding to one particular type of left ventricle. We have thus established a series of rules in a rule base that describe the left ventricle. The region under evaluation will be recognized as such if its parameters fulfill, at best, one of the rules in this rule base. Therefore, we take the maximum strength of the fired rules as the output of the network.

In our case, the membership functions defining the fuzzy sets for the premises used in the nodes of the first layer are isosceles triangular functions, the parameters of which are determined during the training (learning) phase of the recognition system. During the same learning process, the number of membership functions corresponding to each input, and the number of corresponding rules are also determined.

In order to exemplify how recognition of the left ventricle in an image is achieved, let us take the image shown on Figure 1. The first steps of our recognition procedure consist of applying the wavelets based pre-processing algorithm and the neural network based segmentation operation. The result of these operations has already been shown in Figure 3. Next, the resulting image is "exploded," that is, the various tentative interpretations are determined, as shown in Figure 4. In the resulting set of exploded images, the parameters of the regions are evaluated and stored in a table, as depicted in Table 1. This data structure is the input data for the subsequent recognition processing steps.

We can now apply the neuro-fuzzy logic system represented in Figure 7. The membership functions and the rule base have been defined during the training of the system using data from three patients. As already described, the system works with triangularly shaped membership functions for the input nodes.

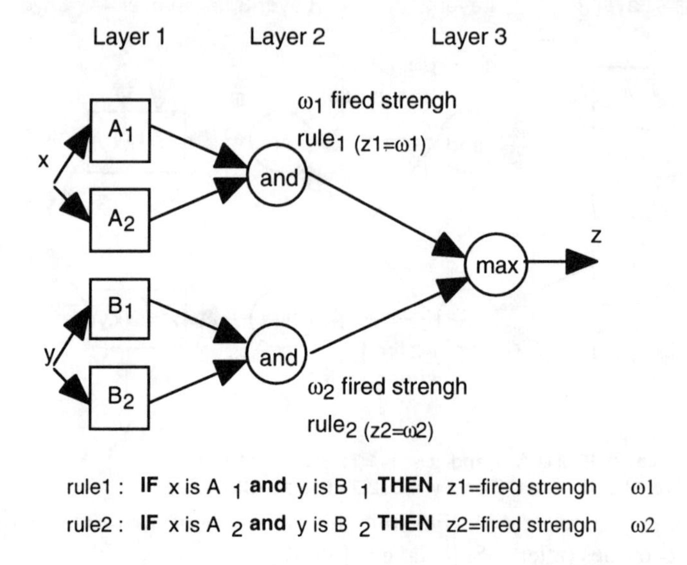

Figure 7. Graphical representation of the equivalent neural network inference system, as implemented.

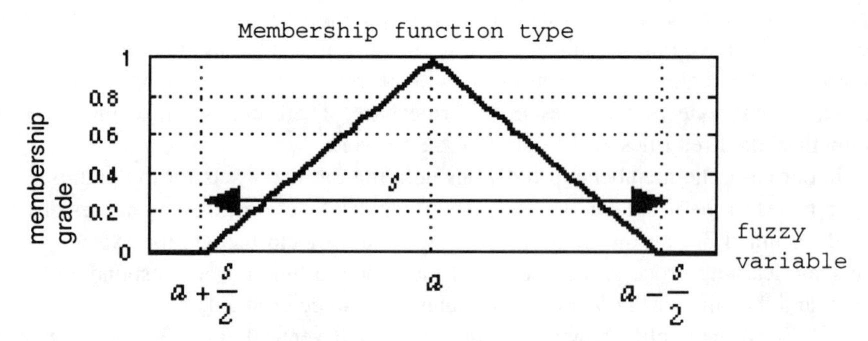

Figure 8. Shape and parameters of the triangular membership functions used.

These symmetric functions are characterized by their center a and their support s, as shown on Figure 8. The complete set of membership functions is represented in Figure 9. Each line shown representing the membership function corresponds to an input parameter of the first layer.

The rule base itself contains 62 rules. However, in the interest of simplicity, we give a representative sample of only six rules likely to be fired during the processing. This is shown using a matrix representation in Table 2.

Table 2. Matrix representation summarizing the parameters of the six rules able to be fired

	fuzzy set associated with label1 with mf	fuzzy set associated with variable2 with mf	fuzzy set associated with variable3 with mf	fuzzy set associated with variable4 with mf	fuzzy set associated with variable5 with mf	fuzzy set associated with variable6 with mf
rule 1	mf11	mf21	mf31	mf43	mf51	mf61
rule 2	mf12	mf21	mf33	mf41	mf51	mf61
rule 3	mf13	mf21	mf33	mf41	mf51	mf61
rule 4	mf13	mf21	mf35	mf42	mf51	mf62
rule 5	mf14	mf22	mf37	mf41	mf52	mf62
rule 6	mf15	mf22	mf34	mf41	mf52	mf62

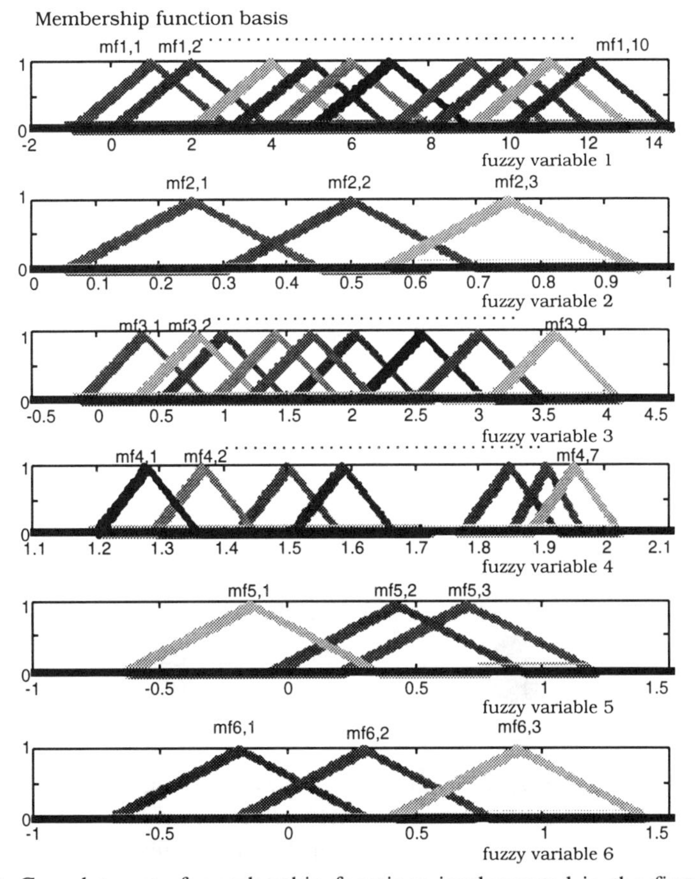

Figure 9. Complete set of membership functions implemented in the first layer of the neuro-fuzzy recognition system.

The (compact) representation of the rules discussed above can be interpreted in the following way for rule #1:

IF parameter1 is The_fuzzy_set_associated_to_variable1_with_mf11
and parameter2 is The_fuzzy_set_associated_to_variable2_with_mf21
and parameter3 is The_fuzzy_set_associated_to_variable3_with_mf31
and parameter4 is The_fuzzy_set_associated_to_variable4_with_mf41
and parameter5 is The_fuzzy_set_associated_to_variable5_with_mf51
and parameter6 is The_fuzzy_set_associated_to_variable6_with_mf61
THEN left_ventricle is the fired strengh of the rule.

After evaluation of the image data, we obtain the fired strengths of the rules used and their associations with the corresponding regions. The max operator is then applied to each processed image to choose the region most similar to the description of the left ventricle encoded in the rules. In Table 3, the computed fired strengths are given, together with the selected region representing the left ventricle (shaded zone). In Figure 10, this same selected region is shown in a square. The other left ventricle hypotheses for the remaining exploded images are shown circled.

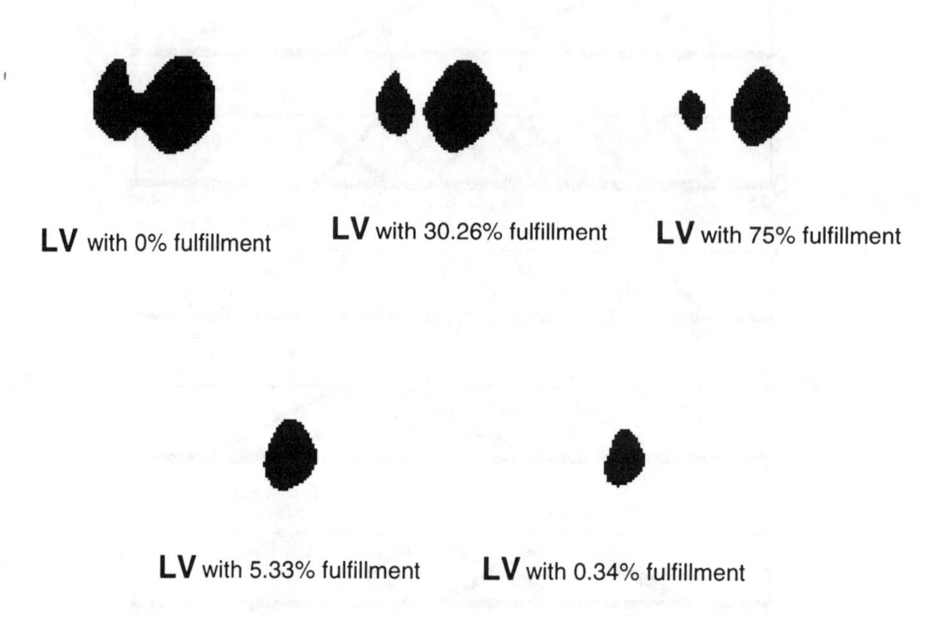

Figure 10. Selected regions in each exploded image resulting from partitioning an actual input image (see also Figure 5). The region representative of the left ventricle is shown in a square.

This result allows us to, first, isolate the right region and, second, to automatically extract the contours of this area, which describes the left ventricle. This is achieved by knowing the gray level value of the finally selected region and its location. Finally, a so-called *zigzag search algorithm* is applied to the retained region in order to contour the whole region. The algorithm is executed until all the perimeter points have been collected. The principle of this final operation and its result are shown in Figure 11.

Table 3. Fired strengths computed and selected representative for the left ventricle (results correspond to the example given in Section 4.1, Table 1, and Figure 5).

Slice	Explode	fired	fired	fired	fired	fired	fired
5.0000	1	0	0	0	0	0	0
5.0000	2	0	0	0	0	0.2241	0.3026
5.0000	3	0	0	0	0	0.7507	0.1137
5.0000	4	0	0	0.0533	0.0524	0	0
5.0000	5	0	0	0.0034	0	0	0

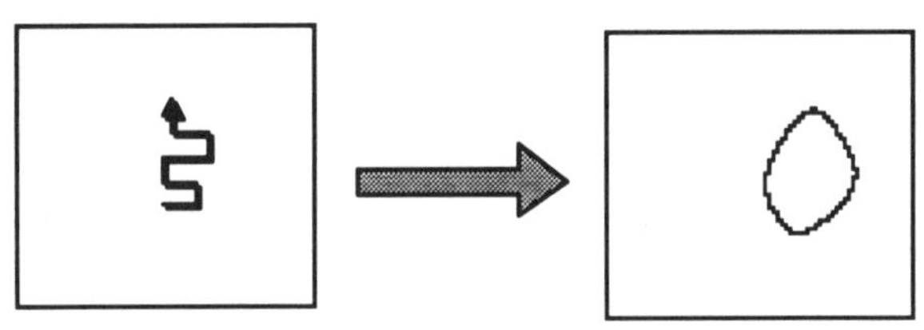

Figure 11. Principle of the final processing step (zigzag search algorithm) and its result.

As a conclusion to this section, a typical example of the final results, obtained for a whole set of diastolic and systolic images of a patient, is shown in Figure 12.

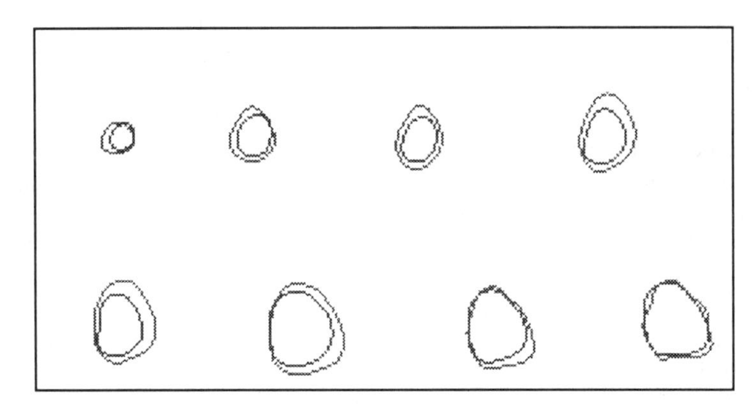

Figure 12. Typical final results for a full set of images of a patient. Diastolic contours are the outer lines, systolic contours are the inner lines.

3.5. Training the recognition system using a neuro-fuzzy technique

The purpose of training the recognition system is twofold. First, it allows us to create and determine the parameters of the membership functions associated with the fuzzy sets used to evaluate each of the input measurements. Second, it enables us to set up the rule base encoding the knowledge needed to recognize the left ventricle. It is a requirement for the system that these tasks be automatically performed. This makes the system comfortable to use. It may be necessary to enlarge the knowledge base to improve its recognition ability, or to correct some particular rules. This will reduce the time needed to tune the system. Furthermore, this enables us to deal with a knowledge base containing a relatively large number of rules and membership functions, a task that may not be easy for a human operator. We can meet our requirement through the parallel combination of two methods. These are the ART leadership algorithm described in Section 3.3., also used during the image segmentation step, and a neuro-fuzzy learning approach inspired from the "FUNNY" system developed by Bersini and Gorrini [27]. In fact, use of the ART-based approach is sufficient to satisfy our requirements. We have, however, chosen to add the neuro-fuzzy learning mechanism to cope with situations for which classification is poor or wrong, even though the necessary rules and membership functions are available. In this fashion, we are able to somewhat optimize the implemented system.

3.5.1. Automated generation of rules and membership functions (ALGORAM)

In this section, we introduce a new approach for automatically producing the membership functions and rules to be stored in a knowledge base. This method makes use of ALGORAM, an acronym for "Automatic Leader Generation Of Rules And Membership functions." The method has been devised in a manner similar to that used for the design of the leadership algorithm described in Section 3.3. We have chosen to partially label the data, as we decided to select several leaders from the input set. Recall that the groups forming the initial partition of the data are known *a priori.* We then select a leader in each group and arrange them into a vector that is said to be labeled.

To gain a better understanding of the proposed method, consider again the data shown in Table 1, summarizing the parameters corresponding to the regions obtained after segmentation of a typical input image, for which we want to build a knowledge base to be used in further recognition and interpretation of the region corresponding to the left ventricle. Recall that each row in this table corresponds to an "exploded" image, extracted from a given slice in the whole image set. For example, there are four "exploded" images for the first, second, and third slices and five "exploded" images for the fifth slice. Each slice can thus be considered to constitute a group of partitioned images. For each slice, or equivalently the associated group of partitioned images, the operator of the system manually designates a leader to start the training phase, pre-labeling the regions of interest. This leader is naturally chosen as being the segmented region, which can be considered to correspond to the left ventricle. If the selection of these initial leaders is poorly performed, the training system will need more iterations to learn to recognize and reject regions that exhibit only weak coherence among them. Consequently, the system generates supplementary membership functions for each input variable to avoid misclassification. This results in a decrease of speed and can be, in some extreme cases, the cause of mistakes when evaluating other patients. Table 1 can then be expanded to include the labels assigned to the leaders, as shown in Table 4.

Table 4. Expansion of Table 1 to include labels for the exploded images in each slice corresponding to the left ventricle.

Slice	Particles	Area	Circularity	*x* Position	*y* Position	Label
1.0000	0.2500	1.0000	1.8066	0.7143	0	0
1.0000	0.2500	0.3613	1.9333	0.7143	-0.2000	1
1.0000	0.2500	0.2514	1.9450	0.7143	-0.2000	0
2.0000	0.2500	0.7948	1.8963	0.7143	-0.1000	1
2.0000	0.2500	0.4306	1.9669	0.7143	-0.2000	0
2.0000	0.2500	0.3006	1.9541	0.5714	-0.2000	0
3.0000	0.2500	0.9480	1.8883	0.5714	0.1000	1
3.0000	0.2500	0.4249	1.9536	0.5714	-0.1000	0
3.0000	0.2500	0.2659	2.0001	0.5714	-0.1000	0
3.0000	0.2500	0.1474	2.0253	0.5714	-0.2000	0
4.0000	0.5000	2.0318	1.8612	0.5714	0.3000	0
4.0000	0.5000	1.6936	1.8860	0.4286	0.3000	0
4.0000	0.2500	1.4017	1.8588	0.4286	0.3000	1
4.0000	0.2500	1.1705	1.8530	0.4286	0.3000	0
4.0000	0.2500	0.7081	1.8192	0.4286	0.2000	0
5.0000	0.5000	2.3555	1.9094	0.5714	0.2000	0
5.0000	0.5000	2.0058	1.8866	0.4286	0.3000	0
5.0000	0.5000	1.6792	1.9201	0.4286	0.3000	1
5.0000	0.2500	1.3671	1.9036	0.4286	0.3000	0
5.0000	0.2500	0.8873	1.8541	0.2857	0.3000	0

In the following, we denote by x each of the input regions to be evaluated, and t_l the partially labeled vector built by grouping together the leaders selected for each slice or partition group. It must, however, be understood that vector t_l will act more as a set of flags providing links to the leaders required by our algorithm than as the expected value to be returned by the recognition system. For example, for the fourth slice, the fourth exploded image could be chosen to tentatively represent the left ventricle. However, in this example, the third image of the partition was preferred, because, in the diastolic phase, the surface of the imaged area is expected to be rather large. However, in the case of a patient with a smaller heart, the input region could better match the fourth region in the fourth slice. It would also make no sense to have the recognition system returning a value of zero. As indicated in Section 3.4.2, the only requirement is that the region tentatively associated with the left ventricle leads to a value larger than those of all the other regions in the slice. Indeed, all the candidate regions for representing the left ventricle should be characterized by high values, whereas regions corresponding to the right ventricle, the atriums, or great vessels should have small values.

The problem to be solved can now be formalized. For that purpose, let us define the variables and parameters as indicated in the table below.

VARIABLE		DESCRIPTION
X		is the set of input regions which can be recognized as representing the left ventricle (candidate regions)
x_{ij}	i = 1,..., m j = 1,..., n	is the collection of m region vectors, each characterized by n parameters
G_k	k = 1,..., o	are the m partition groups obtained after segmentation, or, in other words, the collection of m slices in a given patient image set
sl_{kj}	$k = 1,..., o$ $j = 1,..., n$	is the collection of selected leader regions, each characterized by n parameters
f_i	$i = 1,..., m$	is the collection of labels acting as flags pointing to the selected leader regions
y_i	$I = 1,..., m$	is the collection of the m crisp outputs of the recognition system corresponding respectively to the m input vectors (see Section 3.4.2)
rl_{ki}	$k = 1,..., o$ $j = 1,..., n$	is the collection of returned leaders, each characterized by n parameters
MF		is the set of membership functions
q_j	$j = 1,..., n$	is a vector specifying the number of membership functions to be used for each input
$mf_{jp'}$	$j = 1,..., 3\times n$ $p = 1,...,q_j$	is the collection of membership functions for the recognition system
sg_j	$j = 1,..., n$	is a constant vector containing n sigma values to be used as a support to built a given membership function
RU		is the set of rules
ru_{hj}	$h = 1,..., r$ $j = 1,..., n$	is the collection of r rules characterized by n parameters to be used for recognition

As an example, according to the set of images corresponding to the data collected in Tables 1 and 4, the parameters m, n, and o are $m = 22$, $n = 6$, and $o = 5$. The values of the parameters p, q, and r will be fixed during the training phase as shown later. The training and learning process now proceeds as follows.

a. Automated generation of rules and membership functions for the selected leader in the first slice of the image sequence. We first build triangularly shaped membership functions such that their parameters fulfill the relations $a_{1j} = sl_{1j}$. The support of these functions is given by the sigma vector *sg*. Choosing small values of sigma for each parameter leads to an increase of the knowledge base since more membership functions and rules must be used to classify the data. Membership functions are finally stored using the following format:

$m_{1p} = sl_{1j} - sg_j/2,\ j = 1,\ldots,n$ and $p = 1, 4, \ldots, (3\times n) - 2$

$m_{1p} = sl_{1j},\ j = 1,\ldots, n$ and $p = 2, 5, \ldots, (3\times n) - 1$

$m_{1p} = sl_{1j} + sg_j/2,\ j = 1,\ldots,n$ and $p = 3, 6, \ldots, (3\times n)$

Accordingly, the rules matching these membership functions are also created and the vector **q** is also updated. That is

$ru_{1j} = 1$ and $q_j = 1,\ j = 1, \ldots, n$

b. Verification step. This step checks if the leader in the next slice of the image sequence leads to firing. Indeed, it is possible that, in this slice, the region corresponding to the left ventricle has just slightly changed. In this case, it will be recognized using the same rules and membership functions as for the leader in the preceding slice. If the changes are too important, some of the parameters are no longer in agreement with the currently defined membership functions, and new functions and rules have to be generated, to take into account the data which have not led to firing.

c. Addition of rules and membership functions for the parameters which have not led to firing. In most such situations, only one or two parameters did not produce firing for a newly selected leader. Thus, we need only to add the necessary supplementary information. This avoids rewriting a whole vector of membership functions and the resulting duplication of effort. Also, the vector q is updated, incrementing component q_j by one unit if j is the index of the input with no firing. The update of the membership functions is slightly different from their automated generation, as shown below:

$m_{q(j)p} = sl_{q(j)j} - sg_{q(j)}/2,$
where j corresponds to data unable to fire and $p = (3\times j) - 2$
$m_{q(j)p} = sl_{q(j)j}$,
where j corresponds to data unable to fire and $p = (3\times j) - 1$
$m_{q(j)p} = sl_{q(j)j} + sg_{q(j)}/2,$

where j corresponds to data unable to fire and $p = (3\times n)$

The corresponding new rules are created, based on the existing rules. Even if a rule is not fired, some of its parameters are evaluated. The fired strengths of the rules are thus recalculated, with the difference being that, in this case, the product implication operator is replaced by the algebraic sum. This enables us to select the rule with the greatest fired strength. This rule in turn points to a region with characteristics similar to that of the leader inducing the updates. The rule is simply modified as follows:

$ru_{h+1j} = ru_{1j},\ j=1,\ldots,n$
$ru_{h+1j} = q_{j},$ where j corresponds to data unable to fire

d. Verification of classification quality. During this step, we verify that the leader returned by the recognition system is the same previously selected leader. The returned leader parameter vector rl_{kj} is obtained as follows:

$$rl_{kj} = x_{uj}, \text{ with } y_u > y_v,, u \bullet v \text{ and } x_{uj} > x_{vj}, \in G_k$$

To assess the similarity between the two parameter vectors, we compute the Euclidean distance between the selected leader parameter vector and that of the returned leader in the respective partition group G_k :

$$D_k = //sl_{kj} - rl_{kj} //$$

If the distance above is equal to zero, we can proceed with the treatment of the next leader to evaluate and verify the quality of classification. In this case, the two groups G_k and G_{k+1} are combined into a called super-group. In order to assign a unique representative region to this super-group, the parameter vector rl_{kj} or rl_{k+1j} leading to the largest fired strength is selected. If the distance is not zero, then the rule base has to be modified and possibly new membership functions added, as described in the next section.

e. Addition of rules and eventually of membership functions to avoid misclassification. It can be assumed that the output value of the recognition system for a new leader will not be one. This is due to the fact that, contrary to the case of the leader selected in the very first slice, no new rules for which each premise is associated with a membership function, centered on the value of each parameter characterizing the new leader have been created. To avoid misclassification and to obtain an output value for the new leader that is larger than the value of any other region in the partition group, the rule with the lowest firing strength has to be modified. This is achieved by replacing this premise by a new one, with an associated membership function centered on the value of the corresponding leader input parameter.

If we denote ru_h as the rule with the highest strength and ru_{hj} the premise *j* of the rule with the lowest strength, we are able to define $J = //sl_{kj} - m_{up}//$, as the cost function for partition group *k*, where *j* is the index of the worst fulfilled parameter, $p = (3 \times j) - 1$ and $u = 1,\ldots, q_j$. If there exists a value of *u* such that *J* is below a given threshold, then the corresponding rule is modified as follows:

$ru_{h+1j} = u$, where *j* is the index of the least similar parameter.

If no value of *u* fulfills the threshold condition, the rule has to be associated with a newly created membership function as follows, after having increased q_j by one unit:

$m_{q(j)p} = sl_{q(j)j} - sg_{q(j)}/2$, where j is the index of the least similar parameter and $p = (3 \times j) - 2$

$m_{q(j)p} = sl_{q(j)j}$, where j is the index of the least similar parameter and $p = (3 \times j) - 1$

$m_{q(j)p} = sl_{q(j)j} + sg_{q(j)}/2$

where j is the index of the least similar parameter and $p = (3 \times n)$

$ru_{h+1j} = q_j$, where j is the index of the least similar parameter.

f. Return to step b. (New iteration of the process)

In Table 5 and Figure 13, the operating mode of the training/learning process is shown. The numerical values reported correspond to the example already introduced in Table 4. After evaluation by the neuro-fuzzy inference system, the rules corresponding to a leader which has been fired are stored. According to the example above, the rules having been fired are the rules 1, 2, 3, 7, and 8. The corresponding membership functions are shown in the same order in Figure 13.

3.5.2. Adjustment of membership functions using a descent method (FUNNY)

Nomura, Hayashi, and Wakami [30] have proposed an interesting method enabling the fuzzy inference systems to automatically tune their parameters ("self-tuning") during a learning/training process. This technique is particularly suited for approaches using Sugeno zero order models. In their contribution, these authors suggest a procedure to correct the parameter values of the membership functions associated with the premises and conclusion of a given rule. The procedure is based on the minimization of an objective function, which is the error E between a reference output y^r and the actual output y.

Recall the form of a zero order Sugeno fuzzy system:

IF $x_{1,1}$ is $m_{1,p}$ and $x_{1,2}$ is $m_{2,p}$ and $x_{1,j}$ is $m_{1,p}$ and ... and $x_{1,n}$ is $m_{3*n,p}$
THEN y is b_h for the rule number h.

The error E is given by

$$E = \frac{1}{2}\left(y - y^r\right)^2$$

Table 5. Learning process for the example given in Table 4. Corresponding problems and their solutions as determined automatically by our training process; *mf* stands for membership function.

Parameter values	Problem to be solved	Solution
mf11 mf21 mf31 mf41 mf51 mf61	START	-
mf12 mf21 mf32 mf41 mf51 mf61	2nd image badly Classified! (on 1th Patient data set used)	Addition of 2 mf and 1 rule to avoid misclassification
mf13 mf21 mf32 mf41 mf51 mf62	3rd Image badly Classified! (on 1th Patient data set used)	Addition of 2 mf and 1 rule to avoid misclassification
mf13 mf21 mf33 mf41 mf51 mf62	4th Image: no rule fired! 1mf does not exist	Addition of 1 mf and 1 rule
mf14 mf21 mf33 mf41 mf51 mf62	5th Image: no rule fired! 1mf does not exist	Addition of 1 mf and 1 rule
mf14 mf22 mf33 mf41 mf51 mf62	5th Image: no rule fired! 1mf does not exist	Addition of 1 mf and 1 rule
mf15 mf21 mf33 mf42 mf51 mf62	4th Image badly Classified! (on 1th Patient data set used)	Addition of 2 mf and 1 rule to avoid misclassification
mf14 mf22 mf33 mf41 mf52 mf62	5th Image badly Classified! (on 1th Patient data set used)	Addition of 1 mf and 1 rule to avoid misclassification

If a is the center of the triangle and s its support, then the learning process, or the adjustment or evolution of the parameters, can be described by the following equations:

$$a_{hj}(t+1) = a_{hj}(t) - k_a \frac{\partial(E)}{\partial a_{hj}}$$

$$s_{hj}(t+1) = s_{hj}(t) - k_s \frac{\partial(E)}{\partial s_{hj}}$$

$$b_h(t+1) = b_h(t) - k_b \frac{\partial(E)}{\partial b_h}$$

where k_a, k_s, k_b are adaptation learning factors and t the number of iterations.

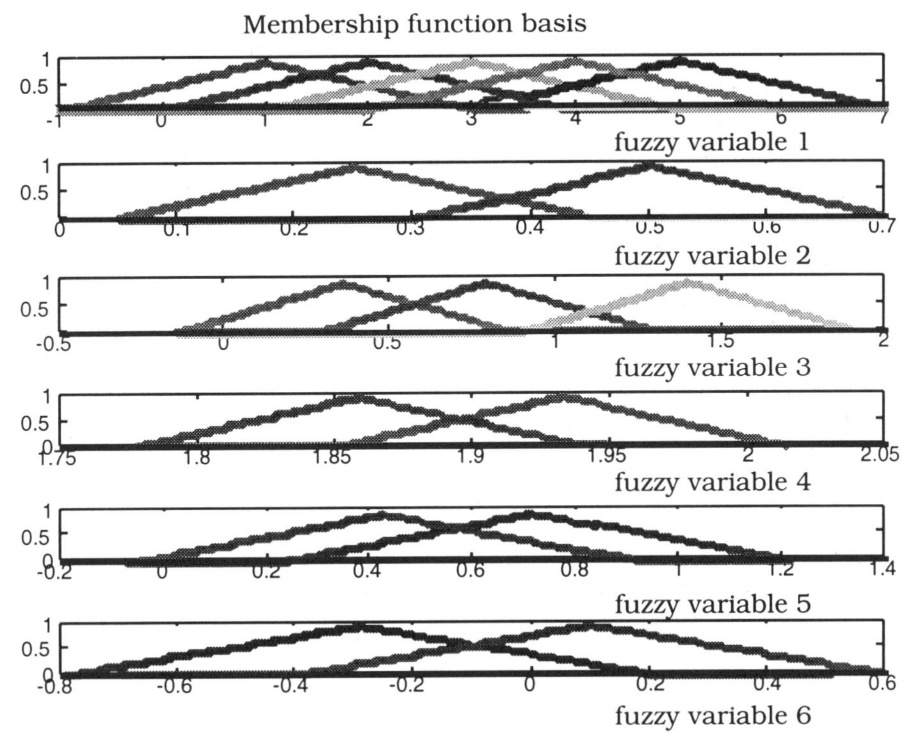

Figure 13. Membership functions created during the training process. They correspond to the example given in Table 4 and to the data summarized in Table 5.

In this method, a given membership function can be adjusted according to the local error E associated with each fuzzy rule and independent of its semantic meaning. This sometimes leads to results which make no sense. An example of the inconsistencies which can occur is given by the membership function associated with the fuzzy set "`small.`" After adaptation of this function, it can happen that the associated range of values is larger than those of the fuzzy set "large."

To remove this drawback, Bersini and Gorrini [27] have proposed an improvement of the method. They suggest linking all rules referring to the same membership functions before proceeding with the adjustments and, in this way, keep a common semantic meaning for all rules. Further, using a gradient-based approach, Bersini and Gorrini propose to express the learning formulas as follows:

$$\frac{\partial(E)}{\partial a_{rj}} = \frac{2k_a\left(y_i^r - y_i\right)\left(\sum_{h=1}^{tmf_{rj}} \omega_h b_h - y_i \sum_{h=1}^{tmf_{rj}} \omega_h\right)\operatorname{sgn}\left(x_{ij} - a_{rj}\right)}{\sum_{h=1}^{tr} \omega_h s_{rj} mf_{rj}\left(x_{ij}\right)}$$

$$\frac{\partial(E)}{\partial s_{rj}} = \frac{k_s\left(y_i^r - y_i\right)\left(\sum_{h=1}^{tmf_{rj}} \omega_h b_h - y\sum_{h=1}^{tmf_{rj}} \omega_h\right)\left(1 - mf_h\left(x_{ij}\right)\right)}{\sum_{h=1}^{tr} \omega_h s_{rj} mf_{rj}\left(x_{ij}\right)}$$

$$\frac{\partial(E)}{\partial b_r} = \frac{k_b \sum_{h=1}^{tb_r} \omega_h \left(y_i^r - y_i\right)}{\sum_{h=1}^{tr} \omega_h}$$

where tmf_{rj} is the total number of rules including mfrj in their premises, *tr* the total number of rules, ω_k the fired strength of the hth rule, and *sgn(x)* the sign of x. In this approach, which defines the output of a rule to be its firing strength, we have slightly modified the learning equations to be used as indicated below:

$$\frac{\partial(E)}{\partial a_{rj}} = \frac{2k_a\left(y_i^r - y_i\right)\left(\sum_{h=1}^{tmf_{rj}} \omega_h^2 - y_i\sum_{h=1}^{tmf_{rj}} \omega_h\right)\operatorname{sgn}\left(x_{ij} - a_{rj}\right)}{\sum_{h=1}^{tr} \omega_h s_{rj} mf_{rj}\left(x_{ij}\right)}$$

$$\frac{\partial(E)}{\partial s_{rj}} = \frac{k_s\left(y_i^r - y_i\right)\left(\sum_{h=1}^{tmf_{rj}} \omega_h^2 - y_i\sum_{h=1}^{tmf_{rj}} \omega_h\right)\left(1 - mf_{rj}\left(x_{ij}\right)\right)}{\sum_{h=1}^{tr} \omega_h s_{rj} mf_{rj}\left(x_{ij}\right)}$$

3.5.3. Combining the automated generation of rules and membership functions and the adjustment of their parameters in a parallel implementation (FUNNY-ALGORAM)

The main objective of the combination of these two techniques is to achieve an optimized system. If we always create a membership function every time a piece of data is badly classified or a rule could not lead to firing, the system rule base may become too large. Applying the FUNNY algorithm simultaneously to the ALGORAM method enables us to adjust the membership functions. This allows an input parameter to lead to a larger or smaller value, depending on the modified location of the membership function center after application of the FUNNY algorithm. To achieve this, we have added (by verifying if a leader is well classified) an intermediate step in our procedure ALGORAM:

```
a. Create rule and membership function for first leader.
b. Check if next leader leads to firing.
```

```
   Add rule and membership function for parameters
   which disabled firing.
c. Check if leader is well classified.
   Apply the FUNNY algorithm to optimize all membership
   functions.
   If application fails, add rule and possibly
   membership function to avoid misclassification.
d. Go to step b.
```

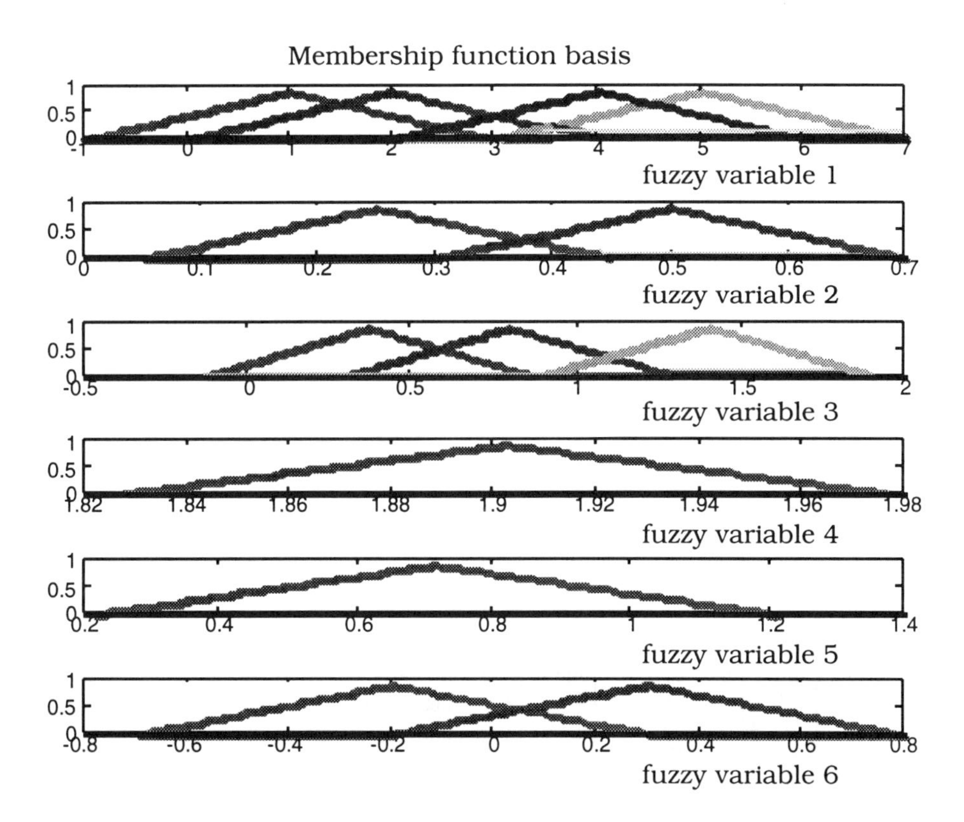

Figure 14. Membership functions created during execution of the parallel version of the training process. They correspond to the example given in Section 3.5.1 and to the data summarized in Table 4.

Table 6. Learning process for the example given in Table 4. Corresponding problems and their solutions as determined automatically by our parallel implementation of FUNNY-ALGORAM; *mf* stands for membership function and epoch for an iteration cycle.

Parameter values						Problem to be solved	Solution
mf11	mf21	mf31	mf41	mf51	mf61	START	
mf12	mf21	mf32	mf41	mf51	mf61	2nd Image badly Classified! (on 1th Patient data set used)	Addition of 2 mf and 1 rule to avoid misclassification after 8 epochs
						3rd Image badly Classified! (on 1th Patient data set used)	All data has been properly classified after 2 epochs
mf12	mf21	mf33	mf41	mf51	mf61	4th Image: no rule fired! 1 mf does not exist	Addition of 1 mfand 1 rule
mf13	mf21	mf33	mf41	mf51	mf61	5th Image: no rule fired! 1 mf does not exist	Addition of 1 mf and 1 rule
mf13	mf22	mf33	mf41	mf51	mf61	5th Image: no rule fired! 1 mf does not exist	Addition of 1 mf and 1 rule
mf14	mf21	mf33	mf41	mf51	mf62	4th Image badly Classified! (on 1th Patient data set used)	Addition of 1 mf and 1 rule to avoid misclassification after 8 epochs
mf13	mf22	mf33	mf41	mf51	mf62	5th Image badly Classified! (on 1th Patient data set used)	Addition of 1 rule to avoid misclassification after 8 epochs

Table 6 shows how this modification improves the behavior of ALGORAM. After evaluation by the trained neuro-fuzzy system, the rules corresponding to a leader which have been fired are held back. According to the above example, the rules which have fired are 1, 2 (×2), 6, and 7. The corresponding created membership functions are shown in the same order in Figure 14.

Other methods that automatically create a rule base, such as, for example ANFIS, or the subtractive algorithm, have also been implemented and tested. We believe that our implementation has several advantages with respect to these other two methods. ANFIS first determines all the permutations between membership functions in order to create the rules, and then adjusts the membership functions. This method is cumbersome and slow when we input, for example, the six parameters introduced in Section 3.4.1 (see also Table 1). We also found that the subtractive method is not well adapted to our application, since we have to provide an output value for each input. Even if we attempt to calculate fulfillment values for the regions (or assign the value 0 to regions other than the leader, which makes no sense), the membership functions and rules created by these approaches are more numerous than in our approach.

4. *IN VITRO* EXPERIMENTS AND APPLICATION TO MEDICAL CASES

To validate the new approach we implemented and to estimate its performance when applied to actual medical image data, we have performed two types of experiments. First, we tested, adjusted, and validated the whole procedure using data sets of images of phantoms modeling real acquisition conditions. Second, the procedure has been applied to image data acquired under medical conditions, to assess its suitability and performance in medical routine.

4.1. Experiments with phantoms

We cannot generally assert that the myocardial volume corresponds exactly to the sum of the region areas found by our algorithm to correspond to the left ventricle in the set of acquired slices (recall that a pixel in a given slice corresponds to an elementary volume or voxel). In other words, the question is to determine to what extent the extracted contours actually represent the left ventricular cavity depicted in a slice. In addition, the ability of our algorithm to separate the two ventricular cavities, taking noise into account, must also be characterized. In order to evaluate both the correctness and performance of our method, we built a series of phantoms that enable us to simulate the ventricular volumes. Since both the shape and dimensions of our artificial "ventricles" are known, we are able to evaluate both the accuracy of the measured left ventricular volumes and the ability to separate the two ventricles, and can assess the robustness of our approach with respect to noise.

The ventricular phantom was built with two bottles of equal diameter and volume (4.2 cm and 75 ml, respectively), containing 4 mCi/cc of 99m-Tc. The distance d between the two bottles (which represents the interventricular septum) can be varied between 0 and 1.5 cm. The experimental work on this phantom has been carried out in two phases. During the first phase, the diameters of the bottles have been measured as a function of the distance d. The principle of this experiment is illustrated in Figure 15. For the second phase, the bottles were submerged in water, as shown in Figure 16, the aim being to simulate the diffusion induced by the thoracic tissues. SPECT images of the phantom have been acquired and slices have been reconstructed after backprojection, using the same parameters as those used for the patient data in Section 4.2.

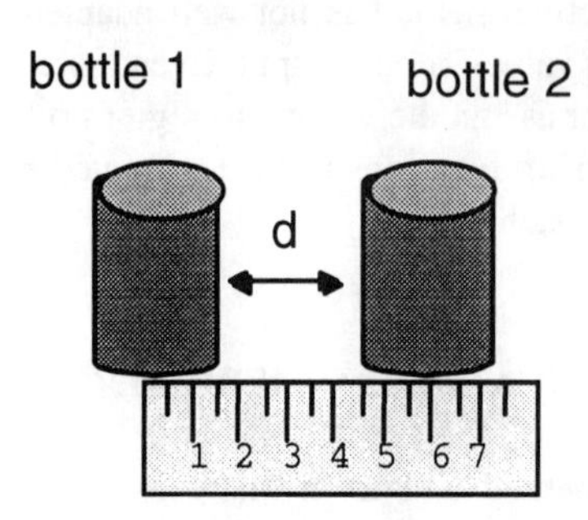

Figure 15. Principle of the first experiment performed to measure the diameter of the bottles separated by a varying distance *d*.

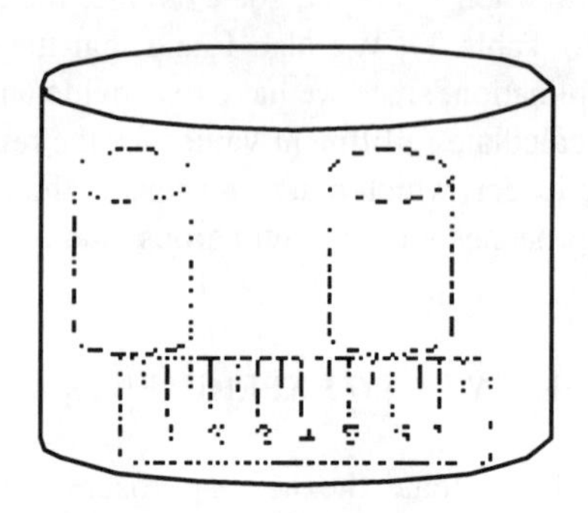

Figure 16. Principle of the second experiment performed under conditions simulating real acquisition conditions (diffusion), to measure the diameter of the bottles separated by a varying distance *d*.

Tables 7, 8 and 9 summarize the experimental values. The corresponding situations evaluated (test cases) are defined as follows:

case no. 1: no diffusion in the images
case no. 2: diffusion present (with water in the tank)
case no. 3: no diffusion, but with 5% of added Gaussian noise in the images
case no. 4: no diffusion, but with 15% of added Gaussian noise in the images
case no. 5: no diffusion, but with 25% of added Gaussian noise in the images

Table 7. Measured diameter for the first bottle

Real distance *d* between the bottles [cm]	Diameter case no. 1 [cm]	Diameter case no. 2 [cm]	Diameter case no. 3 [cm]	Diameter case no. 4 [cm]	Diameter case no. 5 [cm]
0	4.5	3.9	4.5	4.5	4.5
0.2	4.5	3.9	4.2	4.2	4.9
0.5	4.5	3.9	4.5	4.5	4.2
0.7	4.8	3.9	4.2	4.9	4.5
1.0	4.8	4.2	4.9	4.9	4.2
1.5	4.5	3.9	4.5	4.5	4.5
Mean	4.62	3.96	4.47	4.58	4.47
Standard deviation	0.16	0.13	0.26	0.27	0.26

Table 8. Measured diameter for the second bottle

Real distance *d* between the bottles [cm]	Diameter case no. 1 [cm]	Diameter case no. 2 [cm]	Diameter case no. 3 [cm]	Diameter case no. 4 [cm]	Diameter case no. 5 [cm]
0	4.2	3.3	4.2	4.2	4.5
0.2	4.2	3.3	3.9	4.2	4.5
0.5	4.2	3.6	4.2	4.2	4.2
0.7	4.2	3.2	4.2	4.2	4.2
1.0	4.5	3.6	4.5	4.5	3.9
1.5	4.5	3.6	4.5	4.5	4.2
Mean	4.32	3.48	4.25	4.3	4.3
Standard deviation	0.16	0.16	0.22	0.16	0.26

We observe that, for the five test cases investigated, our procedure recognizes the circular structure of the two bottles independent of the distance *d* separating them. More important, the diameters computed remain relatively constant in all instances, demonstrating that our algorithm is suitably robust with respect to the noise. In case no. 2 with images including diffusion, the diameters found are slightly smaller. The slight difference in the measured mean diameters for the two bottles can be partially related to the non-symmetrical position of the two bottles in the water tank, which modifies the diffusion of the gamma rays. Further, the influence of diffusion can also be observed in the apparent lower measured diameter when the two bottles are immersed in the water tank.

Table 9. Measured distance between the bottles
(due to the small amount of data available, correlation coefficients are only indicative).

Real distance *d* between the bottles [cm]	Measured diameter case no. 1 [cm]	Measured diameter case no. 2 [cm]	Measured diameter case no. 3 [cm]	Measured diameter case no. 4 [cm]	Measured diameter case no. 5 [cm]
0	0	0	0	0	0
0.2	0.3	0.6	0.65	0.6	0
0.5	0.6	0.6	0.65	0.6	0.6
0.7	0.6	1.0	1.0	0.6	0.6
1.0	1.0	1.3	1.0	0.6	1.6
1.5	1.3	1.9	1.3	1.31	1.6
correlation coefficient	0.98	0.97	0.92	0.87	0.84

Although the measures are correct in most cases, one can observe that the measured diameters are over-estimated and accuracy decreases when the amount of noise added is too high. This is particularly critical when the distance simulating the ventricular septum is short and the level of noise high (Table 9, case no. 5). In all other cases, identification of the two bottles is correct when the distance between the bottles decreases. One must also take into account the spatial resolution of the gamma camera used, which for this series of experiments was 0.6 mm/pixel. This value can have an impact on the results when the separation between the bottles is small. The correlation coefficients found for the measured distances between the bottles and given in Table 9 can be considered to be fairly good.

4. 2. Clinical Test Cases

For these experiments, we randomly selected twelve patients, and for each of them, we routinely acquired a set of G-SPECT images. Each data set has then been processed with our procedure and the results compared with those obtained using two other techniques routinely employed in our laboratory. The first alternative is a semi-automatic MRI technique where the contours are manually traced (MRI system GE MRMax 0.5T, sequence gradient echo TE = 12ms, TR = 39ms, FLIP = 20°, Nex = 2). The second one is the protocol used every day in our laboratory. For this latter method, thresholding at 40% of the maximum gray value found in the set of images enables us to visualize the diastolic and systolic contours, which can then be interpreted by the physician. However, an operator has to initialize image processing by manually selecting a region of interest corresponding to (including) the left ventricle and, if the two regions corresponding to the ventricles are not isolated or if some contours do not border completely the regions of interest during processing, the errors have to be manually corrected.

Table 10. Calculated LVEF values using the results of the three approaches applied.

PATIENT Number	LVEF (%) MRI-based method	LVEF (%) Every day protocol	LVEF (%) Proposed method
1	53	59	55
2	59	57	53
3	68	68	42
4	59	59	67
5	58	61	48
6	77	74	62
7	55	53	48
8	53	46	54
9	67	70	51
10	7	13	1 *
11	62	73	53
12	43	44	48

For all three methods, the ejection fraction has been calculated. The results are summarized in Table 10. The case marked with a star (*) corresponds to a patient with an important dilated and hypokinetic cardiomyopathy. In this case the very low value of the LVEF obtained by the proposed method is doubtful and may correspond to a limit of the method.

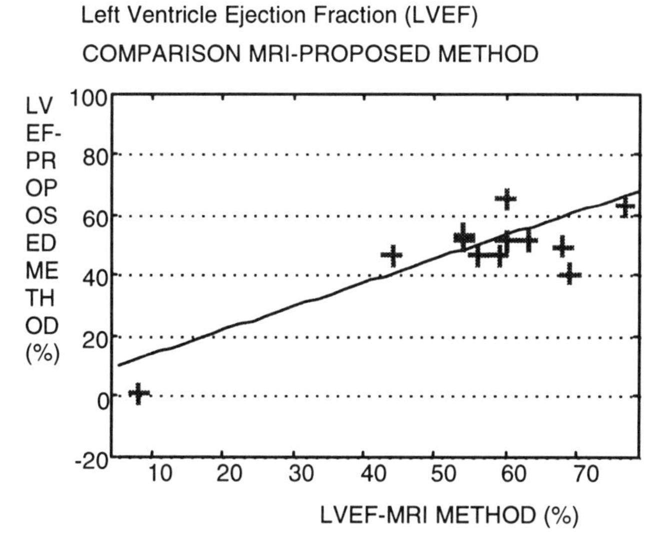

Figure 17a. Comparison of the LVEF values using the three approaches. These figures directly led to the computation of the regression coefficients *r*.

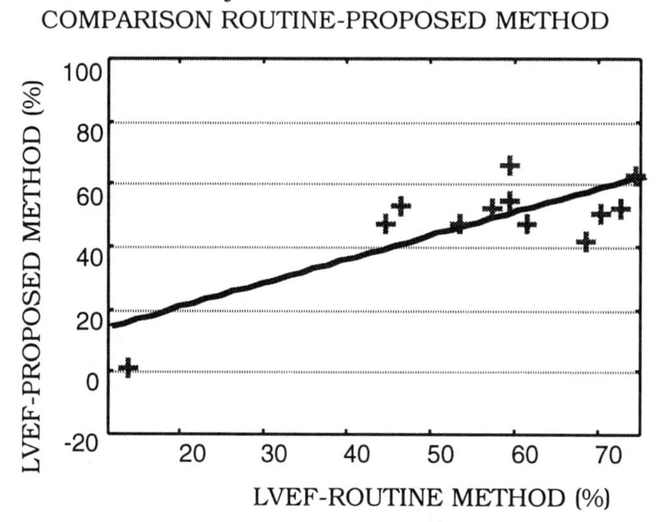

Figure 17b. Comparison of the LVEF values using the three approaches. These figures directly led to the computation of the regression coefficients *r*.

To assess the quality of the results achieved with our automated approach, we calculated the regression coefficients *r* between the results in our procedure and the results obtained by the other two methods (see Figure 17), as an indicative measure to be used for comparison. When compared to the MRI based method, we obtained r = 0.84. This coefficient takes a value of 0.78 when compared to the routine method used in our laboratory.

Due to the limited size of the set of evaluated image data, these values for the regression coefficients r can only be considered acceptable. Also, only the data corresponding to three patients were used to train the recognition system. It is however noteworthy that our implementation was able to automatically select the region corresponding to the left ventricle and to extract the associated contours to compute the LVEF. When comparing our results with the measurements obtained with the two other techniques (MRI and standard nuclear procedure), one must bear in mind that there is no "gold standard" method in medical imaging, which can give the exact value of the LVEF. That means that the correlation coefficients can be compared only in a relation to each other.

4. 3. Implementation issues

We have implemented our method on a Macintosh computer equipped with a 603 PowerPC processor running at 75 MHz. The algorithms have been developed using the programming language of Matlab. The public domain software package Image from the NIH [24] has been used to compute the parameters related to the shape and the location of the various regions extracted from the images under investigation. The computing time for processing a whole patient data set is forty minutes. We expect to speed up execution with a new implementation using the **C** or **C++** programming language.

5. CONCLUSIONS

In this chapter, a new method for fully automatically contouring blood pool myocardial gated SPECT images has been presented. The method was specifically developed to determine the left ventricle cavity surfaces required for calculating the Left Ventricular Ejection Fraction (LVEF). The LVEF is indeed an essential diagnosis parameter in cardiology. The approach is based on the sequential application of three steps. The first step is a wavelet-based image pre-processing algorithm, which can both reduce the noise level in the images and enhance the image contrast to favor separation of the regions corresponding to the two myocardial ventricles. At this stage of the processing, it is particularly crucial to find a point belonging to the border between the regions corresponding, respectively, to the left ventricle and the septum. This point initializes the second step, which performs the segmentation of the images using the neural network leadership algorithm. This step delivers classes built with Gaussian functions, with the initial first class centered around the border point returned by the wavelet-based processing algorithm. The third step selects the region which corresponds to the left ventricle. This determination is made using a fuzzy logic-based decision implemented as a three-layer neural network. This procedure requires training to set up the necessary knowledge base. For that purpose, we developed a new tool called ALGORAM, which can automatically generate rules and membership functions. This allows us to automatically create and/or enlarge both the rule and membership function bases. Combining this tool with an adaptive tuning scheme optimizes the system performance.

The proposed technique has been implemented and applied to both images of phantoms and patients. Results for the phantom images allow us to validate our approach. Further, they show that our method is robust when applied to images corrupted with noise, and that it is able to distinguish the two simulated myocardial ventricles. For the real patient data sets, the LVEF has been calculated and compared to a MRI based method and to the protocol used routinely in our laboratory. Even though the measures delivered by our approach are slightly underestimated with respect to the results obtained with the other two methods, the three sets of results are all in an acceptable overlapping range of values. Therefore, these results validated our method pertaining to the real-life image data. One can also note that the comparison is made using results from methods that are not yet considered a standard in the medical field. Further, even when the measures are determined by experienced operators, there are almost always discrepancies in the results obtained by different operators evaluating the same set of images. Thus, as the results of our method approach those of the other two methods rather well and, as the measures are obtained automatically, it is a good choice to avoid variations in the results to be used for diagnosis. This is further enhanced by the fact that the method is able to segment validly the region corresponding to the left ventricle, and to extract its contour. Finally, the recently implemented software module ALGORAM adds more autonomy to the method and simplifies enlargement of the knowledge base.

However, the current version of the implementation still leaves partially open some questions related mainly to the suitability of the approach in a medical routine. In particular, we are currently collecting a rather exhaustive set of data corresponding to representative clinical cases using both MRI and SPECT acquisition modalities. This will enable us to enlarge the knowledge base and improve the comparison between different techniques. A larger and more complete knowledge base will certainly favor the evaluation of data related to severe pathological cases, for which the data is strongly distorted. This is, for example, the case of some pathologies where the systolic contours are as large as the diastolic ones. Future work will also investigate the influence of knowledge base size on performance. Indeed, if the knowledge base is too large, it could lead to confusion. Another related point is to find the optimal size, as manipulating a large rule base slows down computing speed. We believe that an extension of ALGORAM to "intelligently" destroy unnecessary rules could decrease execution time. Also, the number of measured parameters evaluated in the fuzzy system could be decreased to speed up the training process. This raises the question of the minimal number of parameters needed and of their appropriateness. For example, some of the parameters can be replaced by one unique spatio-temporal parameter maintaining consistency between the slices. Enhancing the segmentation step to have a minimum number of regions to evaluate is another interesting option.

Acknowledgments

The first author would like to thank Prof. Dr. H.-D. Kochs, Head of the Technical Computing Laboratory, and Dr. S. Baginski, for their support and for the facilities made available during the development of the training method of the knowledge base at the Gerhard Mercator University, Duisburg, FRG, during a research stay partially funded by the DAAD (grant n# A/98/34202). This work is also supported by the following institutions: CONACYT (Consejo Nacional de Ciencia y Tecnologia, Mexico D.F. Mexico), SFERE (Société Française d'Exportation de Ressources Educatives), and Apple Computer, France. The authors also would like to thank Dr. Philippe Choquet, Dr. Barbu Dumitrescu, Dr. Luc Mertz, and Dr. Philippe Germain for their help and support.

APPENDICES

1. Automatic Determination of Diastolic and Systolic Images

Acquisition of images is synchronized by the R wave of the ECG (gated-SPECT image acquisition). This means that the eight phases of the cardiac cycle are acquired between two successive R waves. This operation is repeated thirty times and enables us to finally obtain a set of 240 images corresponding to 30 acquisitions. The volume variation with respect to time is sampled over time (at the instants corresponding to the acquisition of the eight phases of the cardiac cycle). In other words, the specific acquisition system does not allow deriving the underlying continuous smooth function, but only an approximation thereof. The set of 240 images is then arranged into eight groups. The first group contains all the first phase images of the 30 acquisition steps, the second group the next phase images in the different acquisitions, and so on. One of these groups contains the information related to the diastolic phase and another that of the systolic phase. These two particular groups have to be determined. Also, the sum of all the images in a given group leads to accumulated radioactivity images (thus, 8 radioactivity images are obtained). Further, the sum of the intensity of all the pixels in a given radioactivity image finally gives the total radioactivity coefficient for that image. The coefficient of maximal value can be shown to be related to the diastolic group and the coefficient of minimal value to the systolic group. Tomographic backprojection reconstruction is then applied to the two groups selected this way, first to the group corresponding to the diastolic phase, second to the one associated with the systolic phase.

To adjust the rotation center, the diastolic image reconstructed according to the procedure above is presented in a horizontal plane to an operator and this operator manually establishes a line passing through the apex and the middle of the myocardium. A visualization software package allows rotation of the displayed image to supperpose this line with the Horizontal Long Axis. The rotation center is then positioned by the operator. In a second step, the lateral limits of the myocardium are established to reconstruct the Vertical Long Axis images. Further reconstruction limits regarding the apex and the base of the myocardium are established to reconstruct the Horizontal Short Axis images. These reconstruction parameters for the diastolic images are stored and automatically applied during the reconstruction of the systolic images, enabling us to take care of the adjustment of the rotation center. The reconstruction software then arranges the whole set of diastolic and systolic images into a new group containing subsets of HSA, HLA and VLA images. Each subset contains alternatively one diastolic image followed by one systolic image. We chose to rearrange these subsets into one diastolic and one systolic image. Also, only HSA diastolic and systolic images are used to compute the LVEF, even though HLA data could be used to ameliorate the estimate of the LVEF. However, as a result, only a slight improvement can be observed in this case.

2. Trust limits of the estimated regression coefficients

To complement the statistical data given in Section 4 for the LVEF calculated using three different methods, the regression coefficient between the routine method and the MRI method has been estimated to be 0.92. Calculation of the trust limits for the estimated regression coefficients are given below:

$Prob(0.51<r<0.954) = 0.95$ for MRI method-Proposed method
$Prob(0.37<r<0.935) = 0.95$ for Routine method-Proposed method
$Prob(0.73<r<0.976) = 0.95$ for Routine method-MRI method

There is, indeed, better agreement between the Routine method and the MRI method. However, as stated in Section 4, there is no standard method available. One can not expect that our results to exactly approach those of the other methods. However, it is acknowledged that training of the system with an enlarged set of patient data would improve the quality of the results and allow us to deal with situations where operator based contouring decisions could lead to inconsistencies.

No correction factor has been applied to the LVEF calculation, even if it is known that there is a systematic underestimation of the volumes in the presence of diffusion. This can be seen when analyzing the phantom of the left ventricle. However, for real data, this error is difficult to estimate as an accurate model is not available. Further, the estimation error is systematic and thus leads to underestimated volumes. As a result, the LVEF being a ratio of volumes smaller than 1, this quantity is slightly underestimated. Note that there is a partial compensation of the error, with the underestimation affecting both terms of the ratio. Applying a correction factor for the diastolic and systolic volumes only slightly ameliorates the LVEF values.

ACRONYMS

ALGORAM	Automatic Leader Generation Of Rules And Membership functions
ART	Adaptive Resonance Theory
ANFIS	Adaptive Neuro-Fuzzy Inference System
G-SPECT	Gated - Single Photon Emission Computed Tomography
FUNNY	FUzzY and Neural Networks
HSA	Horizontal Short Axis
HLA	Horizontal Long Axis
VLA	Vertical Long Axis
LVEF	Left Ventricle Ejection Fraction
SPECT	Single Photon Emission Computed Tomography

REFERENCES

[1] Zadeh, L., Fuzzy sets, *Information and Control*, 8, 338-356, 1965.

[2] Boegl, K. *et al.*, New approaches to computer-assisted diagnosis of rheumatologic diseases, *Radiologe*, 35 (9), 604-610, 1995.

[3] Pietka, E., Computer-assisted bone age assessment based on features automatically extracted from a hand radiograph, *Comput. Med. Imaging Graph*, 19 (3), 251-259, 1995.

[4] Shiomi, S., *et al.*, Diagnosis of chronic liver disease from liver scintiscans by fuzzy reasoning, *J. Nucl. Med*, 36 (4), 593-598, 1995.

[5] Phillips, W., *et al.*, Automatic magnetic resonance tissue characterization for three-dimensional magnetic resonance imaging of the brain, *J. Neuroimaging*, 5 (3), 171-177, 1995.

[6] Phillips, W., *et al.*, Application of fuzzy c-means segmentation technique for tissue differentiation in MR images of a hemorrhagic glioblastoma multiforme, *Magn. Reson. Imaging*, 13 (2), 277-290, 1995.

[7] Brandt, M., *et al.*, Estimation of CSF, white and gray matter volumes in hydrocephalic children using fuzzy clustering of MR images, *Comput. Med. Imaging Graph*, 18 (1), 25-34, 1994.

[8] Bezdek, J., *et al.*, Medical image analysis with fuzzy models, *Stat. Methods Med. Res.*, 6 (3), 191-214, 1997.

[9] Lippmann, R., An introduction to computing with neural nets, *IEEE ASSP Magazine*, 4, 4-22, 1987.

[10] Lin, J.-S., *et al.*, Application of artificial neural networks for reduction of false-positive detections in digital chest radiographs, *Proc. Annu. Symp. Compt. Appl. Med. Care*, 434-438, 1993.

[11] Lin, J.-S., *et al.*, Segmentation of multispectral magnetic resonance image using penalized fuzzy competitive learning network, *Compt. Biomed. Res.*, 29 (4), 314-326, 1996.

[12] Germano, G., *et al.*, Automatic quantification of ejection fraction from gated myocardial perfussion SPECT, *J. Nucl. Med.*, 36, 2138-2147, 1995.

[13] Chin, B. B., *et al.*, Right and left ventricular volume and ejection fraction by tomographic gated blood-pool scintigraphy, *J. Nucl. Med.*, 38, 942-948, 1997.

[14] Meyer, Y., Les ondelettes: algorithmes et applications, Armand Colin, Paris, France, 1994.

[15] Strang, G. and Nguyen T., Wavelets and filter banks, Wellesley-Cambridge Press, Wellesley Ma., U.S.A., 1996.

[16] Carpenter, G. A. and Grossberg, S., A massively parallel architecture for a self-organizing neural pattern recognition machine, *Computer Vision, Graphics and Image Processing*, 37, 54-115, 1987.

[17] Carpenter, G. A. and Grossberg, S., The ART of adaptive pattern recognition by a self-organizing neural network, *Computer*, 21 (3), 77-88, 1988.

[18] Kaparthi, S., Nallan, C. S., and Cerveny, R. P., An improved neural network leader algorithm for part-machine grouping in group technology, *European Journal of Operational Research,* 69, 342-356, 1993.

[19] Mallat, S. and Zhong, S., Characterization of signals from multiscale edges, *IEEE Trans. PAMI*, 11, 710-732, 1992.

[20] Mallat, S. and Hwang, L., Singularity detection and processing with wavelets. *IEEE Trans. PAMI,* 14, 617-643, 1992.

[21] Mallat, S., A theory for multiresolution signal decomposition: the wavelet representation, *IEEE Trans. PAMI*, 11, 674-693, 1989.

[22] Hajj, H., Nguyen, T., and Chin, R., On multiscale feature detection using filter banks, *Proceedings of the Asilomar Conference on Signals, Systems and Computers,* 29, 1996.

[23] Bensaid, A. M., Bezdek, and J. C., Partial Supervision based on point-prototype clustering algorithms, *Proceedings of EUFIT '96*, 2, 1402-1406, 1996.

[24] Rasband W., NIH Image Public Domain Program. *U. S. National Institutes of Health, http:/rsb.info.nih.gov/nih-image/*, 1997.

[25] Patino, L., Mertz, L., Hirsch, E., Dumitrescu, B., and Constantinesco, A., Contouring blood pool myocardial gated SPECT images with a neural network leader segmentation and a decision-based fuzzy logic, *Proceedings of FUZZ IEEE '97*, Barcelona, Spain, 2, 969-974, 1997.

[26] Patino, L., Mertz, L., Hirsch, E., and Constantinesco, A., Segmentation and contouring of blood pool myocardial SPECT images with wavelet-fuzzy constraints, *Proceedings of EUFIT '96,* Aachen, Germany, 3, 2086-2090, 1996.

[27] Bersini, H. and Gorrini, V., FUNNY (FUzzY or Neural Net) methods for adaptive process control, *Proceedings of EUFIT '93*, 2, 55-61, 1993.

[28] Berenji, H. R. and Khedkar, P., Learning and tuning fuzzy logic controllers through reinforcements, *IEEE Trans. Neural Networks*, 3, 724-740, 1992.

[29] Berenji, H. R., *et al.*, Space shuttle attitude control by reinforcement learning and fuzzy logic, *Proceedings of IEEE Int. Conf. on Neural Networks*, 1396-1401, 1993.

[30] Nomura, H., Hayashi, I., and Wakami, N., A learning method of fuzzy inference rules by descent method, *Proceedings of IEEE Int. Conf. on Fuzzy Systems*, 203-

210, 1992.

[31] Jang, J.-S., ANFIS: Adaptive neuro fuzzy inference systems, *IEEE Trans. Systems, Man & Cybernetics*, 23, 665-685, 1993.

[32] Sulzberger, S. M., *et al.*, FUN: Optimization of fuzzy rule based systems using neural networks, *Proceedings of IEEE Int. Conf. on Neural Networks*, 312-316, 1993.

[33] Nauck, D. and Kruse, R., A fuzzy neural network learning fuzzy control rules and membership functions by fuzzy error backpropagation, *Proceedings of IEEE Int. Conf. on Neural Networks*, 1022-1027, 1993.

[34] Nauck, D., Klawonn, F., and Kruse, R., Das NEFCON Modell, in *Neuronale Netze und Fuzzy-Systeme*, Vieweg, Braunschweig, Germany, Chap. 19, 1996.

[35] Jang, J.-S., Sun, C.-T., and Mizutani, E., *Neuro-Fuzzy and Soft Computing*, Prentice Hall, New Jersey, U.S.A., 1997.

[35b] Jang, J.-S., ANFIS architecture, in Jang, J.-S., Sun, C.-T., and Mizutani, E., *Neuro-Fuzzy and Soft Computing*, Prentice Hall, New Jersey, U.S.A., chap. 12, 1997.

[36] Baginski, S., personal communication, *Technische Informatik, Gerhard Mercator Universität, Duisburg, FRG, http://mti.uni-duisburg.de/~baginski/*, 1998.

Chapter 5

Unsupervised Brain Tumor Segmentation Using Knowledge-Based Fuzzy Techniques

Matthew C. Clark, Lawrence O. Hall, Dimitry B. Goldgof, Robert Velthuizen, Reed Murtagh, and Martin S. Silbiger

A system that automatically segments and labels complete glioblastoma-multiforme tumor volumes in magnetic resonance images of the human brain is presented. The magnetic resonance images studied consist of three feature images (T1-weighted, proton density, T2-weighted). The knowledge-based system integrates multispectral analysis, image processing, and fuzzy edge-detection methods. The heuristics in the knowledge base were designed to be independent of a particular magnetic resonance scanning protocol. An unsupervised clustering algorithm provides the initial segmentation. The segmented image, along with cluster centers for each class, is provided to a rule-based expert system, which extracts the intra-cranial region. Multispectral histogram analysis isolates suspected tumor from the rest of the intra-cranial region. Region analysis then removes spatial components that do not contain tumor. Finally, fuzzy edge detection is used to determine a final threshold in T1 space. This system has been trained on twelve volume data sets and tested on twenty-one unseen volume data sets acquired from a single magnetic resonance imaging system. The knowledge-based tumor segmentation was compared with radiologist-verified "ground truth" tumor volumes and results generated by a number of supervised methods. The results of this system generally correspond well to ground truth, both on a per slice basis and, more important, in tracking total tumor volume during treatment over time.

1. Introduction

According to the Brain Tumor Society, approximately 100,000 people in the United States will be diagnosed with a primary or metastatic brain tumor within the next 12 months [16]. One of the primary diagnostic and treatment evaluation tools for

0-8493-9806-1/98/$0.00+$.50

brain tumors has been magnetic resonance (MR) imaging. MR imaging has become a widely-used method of high quality medical imaging, especially in brain imaging where MR's soft tissue contrast and non-invasiveness are clear advantages. MR images can also be used to track the size of a brain tumor as it responds (or doesn't) to treatment. A reliable method for segmenting tumor would clearly be a useful tool [31], [32], [57]. Currently, however, there is no method widely accepted in clinical practice for quantitating tumor volumes from MR images [38]. The Eastern Cooperative Oncology group [17] uses an approximation to tumor cross-sectional area in the single MR slice with the largest contiguous, well-defined tumor. These manual measurements, however, have shown poor reproducibility and tumor response criteria based on these manual estimations have shown poor correlation with quantitative 2D and 3D metrics [10].

Computer-based brain tumor segmentation has remained largely experimental, with approaches including multi-spectral analysis [19], [29], [50], [51], [54], [55], edge detection [2], [12], [21], [22], [45], [59], [60] neural networks [35], [30], [44], and knowledge-based techniques [13], [25], [28], [42], [37]. Researches reported in [6], [7], [34] showed that a combination of knowledge-based techniques and multi-spectral analysis could effectively detect pathology and label normal transaxial slices. Most of these efforts, however, have dealt with either normal data sets or with neuro-psychiatric disorders possessing MR distribution characteristics similar to normals [9].

Supervised pattern recognition methods have also exhibited problems with reproducibility, due to significant intra- and inter-observer variance introduced over multiple trials of training example selection [9]. Furthermore, because supervision, such as the selection of training examples, can be time consuming and requires "domain expertise" to be effective, supervised methods are unsuitable for clinical use. These limitations suggest the need for a fully automatic method for tumor volume measurement, not only for tracking tumor response to therapy, but in planning future treatment as well [10], [31], [32], [57].

In [61], we considered a system that could automatically segment partial glioblastoma-multiforme tumor volumes in a limited range of slices intersecting the ventricles and upward to the top of the head. This system worked well, but did not consider slices below the ventricles, and had a tendency to significantly overestimate tumor volume. Here, we present a knowledge-based paradigm that combines fuzzy techniques, multispectral analysis, and image processing algorithms, to produce an unsupervised system capable of automatically (no human intervention on a per volume basis) segmenting and labeling complete glioblastoma-multiforme tumor volumes from any transaxial MR image that intersects the brain cerebrum. More important, this can be done over a period of time during which the tumor is treated, allowing a tumor's response to treatment to be tracked. Unlike most other efforts in segmenting brain pathology, this system has also been tested on a large number of unseen images with a fixed parameter (rule) set (built from a set of "training images"), and quantitatively compared with "ground truth" images. This allows tumor response to therapy to be tracked over repeat scans, and aids radiologists in planning subsequent treatment. More important, the system's unsupervised nature avoids the problem of observer variability found in supervised methods, providing complete reproducibility of results. Furthermore, observer-based training examples are not required, making the system suitable for clinical use.

The slices processed here were first determined to contain enhancing pathology during "pre-processing" (pathology detection) in [6], [8]. Knowledge gained during pre-processing allows removal of extra-cranial tissues (air, skin, fat, etc.) from a segmentation created by a fuzzy c-means clustering algorithm [5], [23]. The remaining pixels (really voxels, since they have thickness) form an intra-cranial mask. An expert system uses information from multi-spectral, local statistical analysis and fuzzy edge-detection to iteratively segment the enhancing tumor from the intra-cranial mask. A rule-based expert system shell, CLIPS [20], [46] is used to organize the system. Low level modules for image processing and high level modules for image analysis are all written in the C language and called as actions from the right-hand sides of the rules.

Each slice was classified as abnormal by systems described in [6], [8]. Of the tumor types that are found in the brain, glioblastoma-multiformes (Grade IV Gliomas) are the focus of this work. This tumor type was addressed first because of its relative compactness and tendency to enhance well with paramagnetic substances, such as gadolinium. For the purpose of tumor volume tracking, segmentations from contiguous slices (within the same volume) are merged to calculate total tumor size in 3D. The tumor segmentation matches well with radiologist-verified "ground truth" images, and is comparable to results generated by supervised segmentation techniques.

The remainder of the chapter is divided into four sections. Section 2 discusses the slices processed and gives a brief overview of the system. Section 3 details the system's major processing stages and the knowledge used at each stage. The last two sections present the experimental results, an analysis of them, and future directions for this research.

2. DOMAIN BACKGROUND

2.1. Slices of Interest for the Study

The system described here can process any slice intersecting the brain cerebrum in the transaxial plane [43], [47], starting from an initial slice 7 to 8 cm from the top of the brain and moving up toward the top of the head and down toward the shoulders. This "initial slice," which simply needs to intersect the ventricles, is used as the starting point due to the relatively good signal uniformity within the MR coil used [6]. Each brain slice consists of three feature images: T1-weighted (T1), proton density weighted (PD), and T2-weighted (T2) [57].

An example of a normal slice after processing is shown in Figure 1(a) and (b). Figure 1(c) and (d) show an abnormal slice through the ventricles, though pathology may exist within any given slice. The labeled normal intra-cranial tissues of interest are: CSF (dark gray) and the parenchyma tissues, white matter (white), and gray matter (black). In the abnormal slice, pathology (light gray) occupies an area that would otherwise belong to normal tissues. In the approach described here, only part of the pathology (gadolinium-enhanced tumor) is identified and labeled.

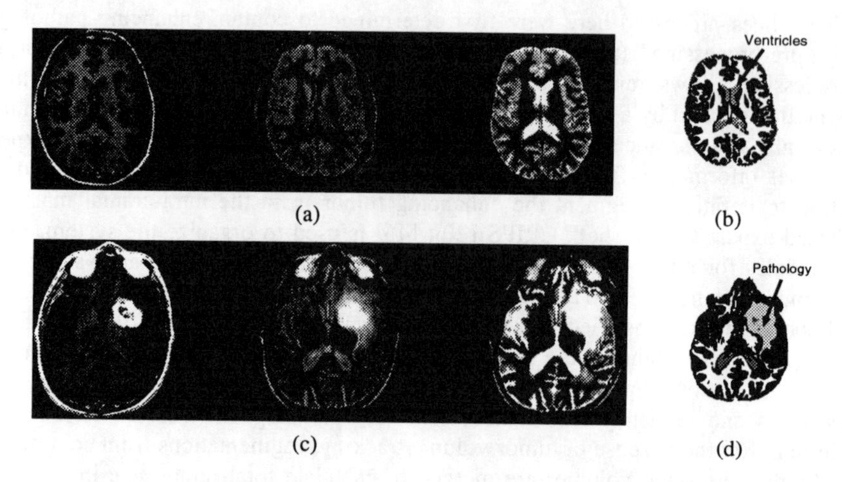

Figure 1. Slices of interest: (a) raw data from a normal slice (T1-weighted, PD-weighted, and T2-weighted from left to right); (b) after segmentation; (c) raw data from an abnormal slice (T1-weighted, PD-weighted, and T2-weighted from left to right); (d) after segmentation. White = white matter; black = gray matter; dark gray = CSF; light gray = pathology in (b) and (c).

A total of 385 slices containing radiologist diagnosed glioblastoma-multiforme tumor were available for processing. Table 4 in Section 4 lists the distribution of these slices across 33 volumes of eight patients who received varying levels of treatment, including surgery, radiation therapy, and chemotherapy prior to initial acquisition and between subsequent acquisitions. A training set of 65 slices was created using tumor size (per slice) and level of gadolinium enhancement as criteria. The heuristics to be discussed in Section 3 were manually extracted from the training subset through the process of "knowledge engineering" and are expressed in general terms, such as "higher end of the T1 spectrum" (which does not specify an actual T1 value). This provides knowledge that is more robust across slices, and avoids dependence on a slice's particular thickness, scanning protocol, or signal intensity, as was the case in [6].

2.2. Basic MR Contrast Principles

One of the key advantages of MR imaging is its ability to acquire multispectral data by rescanning a patient with different combinations of pulse sequence parameters (in our case, repetition time (TR) and echo time (TE)). For example, the MR data used in this study consists of T1, PD, and T2-weighted feature images. A T1-weighted image is produced by a relatively short TR/short TE sequence, a PD-weighted image uses a long TR/short TE sequence, and a long TR/long TE sequence produces a T2-weighted image [36], [15]. For the purpose of brevity, the T1-weighted, PD-weighted, and T2-weighted features will be referred to as T1, PD, and T2, respectively.

Table 1: A Synopsis of T1, PD, and T2 Effects on the Magnetic Resonance Image. TR = Repetition Time; TE = Echo Time.

Pulse Sequence (TR/TE)	Effect (Signal Intensity)	Tissues
T1-weighted (short/short)	Short T1 relaxation (bright)	Fat, Proteinaceous Fluid, Paramagnetic Substances (Gadolinium)
	Long T1 relaxation (dark)	Neoplasms, Edema, CSF, Pure Fluid, Inflammation
PD-weighted (long/short)	High proton density (bright)	Fat, Blood, Fluids, CSF
	Low proton density (dark)	Calcium, Air, Fibrous Tissue, Cortical Bone
T2-weighted (long/long)	Short T2 relaxation (dark)	Iron-containing substances (blood-breakdown products)
	Long T2 relaxation (bright)	Neoplasms, Edema, CSF, Pure Fluid, Inflammation

A particular pulse sequence parameter will provide the best contrast between different tissue types [36] and a series of these images can be combined to provide a multispectral data set. The physics of these pulse sequences are outside the scope of this chapter and their discussion is left to other literature sources [15], [36], [48]. We are primarily concerned with which pulse sequences best delineate specific tissues. After reviewing the available literature, a brief synopsis is shown in Table 1. This synopsis is the starting point for acquired knowledge, which was refined for the specific task of tumor segmentation.

2.3. Knowledge-Based Systems

Knowledge is any piece of information that effectively discriminates between class types [20]. Therefore, tumors will have certain characteristics that are not found in other brain tissues and vice-versa. Two primary sources of knowledge are available in the domain of MR imaging of the brain. The first is tissue signal intensity in feature (or parameter) space, which describes tissue characteristics within the MR imaging system (based on a review of literature [15], [36], [48]. The second is image/anatomical space and includes expected shapes and placements of certain tissues within the MR image. As each processing stage is described in Section 3, the specific knowledge extracted and its application will be detailed.

2.4. System Overview

A strength of the knowledge-based (KB) systems in [6], [7], [34] has been their "iterative processing." Instead of a single classification step, incremental refinement is applied with easily identifiable tissues located and labeled first, allowing a "focus" to be placed on the remaining pixels. To better illustrate the system's organization, we present it at a conceptual level. Figure 2 shows the primary steps in extracting tumor from raw MR data. Section 3 describes these steps in more detail.

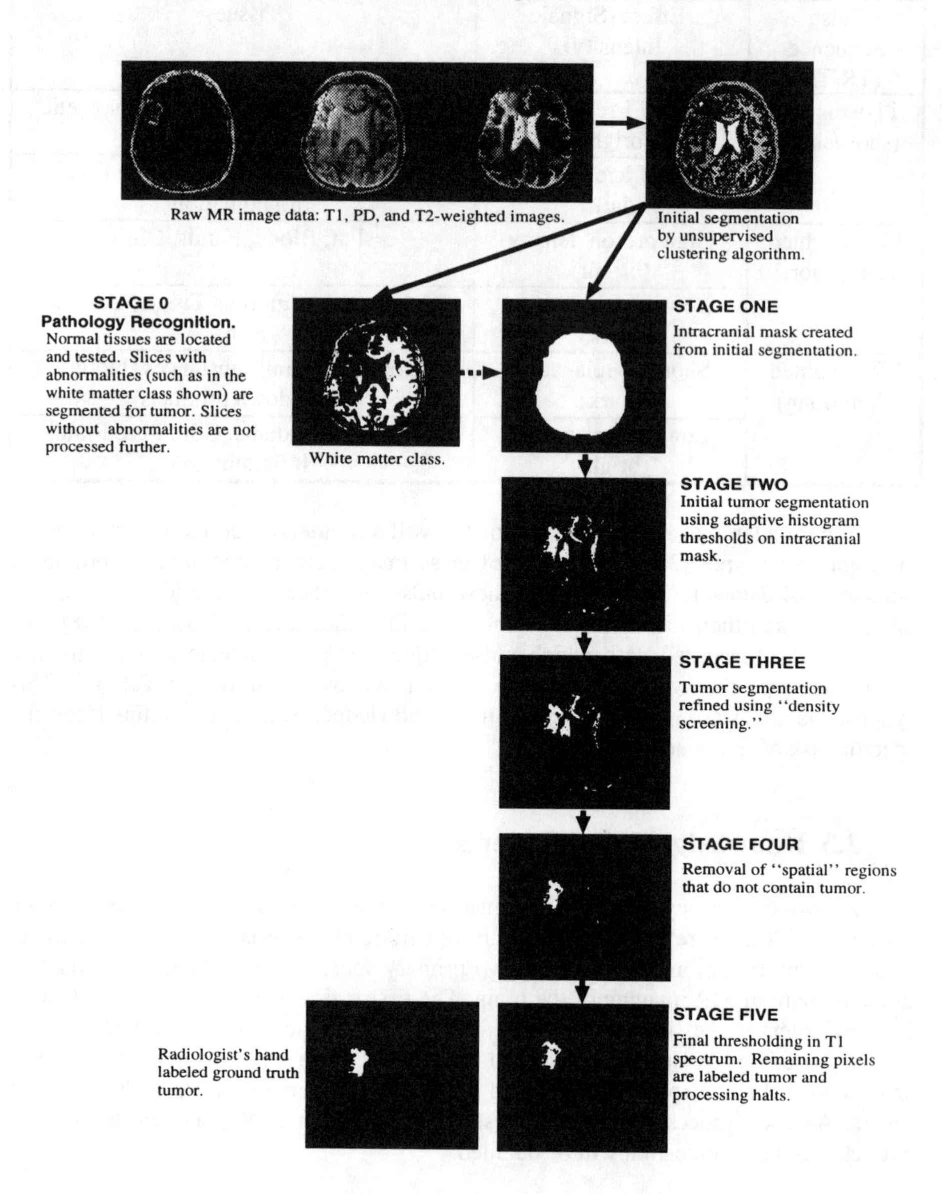

Figure 2. System overview.

The system has six primary steps. First, a pre-processing stage developed in previous works [6], [7], [8], [34] called Stage Zero here, is used to detect deviations from expected qualitative properties within the slice. Slices that are free of abnormalities are not processed by the tumor segmentation system. Otherwise, Stage One extracts the intra-cranial region from the rest of the MR image based on information provided by pre-processing. This creates an image mask of the brain that limits processing in Stage Two to only those pixels contained by the mask..

Applying a combination of adaptive histogram thresholds in the T1 and PD feature images creates an initial tumor segmentation. A "density screening" operation is used in Stage Three to remove additional non-tumor pixels. Density screening exploits the tendency of normal tissue pixels to be more closely grouped together in comparison to tumor pixels.

Stage Four applies a connect-components operation in image space and analyzes each resulting spatially disjoint "region." Those regions found to contain no tumor are removed and the remaining regions are passed to Stage Five for application of a final threshold in the T1 spectrum, using the approximated tumor boundary (determined with a fuzzy edge detector). The resulting image is considered the final tumor segmentation and can be compared with a ground truth image.

3. Classification Stages

By its nature, tumor tissue is much more difficult to model in comparison to normal brain tissues. Therefore, tumor is defined here more by what it is not than what it is. Specifically, the tumor segmentation system operates by removing all pixels considered not to be enhancing tumor, with all remaining pixels labeled as tumor.

3.1. Stage Zero: Pathology Detection

All slices processed by the tumor segmentation system have been automatically classified as containing enhancing pathology and are known to contain glioblastoma-multiforme tumor based on radiologist pathology reports. Because this research is an extension of previous work, knowledge generated during "pre-processing" is available to the tumor segmentation system. Detailed information can be found in [6], [7], [8], [34], but a brief summary is provided.

First, an unsupervised fuzzy c-means (FCM) clustering algorithm [5], [23] is used to segment the slice. The FCM partition is passed to an expert system, which uses a combination of knowledge concerning cluster distribution in feature space and anatomical information to classify the slice as normal or abnormal. Examples of heuristics used to detect pathology include: (1) in a normal slice, the cluster center with the highest T2 value in the intra-cranial region contains primarily CSF; (2) in image space, all normal tissues have roughly the same number of pixels in each brain hemisphere (forming a coarse "symmetry" along the vertical axis). Normal slices will conform to such qualitative "expectations," while abnormal slices will contain deviations, such as the one shown in Figure 2 whose white matter class failed to completely enclose the ventricle area. An abnormal slice, along with the facts generated in labeling it abnormal, are passed on to the tumor segmentation system. Normal slices have all pixels labeled.

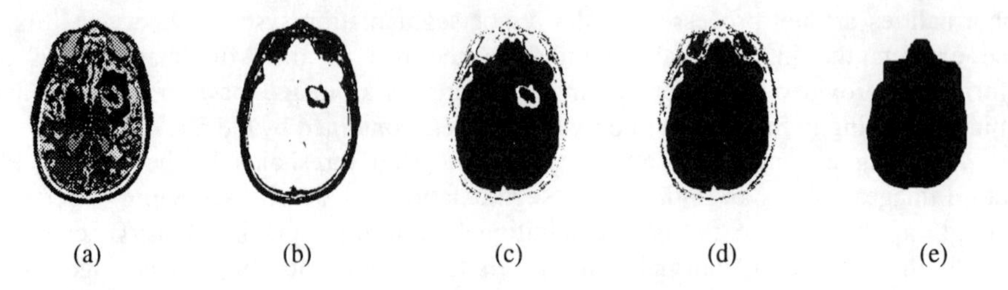

Figure 3. Building the intra-cranial mask. (a) The original FCM-segmented image; (b) pathology captured in Group 1 clusters; (c) intra-cranial mask using only Group 2 clusters; (d) mask after including Groups 1 clusters with tumor; (e) mask after extra-cranial regions are removed.

3.2. Stage One: Building the Intra-Cranial Mask

The first step in the tumor segmentation system is to extract the intra-cranial region from the rest of the MR image. During pre-processing, extra and intra-cranial pixels were distinguished primarily by separating the clusters from the initial FCM segmentation. Extra-cranial clusters are known as Group 1 clusters, while brain tissue clusters are called Group 2 clusters. Enhancing tumor pixels can occasionally be placed into one or more Group 1 clusters with high T1-weighted centroids. These pixels can usually be reclaimed through a series of morphological operations (described below). As shown in Figure 3(b) and (c), however, the tumor loss may be too severe to recover morphologically without distorting the intra-cranial mask.

Group 1 clusters containing significant “Lost Tumor” can be identified, however. During pre-processing, Group 1 and 2 clusters were separated based on the observation that intra-cranial tissues are located within the brain itself, while extra-cranial tissues only surround the brain. A “quadrangle” was developed by Li in [33], [34] to roughly approximate the intra-cranial region. The number of pixels each cluster had within the quadrangle could then be counted. Clusters consisting of extra-cranial tissues will have very few pixels inside this estimated brain, while cluster of intra-cranial tissues will have a significant number. An example is shown in Figure 4.

A Group 1 cluster was considered to contain “Lost Tumor” if more than 1% of its pixels were contained in the approximated intra-cranial region. Because an extra-cranial cluster with no Lost Tumor will have very few pixels within the quadrangle, if any at all, the relatively low value of 1% is sufficient. Pixels belonging to Lost Tumor clusters (Figure 3(b)) are merged with pixels from all Group 2 clusters (Figure 3(c)) and set to foreground (a non-zero value), with all other pixels in the image set to background (value = 0). This produces a new intra-cranial mask similar to the one shown in Figure 3(d).

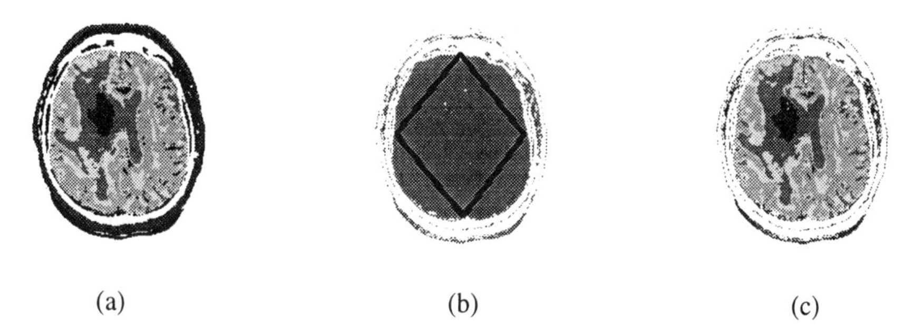

(a) (b) (c)

Figure 4. (a) Initial segmented image; (b) a quadrangle overlaid on (a); (c) classes that passed the quadrangle test.

Because a Lost Tumor cluster is a Group 1 cluster, its inclusion in the intra-cranial mask introduces areas of extra-cranial tissues, such as skin, fat, and muscle (Figure 3(d)). To remove these unwanted extra-cranial regions (and recover smaller areas of lost tumor, mentioned above), a series of morphological operations [26] are applied, which use window sizes that are the smallest possible (to minimize mask distortion) while still producing the desired result.

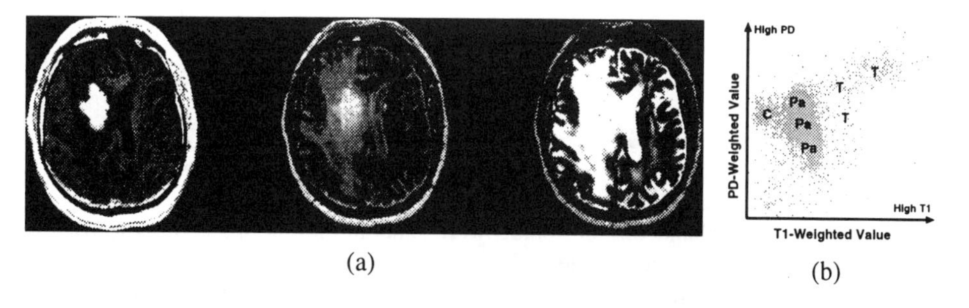

(a) (b)

Figure 5. (a) Raw T1, PD, and T2-weighted data. The distribution of intra-cranial pixels is shown in (b) T1-PD feature space. C = CSF; P = parenchyma tissues; T = tumor.

Applying a 5 × 5 closing operation to the background removes small regions of extra-cranial pixels and enhances separation of the brain from meningial tissues. Then the brain is isolated by applying an eight-wise connected components operation [26] and keeping only the largest foreground component (the intra-cranial mask). Finally, "gaps" along the periphery of the intra-cranial mask are filled by first applying a 15 × 15 closing, then a 3 × 3 erosion operation. An example of the final intra-cranial mask can be seen in Figure 3(e).

3.3. Stage Two: Multi-spectral Histogram Thresholding

Given an intra-cranial mask, there are three primary tissue types: the brain parenchyma (white and gray matter), CSF, and the pathology of interest, which can include gadolinium-enhanced tumor, edema, and necrosis. The first task is to remove as many pixels belonging to normal tissues as possible from the mask.

Each MR voxel of interest has a (T1, PD, T2) location in $\mathbf{R}^3$, forming a feature-space distribution. Pixels belonging to the same tissue type will exhibit similar T1 and T2 relaxation behaviors and water content (PD) [3]. As a result, they also have a similar location in feature space [3]. Figure 5(a) shows the raw signal-intensity images of a typical slice, while (b) shows a histogram for the bivariate features T1/PD with approximate tissue labels overlaid. Some overlap between classes can be observed because the graphs are projections and also due to "partial-averaging" where different tissue types are quantized into the same pixel/voxel.

The typical relationships between enhancing tumor and other brain tissues can also be seen in histograms for each of the three feature images, Figure 6. These distributions were examined and interviews were conducted with experts concerning the general makeup of tumorous tissue, and the behavior of gadolinium enhancement in the three MRI protocols.

From these sources, a set of heuristics were extracted that could be included in the system's knowledge base:

1. Gadolinium-enhanced tumor pixels are found at the higher end of the T1 spectrum.
2. Gadolinium-enhanced tumor pixels are found at the higher end of the PD spectrum, though not with the degree of separation from normal tissues found in T1 space [24].
3. Gadolinium-enhanced tumor pixels were generally found in the "middle" of the T2 spectrum, making segmentation based on T2 values difficult.
4. Tumor pixels with greater enhancement had better separation from non-tumor pixels, while less enhancement resulted in more overlap between tissue types.

Analysis of these heuristics revealed that applying histogram thresholds could remove a significant number of non-tumor pixels in a simple and effective manner, while still preserving the tumor pixels of interest. In fact, the T1 and PD signal intensities having the greatest number of pixels (i.e., the T1 and PD histogram "peaks") were found to be effective thresholds that work across slices, even those with varying degrees of gadolinium enhancement. The T2 feature had no such property that was consistent across all training slices and was excluded. An example of this is shown in Figure 6.

A pixel survived thresholding only if its signal intensity value in a particular feature was greater than the respective intensity threshold. Figures 7(a) and (b) show the results of applying the T1 and PD histogram peak thresholds in Figures 6(b) and (c).

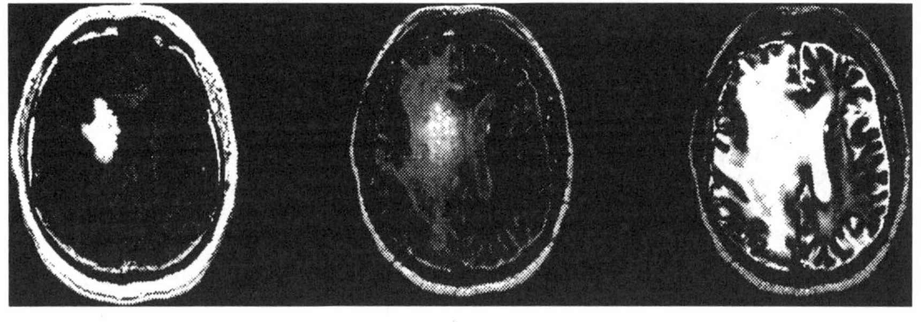

(a) Raw Data

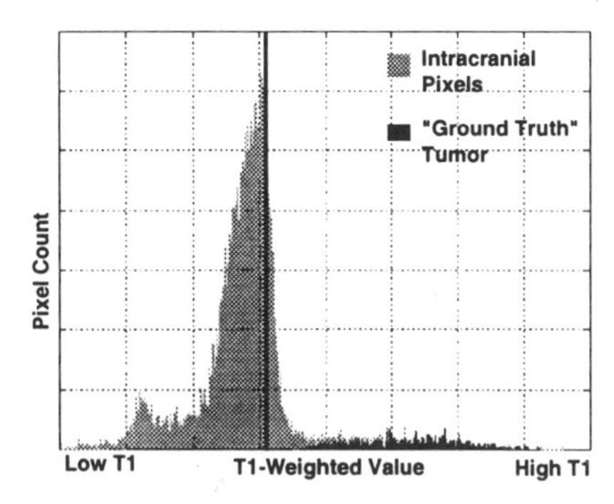

(b) T1-weighted Histogram

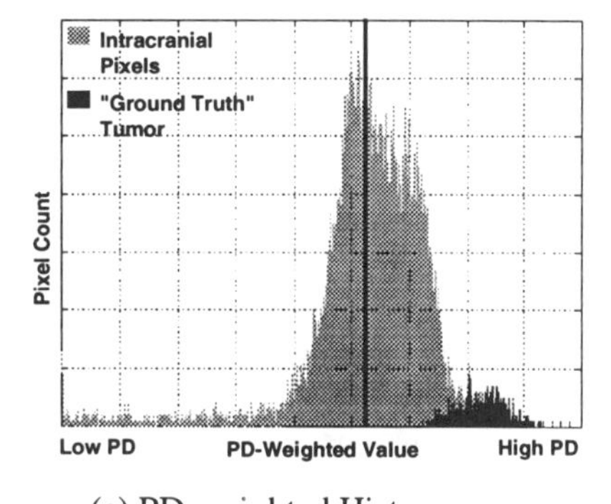

(c) PD-weighted Histogram

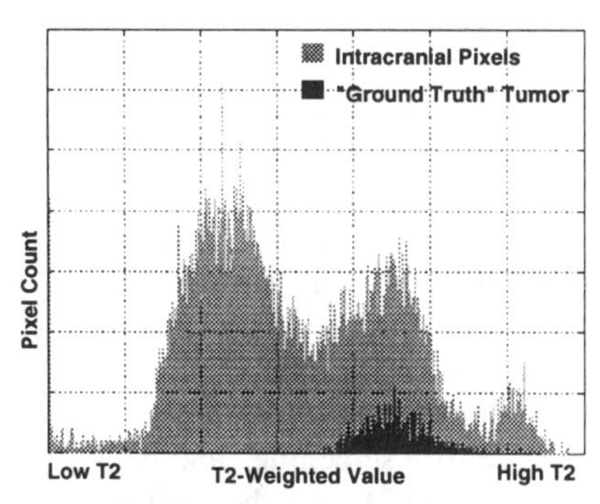

(d) T2-weighted Histogram

Figure 6. Histograms for tumor and the intra-cranial region. Solid black lines indicate threshold in T1 and PD-weighted space.

In both Figures 7(a) and (b), a significant number of non-tumor pixels have been removed by the respective threshold operation, but some non-tumor pixels still remain. Since the heuristics listed above state that gadolinium enhanced tumor has a high signal intensity in both the T1 and PD features, additional non-tumor pixels can be removed by intersecting the two images (where a pixel remains only if it's present in both images). An example is shown in Figure 7(c).

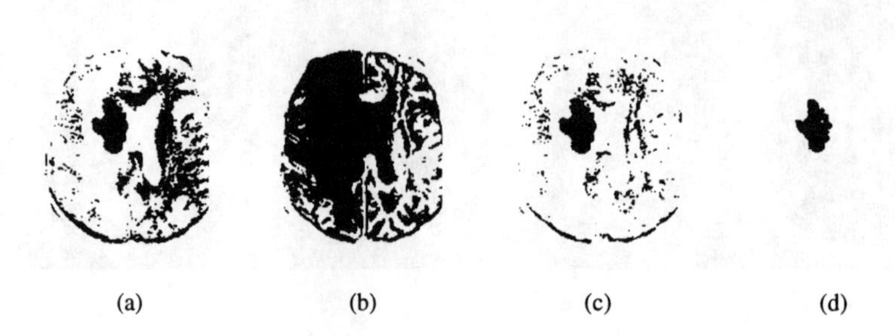

(a) (b) (c) (d)

Figure 7. Multi-spectral histogram thresholding of Figure 6. (a) T1-weighted thresholding; (b) PD-weighted thresholding; (c) intersection of (a) and (b); (d) ground truth.

3.4. Stage Three: "Density Screening" in Feature Space

The threshold operations in Stage Two provide a good initial tumor segmentation, such as the one shown in Figure 7(c). Comparing it with the ground truth image in Figure 7(d), a number of pixels in the initial tumor segmentation that are not found in the ground truth image should be removed. Additional thresholding is difficult to perform, however, without possibly removing tumor as well as non-tumor pixels.

Pixels belonging to the same tissue type will have similar characteristics within the MR coil and signal intensities in the three feature spectrums. Normal tissue types (brain parenchyma and CSF) have a more or less uniform cellular makeup, while tumor can have significant variance due to local degrees of enhancement and tissue inhomogeneity (from partial-averaging of tumor with edema, necrosis, and possibly some parenchyma tissues) [15], [36], [48]. As a result, the distribution of normal tissues in feature space will be relatively concentrated while tumor pixels are more widely distributed [3]. Figures 5(b) and (c) show the different spreads in feature space for normal and tumor pixels. By exploiting this "density" characteristic, non-tumor pixels can be removed without affecting the presence of tumor pixels.

The process is called "density screening" and begins by creating a "quantized" 3-dimensional histogram (T1/PD/T2) of the initial tumor segmentation image. Quantization was performed for two reasons. First, sizes of a three-dimensional histogram can quickly become prohibitively large to store and manipulate. Even a 256^3 histogram (where 256 is the standard number of intensity levels in a gray-scale image) has nearly 17 million elements and in the 12-bit data studied here, slices could have as many as 800 intensity values in a given feature after thresholding. Second, levels of quantization can make the "dense" nature of normal pixels more apparent while still leaving tumor pixels relatively spread out.

For the 12-bit data studied here, 128 bins were used in each feature, which was empirically selected. Using 64 bins blurred separation of tumor and non-tumor pixels in training slices where the tumor boundary was not as well defined. Values similar to 128, such as 120 or 140, did not significantly change the "quantization" effect. Using

MR data with lower or greater resolution (e.g., 256 signal intensities in each feature) may require different quantization levels.

Quantization was performed in each feature by determining the maximum and minimum signal intensity values in the initial tumor segmentation and scaling them into the 128 bins (e.g., the minimum T1 intensity value occupies T1 Bin 1, the maximum T1 intensity value occupies T1 Bin 128). The histograms and scatterplots shown in Figure 8 were created using 128 bins.

From the 3D histogram, three 2D projections are created: T1/PD, T1/T2, and PD/T2. Figure 8(a) shows an example T1/PD projection of the slice from Figure 7(c). A corresponding scatterplot is shown in Figure 8(b). The bins with the greatest number of pixels (the highest "peaks" in Figure 8(a)) are located in the lowest T1/PD corner and are comprised of non-tumorous pixels that should be removed. In contrast, tumor pixels, while greater in total number, are more widely distributed. Thus, their corresponding bins have much lower peaks.

In each 2D projection, non-tumor pixels are removed via a region growing process [27] that starts at the highest peak and "clears" any neighboring bin (including the starting point) whose cardinality (number of pixels in that bin) is greater than a set threshold (T1/PD = 3, T1/T2 = 4, PD/T2 = 3). Figure 8(c) shows the resulting scatterplot after screening. A pixel is removed from the tumor segmentation if it corresponds to any bin that has been "cleared" in any of the three 2D feature-domain projections. Figures 8(d) and (e) are the tumor segmentation before and after the entire density screening process is completed. Note that the resulting image is closer to ground truth.

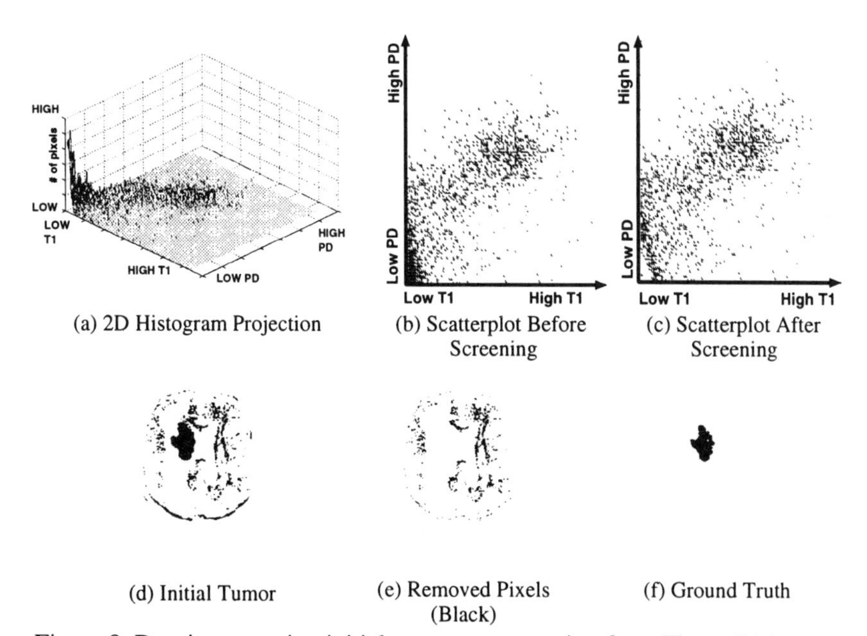

(a) 2D Histogram Projection (b) Scatterplot Before Screening (c) Scatterplot After Screening

(d) Initial Tumor (e) Removed Pixels (Black) (f) Ground Truth

Figure 8. Density screening initial tumor segmentation from Figure 7(c).

The cardinality thresholds were determined by creating a 3D histogram, including 2D projections, of the ground truth tumor image of each training slice, but using the quantization and scaling of the initial tumor segmentation for that training slice. This would be as if a 3D histogram had been created for an initial tumor segmentation, and all non-tumor pixels had been removed, leaving only ground truth tumor pixels. The respective 2D projections of all training slices were examined. It was found that the smallest bin cardinality bordering a bin occupied by known non-tumor pixels made an accurate threshold for the given projection.

3.5. Stage Four: Region Analysis and Labeling

The knowledge extracted and applied in Stages Two and Three only considered pixels individually. Stage Four allows spatial information to be introduced by considering pixels on a region or component level. Applying an eight-wise connected components operation [26] to the refined tumor segmentation generated by Stage Three, allows each region to be tested separately for the presence of tumor. An example is shown in Figure 9.

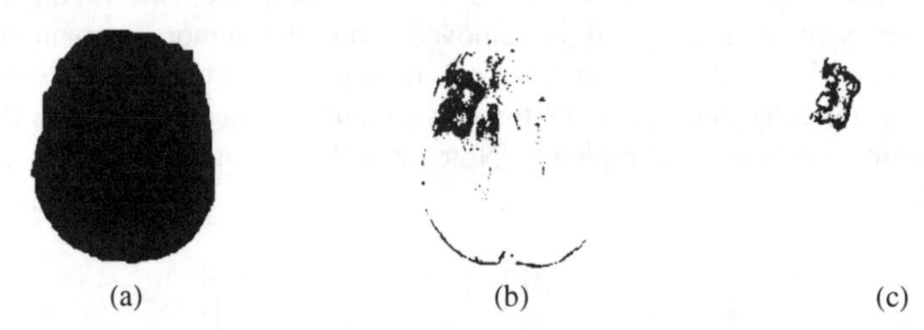

(a) (b) (c)

Figure 9. Regions in image space. After processing the intra-cranial mask (a), (b) is an initial tumor segmentation. Only one region, as shown in the ground-truth image (c), is actual tumor. Region analysis discriminates between tumorous and non-tumorous regions.

After processing the intra-cranial mask shown in Figure 9(a) in Stage Two and Three, a refined tumor segmentation (b) is produced. The segmentation shows a number of regions, but the ground truth tumor in Figure 9(c) shows that only one actually contains tumor. Therefore, regions containing tumor must be separated from regions that do not.

3.5.1. Removing Meningial Regions

Because the blood carries gadolinium to the tumor, other tissues that receive gadolinium-infused blood may also enhance in the MR coil. This includes the meningial tissues immediately surrounding the brain, such as the *dura* of *pia mater*. This may result in regions with a high T1 signal intensity, which may violate the knowledge base's assumption in Section 3.5.2 that regions with the highest T1 value are most likely tumor. Such regions can be identified and removed via anatomical knowledge by noting that, since they are thin membranes, meningial regions should lie along the periphery of the brain in a relatively narrow margin.

Figure 10 shows that an approximation of the brain periphery can be used to detect meningial tissues.

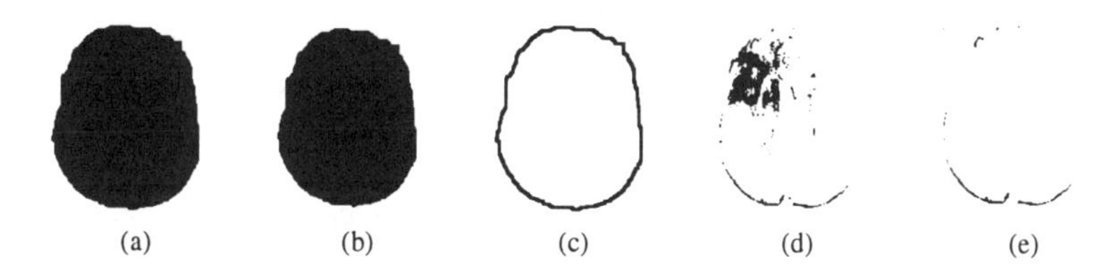

Figure 10. Removing meningial pixels. A "ring" that approximates the brain periphery is created by applying a 7 × 7 erosion operation to the intra-cranial mask (a), resulting in image (b). Subtracting (b) from (a), creates a "ring," shown in (c). By overlaying this "ring" onto a tumor segmentation (d), small regions of meningial tissues (e) can be detected and removed. The unusual shape of the intra-cranial region is due to prior resection surgery.

Applying a 7 × 7 erosion operation to the intra-cranial mask and subtracting the resultant image from the original mask first creates the brain periphery, as shown in Figure 10(a-c). The unusual shape of the intra-cranial region in Figure 10 is due to prior resection surgery. Each component or separate region in the refined tumor mask is now intersected with the brain periphery and is removed as meningial tissue if 50% or more of its pixels are contained in the periphery. Figure 10(d) shows a tumor segmentation which is intersected with the periphery from Figure 10(c). In Figure 10(e), the pixels that will be removed by this operation are shown and they are indeed meningial pixels.

3.5.2. Removing Non-Tumor Regions

After all meningial regions have been removed, trends and characteristics described at a pixel level in Section 3.3 may be applied on a region level to discriminate between regions with and without tumor. Specifically, statistical information about a region, its mean, standard deviation, and skewness in (T1), (PD), and (T2) feature space, respectively, is calculated. By sorting regions in feature space based upon their mean values, rules based on their relative order can be created:

1. Large regions that contain tumor will likely contain a significant number of pixels of highest intensity in T1 and PD space, while regions without tumor likely contain a significant number of pixels of lowest intensity in T1 and PD space.
2. The means of regions with similar tissue types neighbor one another in feature space.
3. The intra-cranial region with highest mean T1 value and a "high" PD and T2 value, is considered "First Tumor," against which all other regions are compared.
4. Other regions that contain tumor are likely to fall within 1 to 1.5 standard deviations (depending on region size) of "First Tumor" in T1 and PD space.

In most cases, the glioblastoma-multiforme tumor is only one spatially compact region with the highest mean T1 value. In some cases, however, tumor growth may have branched into both hemispheres of the brain, causing the tumor to appear disjointed in some slices. In others, the tumor has fragmented as a result of treatment. Moreover, different tumor regions do not enhance equally. Thus, cases can range from a single well-enhanced tumor to a fragmented tumor with different levels of enhancement. In comparison, non-tumor regions are generally more consistent in their makeup than tumorous regions. Therefore, the knowledge base is designed to facilitate removal of non-tumor regions because their composition can be more reliably modeled and detected.

Regions that comply with the first heuristic listed above are the easiest to locate and their statistics can be used to examine the remaining regions. To detect these regions, three image masks are used.

The first mask is created by taking the refined tumor segmentation image and keeping only the highest 20% T1 valued pixels (i.e., if there were 100 pixels in the refined tumor image, the 20 pixels with the highest T1 values are kept). The second mask keeps the 20% highest in PD space. The third mask keeps the 30% lowest in T1 space.

(a) (b) (c) (d) (e)

Figure 11. Using pixel counts to remove non-tumorous regions. Given a refined tumor segmentation after Stage Three (a), spatial regions with a significant number of pixels highest in T1 space (b) or PD space (c) are likely to contain tumor. Regions with pixels lowest in T1 space (d) are unlikely to contain significant tumor. Ground truth is shown in (e).

Each region in the tumor segmentation image is isolated and intersected with each of the three masks. The number of pixels of the region found in a particular mask is recorded and compared with the rules listed in Table 2. An example is shown in Figure 11.

Table 2: Region Labeling Rules Based On Pixel Presence

Region Size	Pixels in intersection with the 3 masks	Action
≤ 5	Any Bottom T1 Pixels AND < 2 Top T1 Pixels	Remove Non-Tumor
≥ 500	> Region Size x 0.06 Top T1 Pixels	Label as Tumor
≥ 5	No Top T1 Pixels AND > Region Size x 0.005 Bottom T1 Pixels AND < Region Size x 0.01 Top PD Pixels	Remove Non-Tumor

Regions that do not activate any of the rules in Table 2 remain unlabeled and are analyzed using the last two heuristics.

Once a region has been positively labeled tumor, the third heuristics indicates that it can be used as a point of reference to search for neighboring tumor regions in feature space. In most cases, the region with the highest T1 mean value can be selected as this point of reference (called "First Tumor"). Occasionally, an extra-cranial region, such as meningial region within the inter-hemispheric fissure, can be selected instead. To prevent this, the selected region is verified via the heuristic that, in addition to a very high T1 mean value, a tumor region will also occupy the highest half of all regions in sorted PD and T2 mean space. For example, if there were 10 regions total, the region being tested must have one of the 5 highest mean values in both PD and T2 space. If the candidate region passes, it is confirmed as First Tumor. Otherwise, it is discarded and the region with the next highest T1 mean value is selected for testing as First Tumor.

Once First Tumor has been verified, the search for neighboring tumor regions can begin. Although tumorous regions can have between-slice variance, the third and fourth heuristics hold for the purpose of separating tumor from non-tumor regions within a given slice. Furthermore, the standard deviations in T1 and PD space of a known tumor region were found to be a useful and flexible distance measure.

Table 3: Region Labeling Rules Based On Statistical Measurements. *Largest* is the largest known tumor region.

(a) Rules Based on Standard Deviation (σ) of "First Tumor"

Region Size	If Region's Mean Values Are:	Action
$\geq$ 10 OR $\leq$ Largest/4	> 1 σ in T1 Space OR > 1 σ in PD Space	Remove
$\geq$ 10 AND $\leq$ Largest/4	> 1.5 σ in T1 Space AND > 1.5 σ in PD Space	Remove

(b) Labeling Rules Based on Region Statistics

$\geq$ 100	$\text{Skewness}_{T1} \leq 0.75$ AND $\text{Skewness}_{PD} \leq 0.75$ AND $\text{Skewness}_{T2} \leq 0.75$	Remove

Table 3(a) lists the two rules that used the standard deviation to remove non-tumor regions, based on the size (number of pixels) of the region being tested. The rule in Table 3(b) serves as a tiebreaker for some regions that remain unlabeled. The term "Largest" is used to indicate the largest known tumor region. In most cases there was only a single tumor region, so the First Tumor region was also the Largest region. In cases with fragmented tumor, however, a larger tumor region would have more pixels (and thus, a more reliable sample) to calculate the mean and standard deviation for the distance measure. Therefore, the system would find Largest by searching for the largest region that was within one standard deviation in both T1 and PD space to the First

Tumor region. After the rules in Table 3 are applied, all regions that were not removed are kept as regions containing tumor.

3.6. Stage Five: Final T1 Threshold

At the end of Stage Four, the regions with no tumor have been removed, but non-tumor pixels may still be found in those regions considered to contain tumor. While enhancing tumor has properties in each of the three available features that have been used as knowledge, discussions with an expert radiologist [39] have indicated that final tumor boundaries are determined by pixel intensities in the T1-weighted image. Thresholds were described in Section 3.3 in a relatively coarse manner because the boundary of enhancing tumor was "obscured" by pixels belonging to non-tumor tissues. With the removal of most of these non-tumor tissues in Stages Two through Four, however, a greater level of focus can be placed and a more precise threshold can be applied.

The threshold is determined using the principle that the spatial boundary between enhancing tumor and surrounding tissues contain pixels whose signal intensities correspond to the tumor/non-tumor boundary in T1-weighted space. One of the most common methods of isolating spatial boundaries between objects of interest (with different intensities) has been edge detection. In our case, edges represent differences in T1-weighted signal intensities, the more distinct boundary a tumor has, the greater its edge strength will be.

Most edge-detection based methods, such as those in [2], [12], [21], [22], [45], [59], [60] attempt to use edges to trace the tumor's contours. This can work well for tumors within distinct boundaries, as shown in Figure 12(d) using a standard Sobel operator, but can have significant problems with more diffuse tumors, shown in Figure 12(i). A variety of edge detection operators have been introduced, such as Canny and Bergholm [27]. Most of these, however, have a number of parameters that are difficult to automatically optimize, especially in a domain where the object of interest can have such wide ranging characteristics [4]. As a result, Dellepiane [2] and Raff and Newman [45] have suggested that edge detection is unlikely to work reliably for complex structures like tumors.

Edge detection may still provide knowledge to be exploited, however, by noting that edges not only approximate the tumor boundary spatially, but can also indicate the approximate signal intensity of that boundary which can be used in a threshold operation. Now detected edges need not be perfect, merely sufficient to indicate the appropriate signal intensity. Edge detection must still be reasonable, however, for the method to work.

To address the problem of detecting edges in tumors with diffuse edges, and to minimize the problem of parameter optimization, the technique introduced there uses a "fuzzy" approach to edge detection presented by Tao and Thompson in [49]. These authors used fuzzy if-then rules that were based on the relationship between each pixel and its eight-wise neighbors. The fuzzy rule captures the concept that a pixel is an edge pixel of it lies on the boundary of two (or more) regions. A particular edge pixel will have similar characteristics (e.g., T1 signal intensity) with one region and dissimilar characteristics with the other region(s). Sixteen structure elements, such as the examples shown in Figure 13, are used in [49] to represent possible edge orientations.

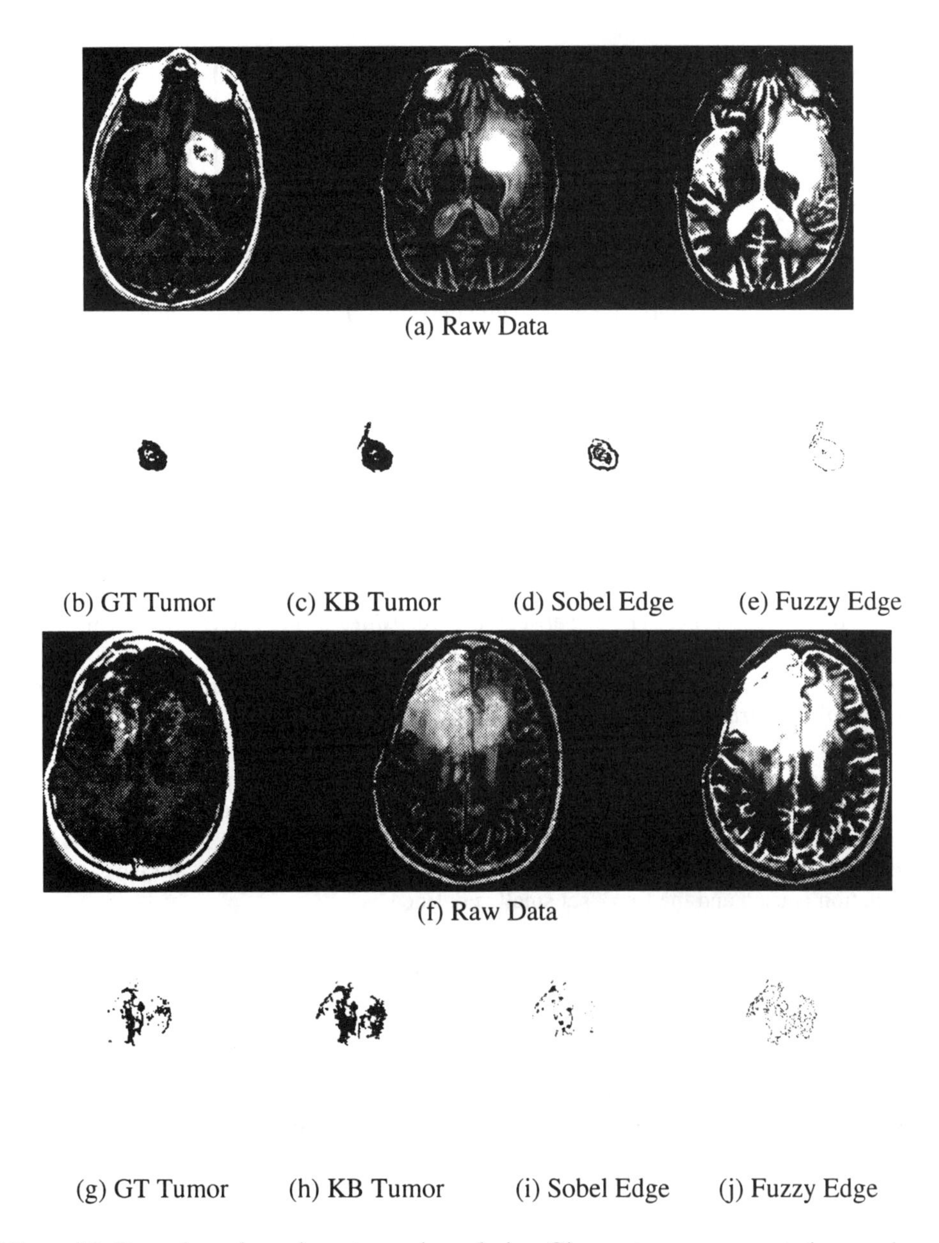

Figure 12. Detecting edges along tumor boundaries. Given a tumor segmentation mask, (b) and (h), produced by Stage Four, edge detection is performed on pixels contained by the mask to find the tumor's boundaries. The results of a Sobel operator are shown in (d) and (i), while (e) and (j) show the results of a fuzzy edge detector described in [49]. The tumor in (a) has distinct edges and both edge detectors work well, although the Sobel operator more closely matches ground truth (b). The tumor in (e), however, performs relatively poorly for both the Sobel and fuzzy edge detectors.

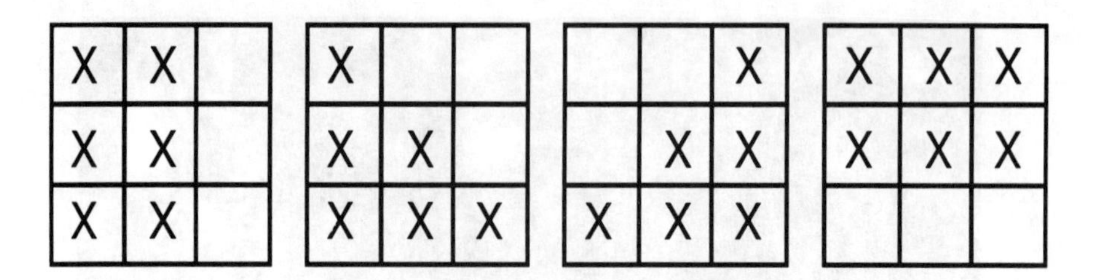

Figure 13. Edge structures for fuzzy edge detection. Edge structures, like the examples shown above, are used to generate the fuzzy if-then rules. Neighbouring pixels covered by an "X" are used to calculate membership in the fuzzy set *small*, while those uncovered are used to calculate membership in the fuzzy set *large*.

For example, the leftmost structure in Figure 13 captures vertical edges, while the rightmost structure captures horizontal edges. Each structure element is used to with a fuzzy if-then rule that is based on the general concept of edges described above:

IF *[the differences (D_x s) between the intensities of the pixels (marked with "X") and the center pixel are small]* AND *[the between the intensities of the pixels (not marked with "X") and the center pixel are large]* THEN *the center pixel is an edge pixel for this structure.*

Thus, an edge pixel is found if its "X" neighbors have similar T1 signal intensities (i.e., the difference is "small") and its neighbors have dissimilar intensities (i.e., the difference is "large"). The authors define the fuzzy memberships small and large as bell-shaped, though they do not specify a particular function. Here, a Gaussian-based function is used and the fuzzy set small is defined as:

$$\mu_{small} = e^{-\frac{Diff^2_{(a,b)}}{2\sigma^2}}$$

where $Diff_{(a,b)}$ is the absolute intensity difference between the center pixel and the eight-wise neighbor (a, b). The fuzzy set large is defined as:

$$\mu_{large} = 1 - \mu_{small}.$$

The actual Gaussian formula is not used to allow a membership of $\mu_{small} = 1.0$ to be returned when $Diff = 0$. Also, in a standard Gaussian function, the value s represents the standard deviation. For defining the fuzzy set, it controls how quickly μ_{small} decreases (and μ_{large} increases) as the intensity difference, Diff, becomes larger. In this preliminary study, s = 2.0, although it could be possible to have rules in the knowledge-base adjust the value according to a tumor's characteristics.

The fuzzy if-then rule described above is used to determine a pixel's "edge potential" (PEP) for a given edge structure (i.e., a particular edge orientation) by calculating the memberships between the center pixel and each of its eight-wise neighbors for the fuzzy set small (if the neighbor is covered by an "*x*" in the edge structure) or large (if the neighbor is uncovered), and returning the minimum membership. For example, using the first structure (PEP1) in Figure 13, its edge potential would be

$$\begin{aligned}\mu_{PEP1}(x, y) = \min\big[&\mu_{small}(Diff_{(x-1,y-1)}), \mu_{small}(Diff_{(x-1,y)}),\\ &\mu_{small}(Diff_{(x-1,y+1)}), \mu_{small}(Diff_{(x,y-1)}), \mu_{small}(Dif_{(x,y+1)}),\\ &\mu_{large}(Diff_{(x+1,y-1)}), \mu_{large}(Diff_{(x+1,y)}), \mu_{large}(Diff_{(x+1,y+1)})\big]\end{aligned}$$

Thus, if the pixel under consideration lies along a vertical edge, PEP1 would be expected to return a relatively high membership. Otherwise, if there is no edge, or it lies in a different orientation, PEP1 will have a low membership value. Given sixteen edge structures, a pixel will have sixteen PEP values calculated. The pixel's final edge membership is set by keeping the membership of the structure that best matched the edge (i.e., the structure with highest membership). Formally

$$PEP(x, y) = \max(\mu_{PEP1}(x, y), \ldots, \mu_{PEP16}(x, y))$$

Once final edge memberships have been calculated for all pixels, the detected edges are "thinned" by removing redundant edge pixels through a local maxima operation. A "pseudo-centroid" of the remaining edge strengths is then calculated and only those edges that are stronger than the pseudo-centroid are kept, producing a final edge image, similar to those shown in Figures 12(f) and (j).

The method proposed by Tao and Thompson was implemented as described in [49], with the addition that the technique considers only pixels contained in an image mask (in this case, the tumor segmentation mask produced at the end of Stage Four). Once the final edge image is produced, the final T1 threshold is calculated by averaging the T1 signal intensities of all pixels contained in the final edge image. During averaging, the signal intensities are not weighted by edge strength as it did not significantly affect the final T1 threshold. Once the T1 threshold is calculated, it is applied to the tumor segmentation image produced at the end of Stage Four (keeping only those pixels whose T1 signal intensity is greater than the T1 threshold). The resultant image is considered the final tumor segmentation and processing halts.

4. Results

A total of 385 slices from 33 volumes across 8 patients diagnosed with glioblastoma-multiforme tumor, who also underwent radiation and chemotherapy treatment, were available for processing. From this data set, 65 training slices, shown in Table 4, were extracted and used to construct the knowledge-base rules.

Table 4: MR Slice Distribution. Parenthesis indicated the number of slices from that volume that were used as training. Rn = Repeat Scan *n*.

	# Slices Extracted from Patient Volume							
Scan	P1	P2	P3	P4	P5	P6	P7	P8
Base	10	13(13)	12	16(1)	9(9)	15	12(7)	7
R1	11(1)	14(14)	12	15(1)	10	12	-	-
R2	11	15(2)	-	15(15)	8(1)	13	-	-
R3	10	-	-	-	8(1)	15	-	-
R4	12	-	-	-	7(1)	15	-	-
R5	-	-	-	-	-	16	-	-
R6	-	-	-	-	-	15	-	-
R7	-	-	-	-	-	14	-	-
R8	-	-	-	-	-	14	-	-
R9	-	-	-	-	-	11	-	-
R10	-	-	-	-	-	14	-	-
R11	-	-	-	-	-	18	-	-
R12	-	-	-	-	-	18	-	-
Total	54(1)	42(29)	24	46(16)	42(12)	187	12(7)	7

After processing by the system, the knowledge-based tumor segmentations were compared with radiologist-verified "ground-truth" tumor segmentations [56], as well as three supervised methods. To measure how well (on a pixel level) a tumor segmentation compares with ground truth, two metrics were used. The first, "Percent Match," is simply the number of true positives divided by the total tumor size. The second, is called a "Correspondence Ratio," and was created to account for the presence of false positives:

$$Correspondence\ Ratio = \frac{True\ Pos. - (0.5 \times FalsePos.)}{Number\ Pixels\ in\ Truth\ Tumor}$$

For comparing on a per volume basis, the average value for Percent Match was generated using

$$Average\ \%\ Match\ = \frac{\sum_{i=1}^{slice \in set} (\%\ match\ )_i\ x\ (number\ ground\ truth\ pixels\ )_i}{\sum_{i=1}^{slice \in set} (number\ ground\ truth\ pixels\ )_i}$$

The average value for the Correspondence Ratio is similarly generated.

Table 5: Tumor Volume Comparison (Pat. = Patient; Vol. = Scan Volume; KB = Knowledge Based; %M = Percent Match; CR = Correspondence Ratio; SDP = Standard Deviation as a Percentage of Total Tumor Volume; # Tr. = Number of Trials; # Obs. = Number of Observers.)

Pat.	Vol.	GT Vol.	KB		kNN				
			%M	CR	%M	CR	SDP	# Tr.	# Obs.
1	Base	7213	0.98	0.87	0.89	0.64	0.07	5	2
	R1	7240	0.98	0.86	0.89	0.52	0.19	5	2
	R2	7470	0.98	0.80	0.88	0.54	0.14	5	2
	R3	6395	0.98	0.84	0.88	0.47	0.40	5	2
	R4	6560	0.97	0.79	0.83	0.32	0.07	5	2
2	Base	12230	0.91	0.73	0.74	0.51	0.09	5	2
	R1	14609	0.76	0.70	0.52	0.29	0.16	5	2
	R2	20924	0.80	0.73	0.49	0.25	0.13	5	2
3	Base	10892	0.92	0.88	0.73	0.46	0.15	5	3
	R1	5971	0.97	0.81	0.72	0.37	0.28	5	3
4	Base	10454	0.84	0.75	0.63	-0.13	0.16	5	2
	R1	10835	0.72	0.69	0.56	-0.14	0.28	5	2
	R2	15788	0.81	0.77	0.69	0.15	0.13	5	2
5	Base	10178	0.94	0.88	0.79	0.51	0.14	4	2
	R1	4657	0.93	0.81	0.68	-0.09	0.32	4	2
	R2	5616	0.94	0.81	0.79	-0.47	0.06	4	2
	R3	9215	0.96	0.90	0.84	0.21	0.14	4	2
	R4	3544	0.90	0.55	0.51	-1.90	0.20	4	2
6	Base	5829	0.96	0.36	0.62	-0.19	0	1	1
	R1	2684	0.98	-0.14	0.88	-0.38	0	1	1
	R2	4354	0.99	0.20	0.96	0.40	0	1	1
	R3	6510	0.97	0.53	1.00	0.97	0	1	1
	R4	8688	0.92	0.63	0.87	0.41	0	1	1
	R5	3079	0.80	0.38	0.81	0.32	0	1	1
	R6	4384	0.96	0.66	0.78	0.47	0	1	1
	R7	4047	0.92	0.53	0.82	0.27	0	1	1
	R8	3935	0.89	0.57	0.80	0.53	0	1	1
	R9	3201	0.90	0.55	0.85	0.61	0	1	1
	R10	3895	0.91	0.61	0.88	0.26	0	1	1
	R11	5878	0.94	0.65	0.67	0.22	0	1	1
	R12	7421	0.86	0.63	0.51	0.21	0	1	1
7	Base	1018	0.81	0.50	0.86	0.13	0	1	1
8	Base	4521	0.99	0.92	0.53	0.30	0	1	1

4.1. Knowledge-Based vs. Supervised Methods

Table 5 shows how well the knowledge-based system performs against the k-nearest (kNN) algorithm ($k = 7$) [11] method for both the Match and Correspondence Ratio metrics. The kNN numbers shown were averaged over multiple trials of ROI selection, meaning that all kNN slice segmentations were effectively training slices. Multiple trials/observations also introduce the question of inter- and intra-observer variability. This is noted in Table 5 by calculating "SDP," the standard deviation as a percentage of tumor volume (for each volume). In some volumes, variability was as high as 40% of tumor volume. In contrast, the knowledge-based system was built from a small subset of the available slices and processed over 300 slices in unsupervised mode with a statistic rule set, allowing for complete repeatability. Table 5 shows that the knowledge-based system has a higher Percentage Match value (indicating it captured more of the ground-truth tumor) in 30 of 33 volumes, and outperformed kNN in Correspondence Ratio in 31 of 33 volumes.

Table 6: Knowledge-Based Tumor vs. kNN. (Pat. = Patient; GT = Ground truth volume; KB = Knowledge-based; kNN SDP = kNN Standard deviation as Percentage of Tumor Volume; Manual kNN = kNN volume after manual tumor extraction.)

Pat.	Scan	GT Vol.	KB Vol.	kNN Vol.	kNN SDP	Manual kNN
1	Base	7213	8561	10022	0.07	6334
	R1	7240	8829	11958	0.19	6794
	R2	7470	9928	11576	0.14	6616
	R3	6395	8132	10870	0.40	5901
	R4	6560	8614	12115	0.07	5690
5	Base	10178	10784	14044	0.14	7938
	R1	4657	5489	10279	0.32	2834
	R2	5616	6773	18603	0.06	3952
	R3	9215	9833	18210	0.14	6729
	R4	3544	5730	19138	0.20	3035

It must be noted, however, that many of the false positives affecting kNNs Correspondence Ratio values are extra-cranial pixels. The kNN method was applied to the whole image and no extraction of the actual tumor was done, which would require additional supervisor intervention, especially in patients with tumor that has "fragmented" as a result of treatment. Some of these kNN volumes had the tumor manually extracted [52], [53, [58], using the "best" segmentation from the multiple trials, and are shown in Table 4.1. Only total tumor volumes were available, so Percent Match and Correspondence Ratios could not be calculated for these volumes. In all 10 volumes listed in Table 4, manually extracted kNN consistently underestimated tumor volume, meaning the method missed more radiologist verified tumor than the knowledge-based method, to a significant degree in some cases.

A comparison was also made against the semi-supervised FCM (ssFCM) algorithm, which was initialized with the same ROI's used to initialize kNN in Table 5 [52], [53], [58]. The resultant ssFCM segmentation was then used to initialize ISG, a commercially available seed-growing tool (ISG Technologies, Toronto, Canada) for supervised evaluation of tumor volumes. The ISG processing also removed any extra-cranial tissues found in the ssFCM segmentation. Results available for the volumes processed by ssFCM and ISG are shown in Table 7. The results reported in Table 7 are a mean over the set trials performed for that volume and, thus, have a standard deviation, also listed. Only final tumor volumes were available from [52], [53], [58] ssFCM and ISG.

Table 7: Knowledge-Based Tumor vs. ssFCM and ISG. Results for Patient 6, 7, and 8 for ssFCM and ISG were unavailable. (Pat. = Patient; GT Vol. = Ground Truth Volume; KB = Knowledge-based; SDP = Standard Deviation as Percentage of Tumor Volume; \# Trial = Number of Trials; \# Obs. = Number of Observers; N/A = Not available.)

Pat.	Scan	GT Vol.	KB Vol.	ssFCM Vol.	ssFCM SDP	ISG Vol.	ISG SDP	# Trials	# Obs.
1	Base	7213	8561	8015	0.07	6067	0.05	5	2
	R1	7240	8829	7757	0.18	5956	0.03	5	2
	R2	7470	9928	7362	0.03	6087	0.03	5	2
	R3	6395	8132	7185	0.09	5361	0.02	5	2
	R4	6560	8614	7332	0.19	5172	0.04	5	2
2	Base	12230	15494	12457	0.06	10027	0.06	5	2
	R1	14609	12874	10916	0.13	7120	0.15	5	2
	R2	20924	19541	13498	0.06	10120	0.06	5	2
3	Base	10892	10855	8018	0.08	N/A	N/A	5	3
	R1	5971	7688	5337	0.03	N/A	N/A	5	3
4	Base	10454	10517	8563	0.05	6258	0.02	4	2
	R1	10835	8614	8901	0.04	6040	0.03	4	2
	R2	15788	15045	14080	0.06	10819	0.17	4	2
5	Base	10178	10784	10334	0.06	7562	0.09	4	2
	R1	4657	5489	3813	0.15	3142	0.24	4	2
	R2	5616	6773	4667	0.09	3483	0.07	4	2
	R3	9215	9833	7852	0.02	7300	0.06	4	2
	R4	3544	5730	3110	0.19	2311	0.13	4	2

In terms of absolute difference, the ssFCM approach #2 was closer to ground truth than the knowledge-base method in 10 out of 18 volumes, with 6 of these cases by more than the standard deviation of the ssFCM volume. In the 8 cases where the knowledge-based method gave better results, however, 7 of them were better than ssFCM by more than the standard deviation. The knowledge-based method performs better against ISG, in 9 out of 16 cases, with all 9 by more than the standard deviation of ISG. Furthermore, ssFCM underestimated total tumor volume in 12 instances, while ISG underestimated tumor volume in all 16 available volumes, which is not helpful for any use involving treatment, since the methods missed more radiologist verified tumor.

4.2. Evaluation Over Repeat Scans

An important use of tumor volume estimation is in tracking a tumor's response to treatment based on its growth/shrinkage. From the 33 volumes available, 25 "transitions" could be tracked (e.g., Baseline Scan to Repeat Scan 1 is one transition).

Examining tumor growth/shrinkage over repeat scans, the knowledge-based method failed to properly track 3 of 25 transitions (12%). The kNN method, without manual tumor extraction, failed on 8 of 25 transitions (32%), while the manually extracted kNN volumes failed in 2 of 10 transitions (20%). The ssFCM method failed on 3 of 13 transitions (23%), while ISG failed on 4 out of 12 (33%). Since the kNN, ssFCM, and ISG volumes are based on multiple trials, it is difficult to assign a specific cause, although the importance of supervised removal of extra-cranial tissues, handled automatically by the knowledge-based system, should be noted. Also, as a percentage, the knowledge-based system had a lower rate of failure than the supervised methods.

For the knowledge-based system, the failure in Patient 6, Repeat Scans 8 to 9, was due to slight over-estimation of tumor volume in Scan 9 where the T1 threshold was set too low. The failure of Patient 4, Baseline to Repeat Scan 1, was due primarily to poor extraction of the intra-cranial region in 3 slices, resulting in significant tumor loss. The third transition failure, Patient 2, Baseline to Repeat Scan 1, was due primarily to the presence of a significant amount of fluid, possibly hemorrhage, in the Baseline scan. This distorted the PD histogram used in Stage Two, resulting in significant overestimation. In Repeat Scan 1, however, not only had the fluid disappeared, but pathology reports noted the slight decrease in gadolinium enhancement. The tumor boundaries were not properly captured in Stage Five and resulted in a T1 threshold that was too high. Thus, the initial overestimation followed by the decreased gadolinium enhancement caused the trend to appear to be tumor shrinkage instead of growth. Figure 14 illustrates the total tumor volumes for the six patients with repeat scans using knowledge-based, kNN, and radiologist-verified ground truth.

Examples of knowledge-based segmentation versus ground-truth are shown in Figure 15 to visually show the knowledge-based system tumor correspondence to radiologist-labeled tumor. Figure 15 (a-c) shows a worst case segmentation, while (d-f) show a best case. Both examples are from slices in the test set.

5. Discussion

A knowledge-based multi-spectral analysis tool that segments and labels complete glioblastoma-multiforme tumor volumes has been presented. The presence of the knowledge base allows "informed" unsupervised segmentation and classification decisions to be made. Combined with the paradigm of iterative refinement, this system has increased power and flexibility. This is in contrast to most other multi-spectral efforts such as [19], [51], [54] which attempt to segment the entire brain image in one step, based on either statistical or (un)supervised classification methods.

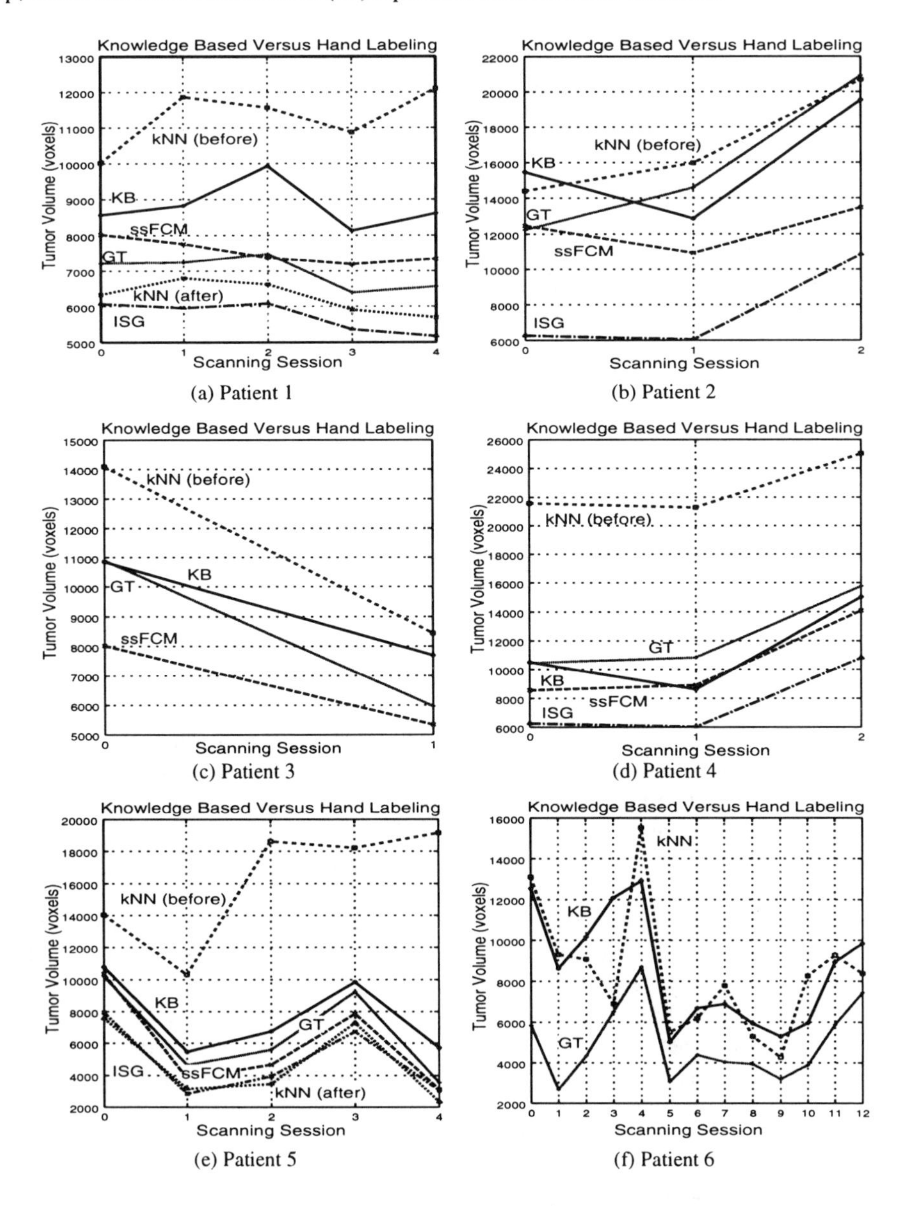

(a) Patient 1

(b) Patient 2

(c) Patient 3

(d) Patient 4

(e) Patient 5

(f) Patient 6

Figure 14. Tracking tumor growth/shrinkage over repeat scans. KB = knowledge-based system. GT = ground truth.

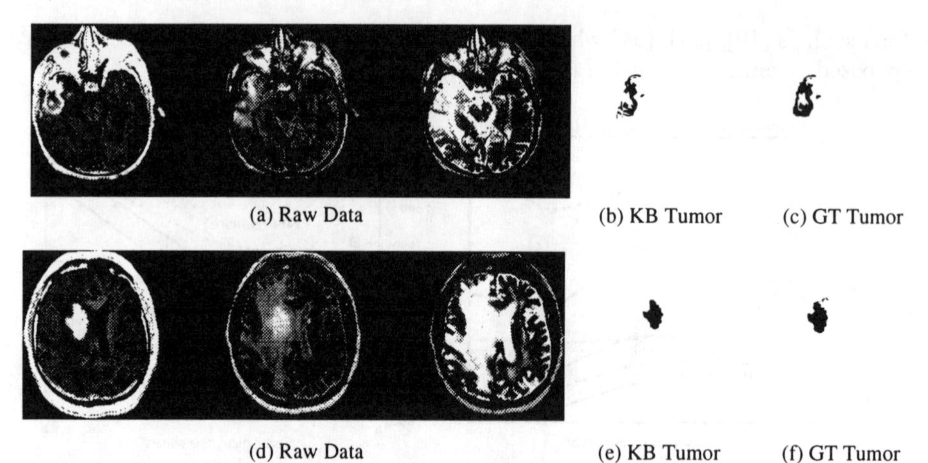

Figure 15. Comparison of knowledge-based tumor segmentation vs. ground truth. Worst case (a-c) and best case (d-f).

The knowledge base was built using a process called "knowledge-engineering," as the heuristics most relevant towards tumor segmentation had to be manually extracted from the available domain information, and implemented into a rule-based system. More important, a relatively small training set was used – sixty five slices over five patients. Yet, the system performed well. A larger training set would most likely allow new and more effective trends and characteristics to be revealed. Thresholds used to handle a certain subset of the training set could be better generalized.

Processed slices had a relatively large thickness of 5 mm. Thinner slices would exhibit reduced partial-averaging and provide better tissue contrast. Although the system relied on distributions in feature space, the heuristics were based on general tissue characteristics and relative relationships between tissues to minimize dependence upon specific feature-domain values. The data volumes studied here were acquired with different scanning parameters, and gadolinium-enhancement has been found to be generally very robust in different protocols and thicknesses [4], [24]. The extent of this robustness has not been rigorously tested, but should acquisition parameter dependence become an issue, a large training base across a wider range of parameters could allow heuristics to adjust the system to a slice's specific parameters. Such information is easily included when processing starts. All patients received various degrees of treatment before and between scans, such as surgery, radiation, and chemotherapy. Despite the changes these treatments can cause, however, such as demyelinization of white matter, no modifications to the knowledge-based system were necessary. Approaches that use specific feature values, like neural networks [1] or anything depending on a specific set of training examples, could have difficulties in dealing with slightly different imaging protocols and the effects of treatment.

As stated in the introduction, no method of quantitating tumor volumes is widely accepted and used in clinical practice [38]. Tumor area is approximated by the Eastern Cooperative Oncology group [17] in the single MR slice with the largest contiguous, well-defined tumor evident. The longest tumor diameter is multiplied by its perpendicular to yield an area. Tumor response to treatment is classified into five categories, ranging from complete response (no measurable tumor remains) to progression, based on changes greater than 25% in the tumor area between scans and visual observations. Clearly, this approach does not address full tumor volume, depends on the exact boundary choices, and the shape of the tumor [17], [31]. By itself, the approach can lead to inaccurate growth/shrinkage decisions [10].

The promise of the knowledge-based system as a useful tool is demonstrated by the successful performance of the system on the processed slices. The final KB segmentations compare well both with radiologist-verified "ground truth" images and supervised methods. More important, it was able to segment tumor without the need for (multiple) human-based ROIs or post-processing, which make supervised methods clinically impractical.

Overall, the knowledge-based approach tended to overestimate the tumor volume. Only two patients (2 and 4) showed noticeable underestimation by the knowledge-based system. In all other cases, the knowledge-based system shows a number of "false positives." Since the system segments tumor by removing only pixels proven not to be tumor, leaving anything that remains being labeled as tumor, a higher level of false positives is not inconsistent with the paradigm. The final T1 threshold is applied using a fixed parameter σ, but the knowledge-based system's errors suggest that creating rules to automatically set σ (making Stage Five more adaptive to tumors with different degrees of enhancement) is worthy of investigation.

It should also be noted, however, that the process of creating ground truth images is very imprecise [40] and has approximately a 5% inter-observer variability in tumor volume [56]. All brain tumors have micro-infiltration beyond the borders defined with gadolinium enhancement. This is especially true in glioblastoma-multiformes, which are the most aggressive grade of primary glioma brain tumors, and no one can tell the *exact* tumor borders, even with invasive histopathological methods [9], [18], [41], which were unavailable. Ground truth images mark the areas of tumor exhibiting the most angiogenesis (formation of blood vessels, resulting in the greatest gadolinium concentration) and represent those pixels which are "statistically most likely" to contain tumor [40], [41]. Such pixels would have the highest level of agreement between radiologists, but they do not guarantee that all tumor has been identified [40]. Therefore, the knowledge-based system may often capture tumor boundaries that extend into areas showing lower degrees of angiogenesis (which would still be treated during therapy) [41].

Future work includes introducing new tumor types, such as lower grade gliomas, as well as complete labeling of all remaining tissues. Also, newer MRI systems may provide additional features, such as diffusion images or edge strength to estimate tumor boundaries, which can be readily included into the knowledge base. The knowledge-base also allows straightforward expansion as new tools are found effective.

In conclusion, the knowledge-based system is a multi-spectral tool that shows promise in completely and effectively segmenting glioblastoma-multiforme tumors without the need for human supervision. It has the potential of being a useful tool for segmenting tumor for therapy planning, and tracking tumor response. Finally, the knowledge-based paradigm allows easy integration of new domain information and processing tools into the existing system when other types of pathology and MR data are considered.

Acknowledgments

This research was partially supported by a grant from the Whitaker Foundation and a grant from the National Cancer Institute (CA59 425-01). Thanks to Dr. Mohan Vaidyanathan for his assistance in the ground-truth work.

REFERENCES

[1] S. Amartur, D. Piriano, and Y. Takefuji: Optimization neural networks for the segmentation of magnetic resonance images. *IEEE TMI 11*, 2 (June 1992), 215-221.

[2] M. Bomans, K. Hohne, U. Tiede, and M. Riemer: 3D segmentation of MR images of the head for 3D display. *IEEE TMI 9* (1990), 177-183.

[3] P. Bottomley, T. Foster, R. Argersinger, and L. Pfeiffer: A review of normal tissue hydrogen NMR relaxation times and relaxation mechanism from 1-100 MHz: Dependency on tissue type, NMR frequency, temperature, species, excision and age. *Medical Physics 11* (1984), 425-448.

[4] R. Bronen and G. Sze: Magnetic resonance imaging contrast agents: Theory and application to the central nervous system. *Journal of Neurosurgery 73* (1990), 820-839.

[5] R. Cannon, J. Dave, and J. Bezdek: Efficient implementation of the fuzzy c-mean clustering algorithms. *IEEE Transaction on Pattern Analysis and Machine Intelligence 8,* 2 (1986), 248-255.

[6] M. Clark, L. Hall, D. Goldgof, *et al*: MRI segmentation using fuzzy clustering techniques: Integrating knowledge. *IEEE Engineering in Medicine and Biology 13,* 5 (1994), 730-742.

[7] M. Clark. L. Hall. C. Li, and D. Goldgof: Knowledge based (re-)clustering. In *Proceedings of the 12th IAPR International Conference on Pattern Recognition* (1994), pp. 245-250. Jerusalem, Israel.

[8] M.C. Clark,: *Knowledge-Guided Processing of Magnetic Resonance Images of the Brain.* Ph.D. thesis, University of South Florida, 1997.

[9] L. Clarke, R. Velthuizen, M. Camacho, J. Heine, M. Vaidyanathan, L. Hall, R. Thatcher, and M. Silbiger: MRI segmentation: Method and application: *Magnetic Resonance Imaging 13,* 3 (1995), 343-368.

[10] L. Clarke, R. Velthuizen, M. Clark, G. Hall, D. Goldgof, *et al:* MRI measurement of brain tumor response: Comparison of visual metric and automatic segmentation. To appear, *Magnetic Resonance Imaging*.

[11] B. Desarthy: *Nearest Neighbor (NN) Norms: NN Pattern Classification Techniques.* IEEE Computer Society Press, Los Alamitos, CA, 1991.

[12] S. Dellepiane: Image segmentation: Errors, sensitivity, and uncertainty. In *Proceeding of the 13th IEEE EMB Society* (1991), vol. 13, pp. 253-254.

[13] S. Dellepiane, G. Venturi, and G. Vernazza: A fuzzy model for the processing and recognition of MR pathological images. In: *IPMI 1991* (1991), pp. 444-457.

[14] S. Dougherty: Discussions held with Sean Dougherty concerning edge detection algorithms in MR images, September 1997.

[15] T.C. Farrar: *An Introduction to Pulse NMR Spectroscopy.* Farragut Press, 1987.

[16] G.B. Feldmann: Brain tumor facts and figures. *The Brain Tumor Society* - http: //www.btfacts.htm, July 3 1997.

[17] L. Feun: Double-blind randomized trial of the anti-progestational agent mifepristone in the treatment of unresectable meningioma, phase iii. *Tech. Rep. SWOG-9005*, University South Florida, Tampa, Fl., Southwest Oncology Group, 1995.

[18] R. Galloway, R. Maciunas, and A. Failinger: Factors affecting perceived tumor volumes in magnetic resonance imaging. *Annals of Biomedical Engineering 21* (1993), 367-375.

[19] G. Gerig, J. Martin, R. Kikinis, *et al.*: Automating segmentation of dual-echo MR head data. In: *The 12th International Conference of Information Processing in Medical Imaging* (IPMI 1991).

[20] J. Giarratano and G. Riley: *Expert Systems: Principles and Programming,* second ed. Boston: PWS Publishing, 1994.

[21] P. Gibbs, D. Buckley, S. Blakband, and A. Horsman: Tumor volume determination from MR images by morphological segmentation. *Physics in Medicine and Biology 41,* 11 (November 1996), 2437-2446.

[22] L. Gong, and C. Kulikowski: Automatic segmentation of brain images: Selection of region extraction methods. In: *SPIE Vol. 1450 Biomedical Processing II* (1991), SPIE, pp. 144-153.

[23] L. Hall, A. Bensaid, L. Clark *et al*: A comparison of neural network and fuzzy clustering techniques in segmenting magnetic resonance images of the brain. *IEEE Transaction on Neural Networks 3,* 5 (1992), 672-682.

[24] R. Hendrick, and E. Haacke: Basic physics of MR contrast agents and maximization of image contrast. *JMRI 3,* 1 (1993), 137-148.

[25] G. Hillman, C. Chang, H. Ying, *et al.*: Automatic system for brain MRI analysis using a novel combination of fuzzy rule-based and automatic clustering techniques. In: *Medical Imaging 1995: Image Processing* (February 1995), *SPIE*, pp. 16-25. San Diego, CA.

[26] A. Jain: *Fundamentals of Digital Image Processing.* Englewood Cliffs, N.J.: Prentice Hall, 1989.

[27] R. Jain, R. Kasturi and B. Schunck: *Machine Vision.* McGraw-Hill, Inc., 1995.

[28] M. Kamber, R. Shingal, D. Collins, G. Francis, and A. Evans: Model-based 3D segmentation of multiple sclerosis lesions in magnetic resonance brain images. *IEEE TMI 14,* 3 (1995), 442-453.

[29] R. Kikinis, M. Shenton, G. Gerig, *et al*: Routine quantitative analysis of brain and cerebrospinal fluid spaces with MR imaging. *JMRI* 2 (1992), 619-629.

[30] E. Kischell, N. Kehtarnavaz, G. Hillman, H. Levin, M. Lily, and T. Kent: Classification of brain compartments and head injury lesions by neural networks applied to MRI. *Neuroradiology 37* (1995), 535-541.

[31] N. Laperrire and M. Berstein: Radiotherapy for brain tumors. *CA - A Cancer Journal for Clinicians 4* (1994), 96-108.

[32] N. Leeds and E. Jackson: Current imaging techniques for the evaluation of brain neoplasms. *Current Science 6* (1994), 254-261.

[33] C. Li: Knowledge based classification and tissue labeling of magnetic resonance images of the brain. Master's thesis, University of South Florida, 1993.

[34] C. Li, D. Goldgof, and L. Hall: Automatic segmentation and tissue labeling of MR brain images. *IEEE TMI 12,* 4 (December 1993), 740-750.

[35] X. Li, S. Bhide, and M. Kabuka: Labeling of MR brain images using boolean neural network. *IEEE TMI 15,* 2 (1996), 628-638.

[36] R.B. Lufkin: *The MRI Manual.* Year Book Medical Publishers, Inc., 1990.

[37] W. Menhardt, and K. Schmidt: Computer vision on magnetic resonance images. *Pattern Recognition Letters 8,* 2 (September 1988), 73-85.

[38] R. Murtagh, S. Phuphanich, N. Imam, L. Clarke, M. Vaidyanathan, *et. al:* Novel methods of evaluating the growth response patterns of treated brain tumors. *Cancer Control* (1995), 293-299.

[39] F. Murtagh: Discussions held with Dr. F. Reed Murtagh, M.D., Dept. of Radiology, University of South Florida, April 1997.

[40] F. Murtagh: Discussions held with Dr. F. Reed Murtagh, M.D., Dept. of Radiology, University of South Florida, October 1, 1997.

[41] F. Murtagh: Discussions held with Dr. F. Reed Murtagh, M.D., Dept. of Radiology, University of South Florida, October 22, 1997.

[42] A. Namasivayam and L. Hall: Integrating fuzzy rules into the fast, robust segmentation of magnetic resonance images. In: *New Frontiers in Fuzzy Logic and Soft Computing Biennial Conference of the North American Fuzzy Information Processing Society- NAFIPS 1996* (1996), pp. 23-27. Piscataway, NJ.

[43] R. Novelline and L. Squire: *Living Anatomy.* Hanley and Belfus, 1987.

[44] M. Ozkan, B. Dawant, and R. Maciunas: Neural-network-based segmentation of multi-modal medical images: A comparative and prospective study. *IEEE TMI 12,* 3 (September 1993), 534-545.

[45] U. Raff and F. Newman: Automated lesion detection and lesion quantitation in MR images using autoassociative memory. *Medical Physics 19* (1992), 71-77.

[46] G. Riley: Version 4.3 CLIPS reference manual. Tech. Rep. JSC-22948, Artificial Intelligence Section, Lyndon B. Johnson Space Center, 1989.

[47] H. Schnitzlein and F.R. Murtagh: *Imaging Anatomy of the Head and Spine: A Photographic Color Atlas of MRI, CT., Gross. And Microscopic Anatomy in Axial, Coronal, and Sagittal Planes,* second ed. Baltimore: Urban & Schwarzenberg, 1990.

[48] D.D. Stark, G. William, and J. Bradley: *Magnetic Resonance Imaging, Second Ed., Volume One.* Mosby Year Book, 1992.

[49] C. Tao and W. Thompson: A fuzzy if-then approach to edge detection. In: *1993 IEEE International Conference on Fuzzy Systems* (1993), IEEE, pp. 1356-1360.

[50] T. Taxt and A. Lundervold: Multispectral analysis of the brain in magnetic resonance imaging. In: *IEEE Workshop on Biomedical Image Analysis* (1994), pp. 33-42. Los Alamitos, CA, U.S.A..

[51] T. Taxt and A. Lundervold: Multispectral analysis of the brain using magnetic resonance imaging. *IEEE TMI 13,* 3 (September 1994), 470-481.

[52] M. Vaidyanathan, R. Velthuizen, L. Clarke, and L. Hall: Quantitation of brain tumor in MRI for treatment planning. In: *Proceedings of the 16th Annual International Conference of the IEEE Engineering in Medicine and Biology Society* (1994), vol. 16, pp. 555-556.

[53] M. Vaidyanathan, R. Velthuizen, P. Venugopal, and L. Clarke: Tumor volume measurements using supervised and semi-supervised MRI segmentation methods. In: *Artificial Neural Networks in Engineering - Proceedings (ANNIE 1994)* (1994), vol. 4, pp. 629-637.

[54] M. Vannier, R. Butterfield, D. Jordan, *et al.*.: Multispectral analysis of magnetic resonance images. *Radiology 154*, 1 (January 1985), 221-224.

[55] M. Vannier, C. Speidel, and D. Rickmans: Magnetic resonance imaging multispectral tissue classification. *News Physiol. Sci. 3* (August 1988), 148-154.

[56] R. Velthuizen and L. Clarke: An interface for validation of MR image segmentations. In: *Proceedings of the 16th Annual International Conference of the IEEE Engineering in Medicine and Biology Society* (1994), pp. 547-548.

[57] R. Velthuizen, L. Hall, and L. Clarke: Unsupervised fuzzy segmentation of 3D magnetic resonance brain images. In: *Proceedings of the IS&TSPIE 1993 International Symposium on Electronic Images: Science & Technology* (1993), vol. 1905, pp. 627-635. San Jose, CA, Jan. 31- Feb. 4.

[58] R. Velthuizen, S. Phuphanich, L. Clarke, L. Hall, *et al.*.: Unsupervised tumor volume measurement using magnetic resonance brain images. *JRMI 5*, 5 (1995), 594-605.

[59] Z. Wu, and R. Leahy: A graph theoretic approach to segmentation of MR images. In: *SPIE Vol. 1450 Biomedical Image Processing II* (1991), pp. 120-132.

[60] Z. Wu, and R. Leahy: Image segmentation via edge contour finding: A graph theoretic approach. In: *IEEE Computer Vision and Pattern Recognition* (1992), pp. 613-619.

[61] M. Clark, L. Hall, D. Goldgof, *et al.*.: Automatic Tumor Segmentation Using Knowledge-Based Techniques. *IEEE: TMI, 17*, 2 (April 1998).

Abbreviations

MR(I): Magnetic Resonance (Imaging)
T1: T1-weighted
PD: PD-weighted
T2: T2-weighted
PEP: pixel edge potential
FCM: fuzzy c-means clustering
ssFCM: semi-supervised FCM
kNN: k-nearest neighbors

Part 2.

Neuro-Fuzzy Knowledge Processing

Chapter 6

An Identification of Handling Uncertainties Within Medical Screening: A Case Study Within Screening for Breast Cancer

Fredrik Georgsson and Patrik Eklund

We describe a top-down approach of how to use artificial intelligence within different processes in screening. Different aspects of the screening are described using a more formal language, making it possible to reason about a screening. This in turn makes it possible to develop computer support within different steps of the screening. The support includes the whole process from the first selection of people to screen in a population, to the final decision of whether this particular person should be selected or not.

The formalism and definition of where in the screening process different kinds of knowledge are needed makes it easy to adapt the formalism to a number of different screenings. In particular, we deal with screenings that involve some sort of image processing. The formalism is applied to a screening for breast cancer.

1. INTRODUCTION

In this chapter, we will present a conceptual framework for medical screening, with a case study focused on breast cancer diagnosis. Our motivation for this work is to enable us to place data and image analysis methods into perspective within the framework, and thereby present requirements and recommendations for development of various analysis methods in this context. The vast majority of research developed in these and related fields normally tends to be entirely method oriented, usually with little indication of how to incorporate respective methods into a corresponding patient care management scenario. For breast cancer, analysis work is almost exclusively focused on images, i.e., mammograms, and analysis methods concentrated on particular signs of cancer, such as micro-calcifications. The literature reveals very few, if any, proposals

0-8493-9806-1/98/$0.00+$.50

of information system architectures, within which analysis methods of various kinds are put into practice.

From the point of view of patient care management, the diagnosis of breast cancer is basically built around a screening process, through which patients are pipelined. Within various steps of the screening process, different methods are applied to patients to effectively and efficiently decide upon a diagnosis, or at least reach a conclusion on consequent actions to be taken in the screening process. Roughly speaking, the screening process is a sequence of methods applied to patients, or actually to the information scope of that patient being available at that particular time. Thus, the result of each method is a refinement of the content of information about the patient. In addition, the result is, in particular, a choice of the next method to be applied, in case the diagnosis task was not finalized. We should note that the screening process must fulfill cost-effectiveness criteria, but a full modeling of these aspects is beyond the scope of this chapter.

For the breast cancer case study, image analysis aspects are also mentioned, even if details related to particular methods are presented elsewhere. Developments within the case study have been done in co-operation with the Mammographic Unit at Umeå University Hospital [4]. The chapter is concluded with a suggestion of a system architecture in which particular methods within the screening process are integrated.

The outline of this chapter is as follows. First, the screening process is described from a medical point of view. Then the different steps in a screening are identified, and within each step, different parameters are identified. Based on this, a formalism is proposed. Finally, the entire theory is applied to a case study in screening for breast cancer.

2. SCREENING

A screening is traditionally defined as the presumptive identification of unrecognized disease or defect by application of tests, examinations, or other procedures that can be applied rapidly [30]. To this definition, we add that there must be an effective treatment for the "disease or defect" [15]. Obviously, there would be no point in diagnosing people as having a "*disease or defect*," if it was not curable. The degree of "disease or defect" must be rather severe to motivate such an effort to locate it. Attention must also be drawn to the fact that the method used must be rapidly applicable and that the method must be accurate [25]. This is one motivation for computer-assisted screening.

2.1. Notations

Given information about a patient at a certain point in time, the objective of a (diagnostic) method is typically either to classify the patient as a *negative* (N) or a *positive* (P) with regard to a particular disease D.

We will typically denote an entire population of people, p, with A_0. At a certain point in time, t, we have a static disease specific subset, D, such that $D \subseteq A_0$. A screening is a process of finding a finite series of As, or more correctly, the methods m

that generate the series, such that $A_0 \supseteq A_1 \supseteq \ldots \supseteq A_n$. We also want to maximize the cardinality of $A_n \cap D$, see Figure 1. Each set A_i is formed from A_{i-1} by the use of a method m_i according to

$$A_i = \{p \in A_{i-1} \mid m_i(p) \notin \{N, P\}\},\ i = 1,2,\ldots,n\,.$$

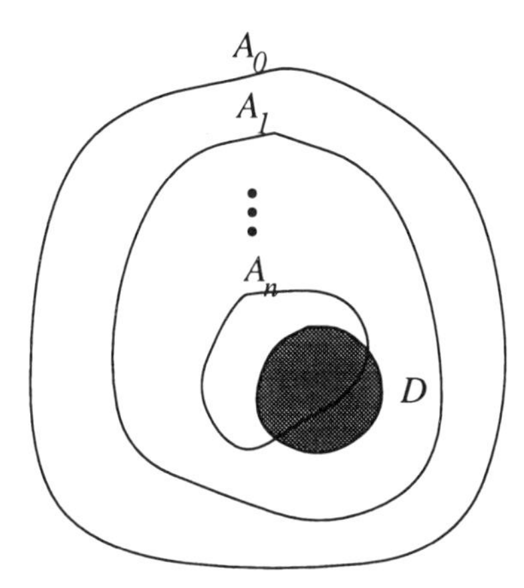

Figure 1: Zooming in on the disease set D by a sequence of A's.

To each method m, a cost-evaluation c is associated. Hence, the total cost of forming A_{i-1} is

$$\text{cost}_i = \text{card } A_{i-1} \cdot c(m_i)$$

as we must apply m_i to each individual in A_{i-1}. This holds under the assumption that there is one m associated with each A. This is not necessarily the case, but in all practical cases, it will be applicable. Consequently the entire cost for forming A_n is:

$$\textbf{const}_n = \sum_{k=1}^{n} \text{cost}_k = \sum_{k=1}^{n} \text{card } A_{k-1} \cdot c(m_k) \qquad (1)$$

An additional constraint on screening is to minimize the cost of applying the series of methods from which we form our *A*s. Studying Equation 1, it is clear that we might either minimize the cost of the method or the size of the sets. Typically, we start with an entire population so it is hard to reduce the size. Consequently, we are forced to choose an inexpensive method. Then, when the size of the A_n decreases, we can afford to use more expensive methods. It is clear that the more we reduce the size of A in the first steps, the more there is to gain. If we choose a series of methods that does a poor job in finding D, we must pay for that by caring for the subjects that belong to D but are not detected by the screening, that is, they do not belong to A_n. This implies that it is only useful to screen for diseases that produce a great cost if left unattended. Therefore, one may be motivated to screen only for lethal diseases. If we are dealing with a highly infectious disease we must take into account the fact that persons belonging to the disease set might spread the disease before the individual suffers from the disease. Obviously, this indicates that we have to deal with a very complicated pattern of epidemiological nature, which again is outside the scope of this work.

The set $A_n \setminus D$ contains the false positives, $D \setminus A_n$ the false negatives, and $A_n \cap D$ represents the accuracy of the screening. Hence, if we choose A_n so large that it covers all of D, we will have no false negatives and a maximal accuracy but to the cost of false positives. If we choose a set $A_n \subset D$, we will have no false positives but a lower accuracy and false negatives. Hence, we will require A_n to be somewhat larger than D. The quantities discussed here are normally given as percentages by considering only their size and normalizing them with the size of A_n.

2.2. The screening program

In the previous section, the screening was formalized. In this section, we will express the screening as an algorithm. This algorithm can be expressed as

while m *not indicating* P *or* N	*1*
$\quad m := \text{select}_D(p)$	2
$\quad$ *if* m *indicates* P *or* N *then break*	3
$\quad p := p \mid \langle m, m(p) \rangle$	4
end while	5

Figure 2: The screening expressed in algorithmic form.

In this notation, a person p is characterized by the data available about the person. Initially this information is $p = \langle \perp, I_0 \rangle$ where $\perp$ denotes the empty method. Here, I_0 is the universal knowledge about the person, a knowledge that is contained in the person itself and not in a journal system of any kind. The person might be viewed as a set of data and, by applying different methods, we open windows into this data (see Figure 3). The aim is to open just as many windows of information as needed to be able to classify the patient as having the disease D or not.

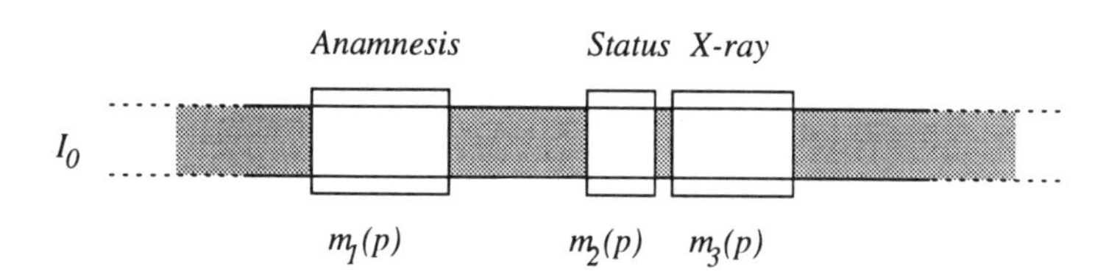

Figure 3: The information about a person is the set of all $m_i(p)$s; see Example 1 for more detail.

The role of a journal system is to store all this information in one place so that it is easy to access.

Example 1. *We are trying to classify a person* p *who is complaining of chest pain. The first method* m_1 *to apply is to check the patient's anamnesis. This means we check all the previous "information windows" that are already known about the patient. Further,* m_2 *is an auscultation (see the following section) performed to obtain the patient's status. Finally, in method* m_3*, a chest X-ray is requested, and the X-ray image shows chronic pulmonary oedema, a possible sign of heart failure.*

In Example 1, we find a sequence of methods that leads to the detection of a possible heart failure before it occurs. However, it is not a screening, since the sequence of methods is applied to the patient only after initial complaints of chest pain. To make it a screening for heart failure, we would have to ask an entire population if they feel any chest pain and then apply m_2 and m_3 to all those that felt chest pain. The method m_1 is, however, not a very good initial method, since many other diseases are characterized by pain in the chest. We also note that the work of the select function is carried out by a physician.

2.3. The methods

In screening, several different types of methods can be applied [15].

Observation. An observation is the most widely used examination form in a screening. It is used especially when screening for lesions on the patient's skin and in cavities. The observation must be performed by a physician, which makes the method rather expensive, unless it may be performed in association with an annual routine visit to a physician.

Palpation. With palpation, a physician may feel changes that are close to the patient's skin or a cavity. Palpation may be used in the screening for breast cancer and to examine patient's lymph nodes and other glands.

Auscultation and Percussion. Auscultation may be used to diagnose early heart diseases along with percussion. Auscultation is when a physician listens to the patient with a stethoscope, or some other sound-amplifying device. For percussion, the physician listens to the sounds generated in the patient after a mechanical influence [17]. These two methods are completely harmless to the patient, but must be performed by a physician.

Internal imaging. Internal images are generated by sending some kind of radiation into the patient, and measuring the response of this exposure [7]. Examples of these techniques are X-ray imaging, Computer Tomography (CT), Magnetic Resonance Imaging (MRI), ultra sound and nuclear medicine (scintograms, Single Photon Emission CT (SPECT), and Positron Emission Tomography (PET)) [13]. Roughly, it can be stated that nuclear medicine and, in some cases, ultrasound are often more suitable if a *physiological* process is of interest. CT, MRI, and X-ray are better for *anatomical* imaging. Internal imaging may be performed by *radiographers* (persons trained to handle the image taking process), who are more cost efficient than *radiologists* (physicians trained to judge medical images), since they require less training. The images are then presented to the radiologist, who may concentrate fully on the task of judging them. This also makes it possible to double judge patients, a task that is difficult when the patient is present, as in observation or palpation. A disadvantage with internal imaging is that the radiation may be ionizing and hence application should be restricted. Ultrasounds and MRI are exceptions to this [17], [19].

Laboratory tests. In a laboratory test a chemical substance is removed from the patient and tested. The substance may be feces, saliva, blood, or scraping of cells. Laboratory tests are also common in the second phase of screening, namely, diagnosing those selected from the screening, that is the persons that belong to A_n. As in the imaging case, we may be able to separate the testing from the judging with laboratory testing.

3. THE SELECT FUNCTION

A major side effect of applying a method to patient information is that it results in further (and refined) information about the patient by adding to the patient information. This new information might contain the diagnostic classification (usually N or P) regarding the disease in question. If so, the patient is removed from further processing by the screening system. Positive classification results in patients being sent to corresponding treatment. In the situation where the diagnostic classification could not be reached, i.e., $\langle \perp, I_0 \rangle$, where there were not sufficient data to make a diagnosis, a proposal for the next method to be applied should be returned.

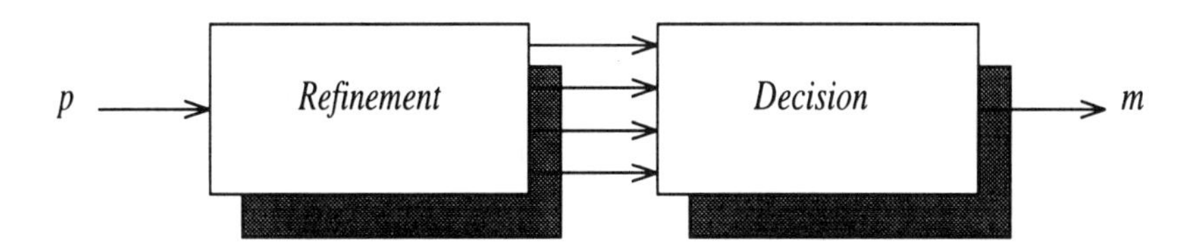

Figure 4: The select function.

Given a patient, the select function produces a selection of a method (see Figure 4) to be applied to the information about that patient, where the objectives of the method are as described above. Traditionally, a domain expert performs the task of the select function, but as we shall see, we can also implement computerized select functions.

As we have seen before (Figure 2, line 2), the select function takes a set of tuples as input. Each tuple consists of a method m and the additional information obtained with this method is denoted $m(p)$. Note that for the initial tuple $\langle \perp, I_0 \rangle$, the select function provides the first method to be applied in the screening process.

The select function should be seen as being composed of a refinement and a decision function, respectively.

The reason for refining the data is that the information may be in different formats such as text, image, or sound, and must be transferred to some numerical representation before further processing in the decision step. Decisions are taken based on refined data as available. Refinements by domain experts are performed based on expertise and experience, and can thus be seen as provided by (human) cognitive systems. A related discussion can be found in [31].

Decisions are based on refined information, and decision techniques are obviously based on entirely different styles of data manipulation.

Example 2. *In Example 1, it was stated that the X-ray images showed signs of chronic pulmonary oedema but there was no indication of how this conclusion was reached. Roughly speaking, one can detect pulmonary oedema by the fact that the heart appears enlarged on a chest X-ray. Thus we must concentrate our efforts, first to localize the heart in the X-ray image and, second, to measure its size.*

3.1. The decision step

The goal of the decision step is to classify patients as either positives or negatives. If it is not possible to make a decisive diagnosis, we at least require that we can determine the next method to be applied so that discrimination becomes more possible. In a decision scenario based on logical considerations, the diagnosis is based upon the information extracted by any of these methods formed by some sort of rule. Linguistic terms are frequently used to distinguish normality from abnormality in situations where quantitative assessments are not available.

By applying a method m to a patient p, we obtain the information $m(p) \in I$. This information is pre-processed and transformed into information that fits the linguistic rules. Hence, the pre-processor may be viewed upon as a function $\pi: I \rightarrow X$ where I is the method specific information and $X = \{x_1, x_2, \ldots, x_l\}$ is the disease specific information obtained from I. The set X is then used to diagnose the patient by rules as

$$R_i: \text{If } x_1 \text{ is } L_1 \text{ and } x_2 \text{ is } L_2 \text{ and ... and } x_l \text{ is } L_l \text{ then } C_i.$$

Here L_j is a linguistic variable for x_j, and C_i is the logical consequence of R_i, where C_i may either be a diagnosis, i.e., $C_i \in \{N, P\}$, or a method m. One example of a system using these kinds of rules may be found in [23]. For the mammography case study, the use of Bayesian networks has been suggested in [20].

Example 3. *We could formalize the statement in Example 2, saying that an enlarged heart is a sign of chronic pulmonary oedema,*

$$\textit{If heart}^{(size)} \textit{ is Large then heart}^{(Diagnose)} \textit{ is chronic pulmonary oedema}$$

But what exactly is meant by "large?" As a rule of thumb, one can say that a heart is enlarged if the ratio between the cardiac and thoracic diameters is greater than 0.7 [7]. *Hence we must understand and be able to detect these entities as well.*

3.2. Disease-specific knowledge

In order to find a fully automatic π-function, we must extract the disease-specific information used by domain experts in their judgment. It is obviously non-trivial to find formal methods to do this, but there are a number of approaches to take. An interviewing approach, expressing the knowledge in linguistic variables is described in [23]. As opposed to interviews, we can also "record" the actions of a large number of experts and try to find dependencies and causalities between variables. An approach involving some kind of interviewing is, in any case, preferable, since we need experts to evaluate and verify extracted facts, and to approve the use of particular parameterized methods. Note that parameterized methods might not lend themselves to useful interpretations, such as in the case of multi-layer neural networks, whereas methods based on Bayesian computing provide validity and soundness of parameter estimations, at least from a statistical viewpoint.

The parameters related to π-functions reside either in the function expression (e.g., within the syntax of a rule base) or in the algorithm specifying how the function is computed (how output is inferred from input). A general criterion of suitability of π-functions is that its parameters have some relevance to the corresponding parameters being used in the domain where the function is to be integrated.

Example 4. *If* $p = \{\langle m_1, I_1 \rangle, \ldots, \langle m_l, I_l \rangle\}$ *is the set of all information obtainable from some internal imaging methods, we must adapt this point oriented information (images) to information about objects of different shape, size, color, or texture, i.e., what is normally meant by image analysis. The object entities are then matched with the disease-specific linguistic rules.*

3.3. The refinement step

Consider the situation where $m(p)$ is the set of all information that may be contained in an image. There are a number of more or less well-defined (sub-) steps that are associated with a function from $m(p)$ to a disease-specific information set (see Figure 5).

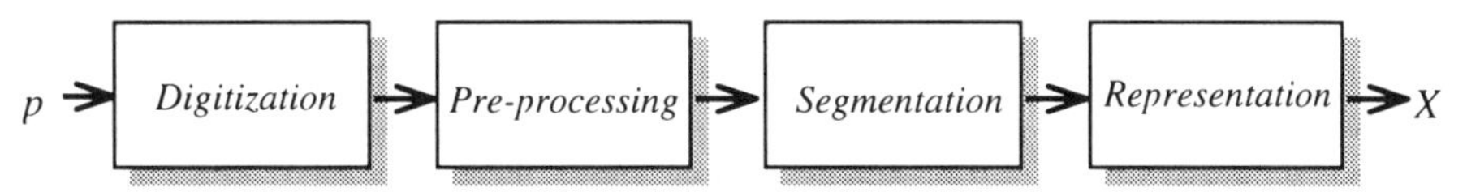

Figure 5: A high-level description of the refinement step.

The first step is to discretize the image to a numerical format. This is either done directly in the acquisition device or by discretizing the information externally. Some acquisition devices may appear to be fully digital, such as a CT, but internally the CT faces basically the same problems with digitalization as a digital X-ray. However, the problem of discretization is usually hidden from these considerations. In order to make a good discretization, there must be an understanding of which types of spatial and intensity resolutions are actually needed. These are very much disease specific, but image-processing knowledge is also needed to know how to reach the disease specific demands [7], [11].

One of the main purposes of pre-processing is to remove noise, but one might view pre-processing as a general enhancement of the image. There might be a very delicate difference between what is regarded as noise and what is regarded as important information. The only way to discriminate between the two is to use disease-specific knowledge.

The segmentation step is in many ways the hardest one and requires a high degree of image analysis knowledge. However, disease-specific knowledge is strongly supportive in acceptable implementations. The typical output from this step is an image in which each object is marked binary. By "*object*" we mean an area in the image showing some property that is used in the diagnosis. This object might differ from the background in intensity or texture or it may be defined by its borders. Some of the techniques usable here are described in [1], [8], [11], [16], [22].

The final step is to represent the objects found in the segmentation step numerically. Identification of those features that are of greatest interest is very much disease specific. Furthermore, it is highly desirable to have the representation match the are the size, shape, and borderlines of the object. For this task, there are several well-known techniques [2], [6], [14], [18], [28].

4. A BREAST CANCER CASE STUDY

In this section we will test all ideas that are put forward in a case study. The case chosen is the screening for breast cancer. Breast cancer as a disease, and for the purpose of this work, is interesting because:

- there are at least two completely different screening methods,
- there are clearly distinguishable risk groups,
- a large economical gain may be expected from a successful screening program,
- screening has been in progress in Sweden for many years; consequently, there is vast experience within this field.

The purpose of this case study is not to construct and implement a system but only to show the possibilities of the formalism derived in the earlier sections. In the following sections, the different kind of knowledge will be displayed in a graph-like manner. The causality indicated by the arcs must of course be quantified by using large data sets. These causalities are not resolved here.

4.1. Minimizing A_0 as much as possible in one step

We would like to minimize the cardinality of A_0 as much as possible in one step, before we apply a more costly screening method. To do this we first analyze what factors are presumed to cause breast cancer. This is done by a literature study [9], [29] and communication with a domain expert (*radiologist*). The facts that are obtained are organized in a graph, see Figure 6.

This information may be used to extract rules for going from A_0 to A_1. To do this, different linguistic variables are formed (see Table 1).

Table 1: The linguistic variables of some breast cancer causes

Age	{*young, middle-aged, old*}
Contraceptives	{*never, some, many*}
Weight	{*under, normal, over*}
Relatives	{*none, some, several*}
Menarche	{*early, normal, late*}
Menopause	{*no, yes*}
Pregnancies	{*none, few, many*}
...	...

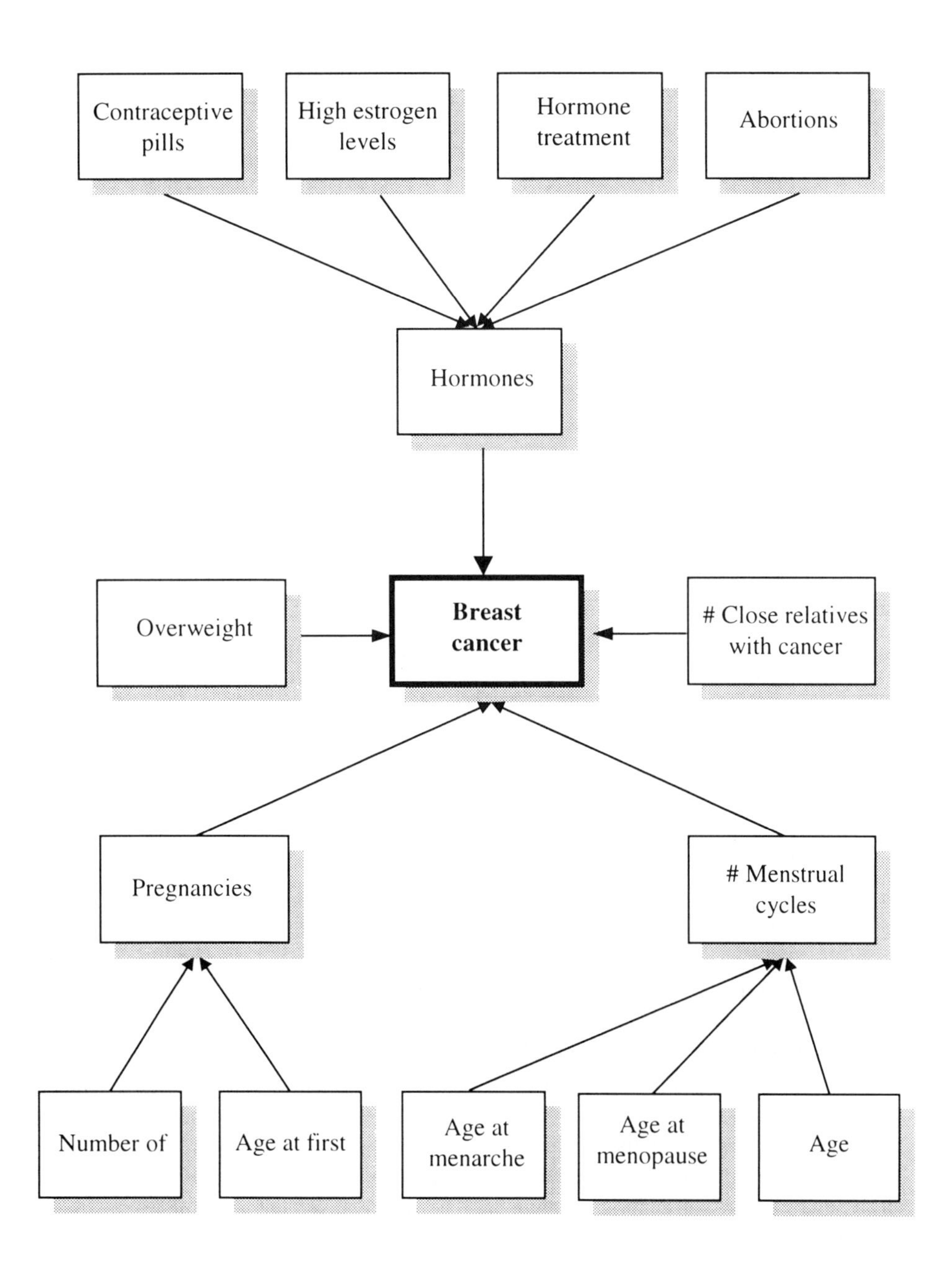

Figure 6: Some believed causes of breast cancer

Using Table 1 and Figure 6, we can form the following example.

Example 5. *A person's membership to* A_1 *may be formalized as*

$$p^{(BreastCancer)} = \Phi\left(p^{(Age)}, p^{(Contraceptives)}, \ldots, p^{(Pregnancies)}\right)$$

Here, Φ *is a logical derivation of the different* $p^{(-)}$s. *This is needed since some of the properties increase the membership to the set "BreastCancer" while some decrease it. Then a crisp* A_1 *is formed by*

$$A_1 = \{p \mid p^{(BreastCancer)} > \kappa\}$$

where κ *is selected given a number of factors, such as risk and cost. A small value of* κ *increases the cardinality of* A_1 *and we are more likely to maximize* $A_n \cap D$, *but the cost of the screening increases.*

In conventional breast cancer screening, at least in Sweden, two different memberships are used in forming A_1 [29], namely sex (female) and age (over 50 years). Indirectly, the geographic location of the person is used since the screening is organized by the county councils and only people living in that specific county are invited to participate in the screening. By introducing computer assistance, it would be possible to allow a more elaborated method for finding A_1. Namely, it is possible to use the whole causality graph shown in Figure 6, or only parts of it.

4.2. Finding the screening method

After forming A_1 from A_0 we would like to continue to zoom toward A_n and, hopefully, the disease set D. To be able to do this we must find suitable screening methods. Traditionally there are two different methods for screening for breast cancer, namely palpation and mammography.

It has been shown, see [9], [32], that palpation gives insufficient information I from which it is not possible to derive sufficient disease specific information X. Hence, palpation is not a good method to use in a screening. It may be possible to combine palpation with other methods providing signs and symptoms as can be seen in Figure 7.

Mammography, on the other hand, gives an I that in turn makes it possible to find an X that enables an accurate diagnosis. The disadvantage of mammography is the complex organization needed to use it as a screening method and the fact that the equipment is rather expensive.

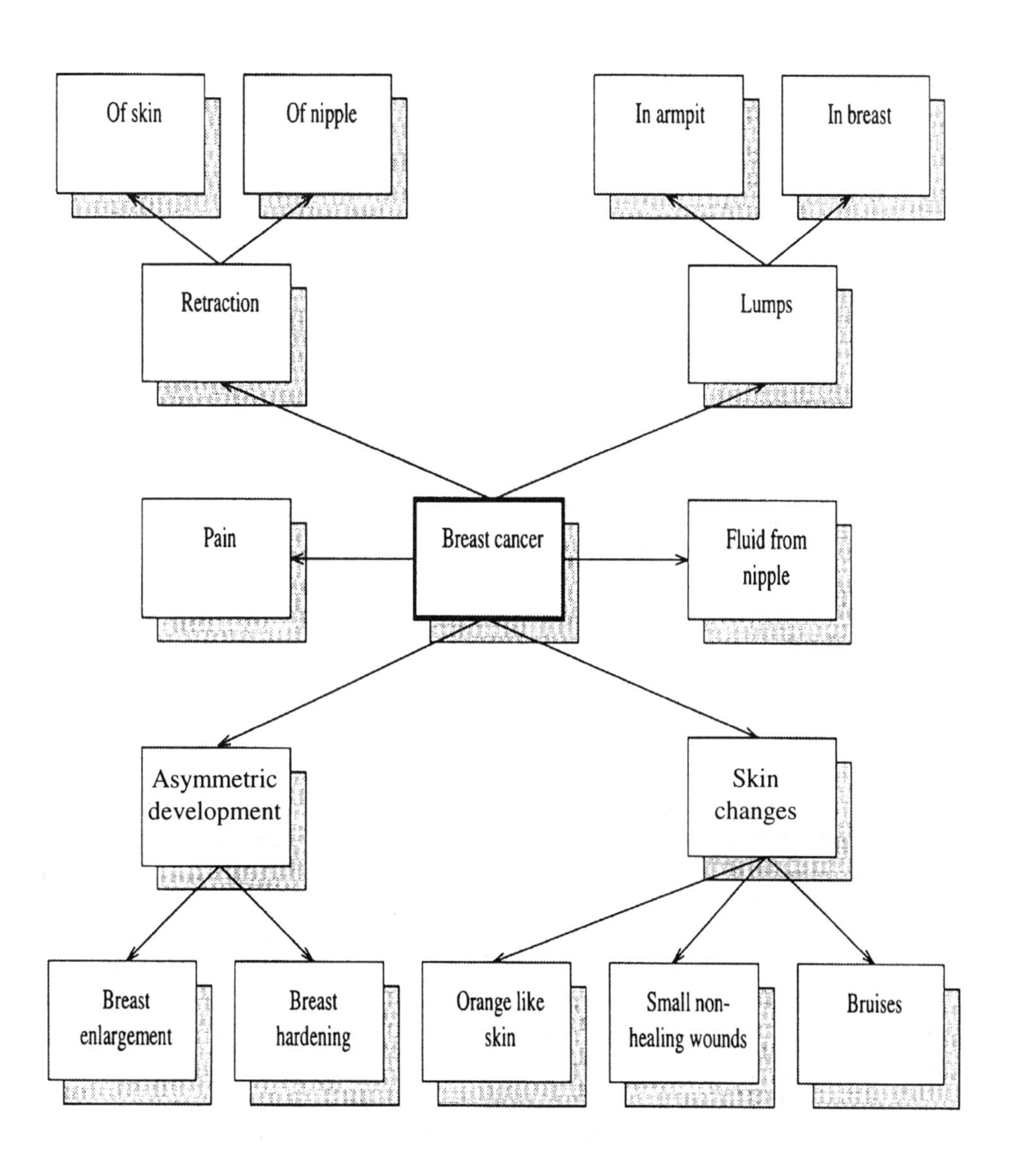

Figure 7: Some non-mammographic signs of breast cancer

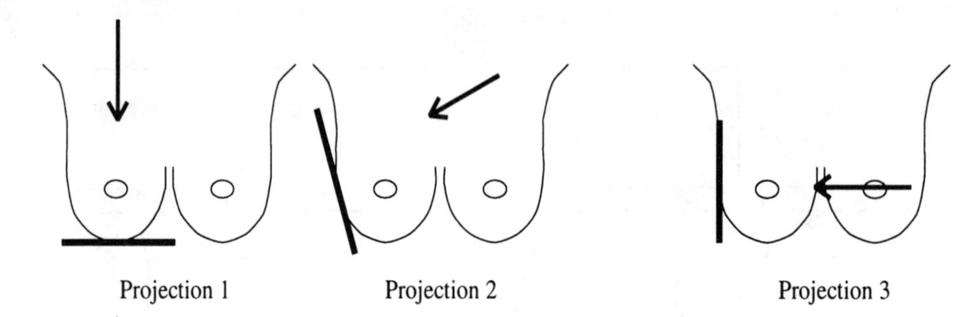

Figure 8: The three different projections used in mammography.

Although we decided to use mammography as a method, there are still several important parameters that must be dealt with. One is what projection to use (see Figure 8 for choices). If we use the wrong projection, I will become insufficient to form a satisfactory X and it will not be possible to achieve a good selection from the screening. In a screening, "projection 2" is the most common, since it maximizes X [29].

In all, much of the process of defining what method to use in a screening must be left to domain experts.

4.3. Defining disease-specific knowledge

The knowledge used by a radiologist when judging a mammogram must be described in the same way as we described the causes of breast cancer. The description is formed by interviewing radiologists and performing a literature study. The result can be seen in Figure 9.

In this figure, it is clear that there are several completely different sets of phenomena that the radiologist uses in the judging process. This indicates that there are subsets within X that must be dealt with. These subsets are "Mass lesions" (x_m), "Calcifications" (x_c), "Parenchymal deformations" (x_p), and "secondary signs" (x_s). All of these subsets are characterized by the fact that they can be described by linguistic variables. Some of these subsets are more important than others in the final diagnosis, and these are the gray nodes in Figure 9. A next step might be to form fuzzy classes for these variables, as was done in [23].

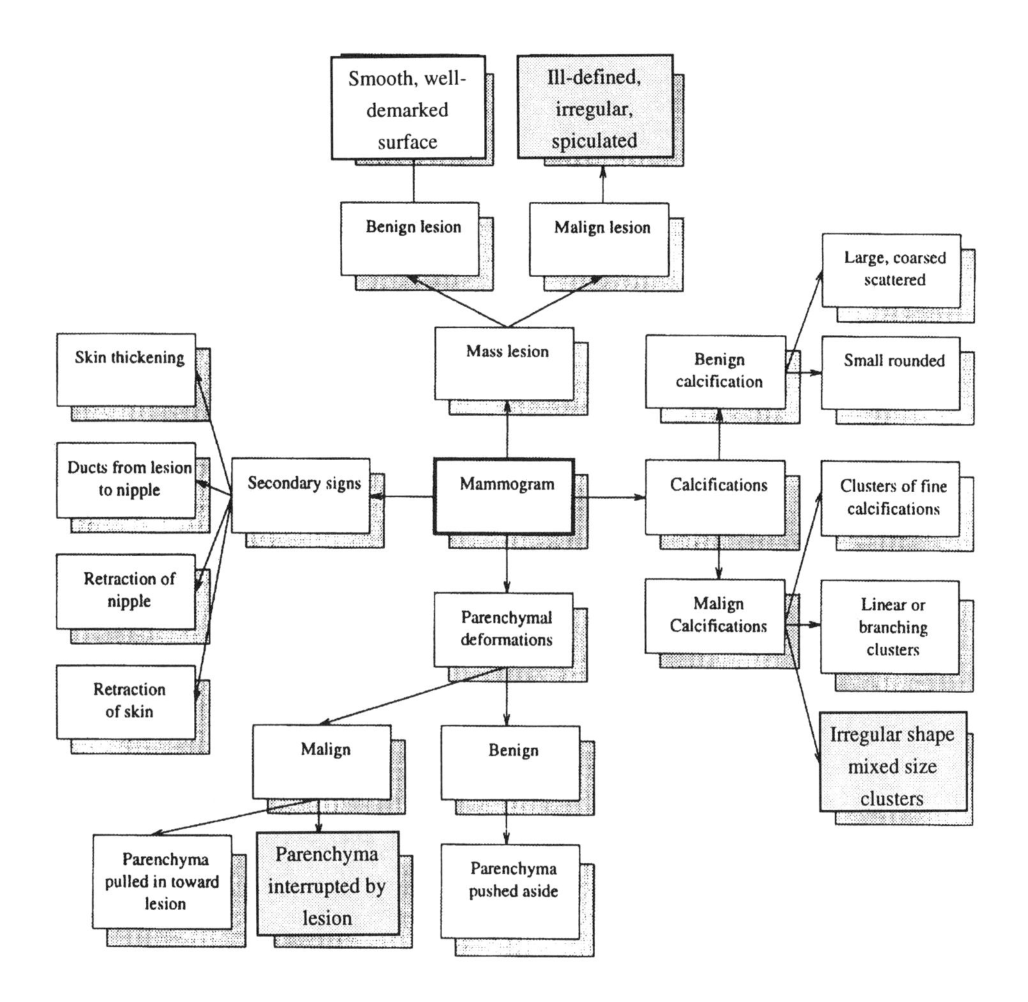

Figure 9: Medical knowledge

Example 6. *Some simple linguistic variables for describing micro-calcifications would be the sets, see Figure 9, {fine, small, large} describing the size and {regular, irregular} describing the shape. A calcification* c*, or as we see it through* π*,* x_c*, would not be a strong indication for malignant change if* c *is large and regular. More formally,*

$$\textit{if } x_c^{(size)} \textit{ is large and } x_c^{(shape)} \textit{ is regular then } x_c^{(Diagnose)} \textit{ is benign.}$$

The radiologist does not use only the properties of objects, but also the objects' placement in the mammogram, i.e., the radiologist uses anatomical knowledge.

Example 7. *If a mass lesion* m *or* x_m *seen through* π*, is believed to be benign by its appearance it would still cause suspicion if it was located in the upper quadrant of the breast* [4]. *To formalize this, we form a set of linguistic variables {upperQ, lowerQ} and obtain the following rules:*

$$R_1: \text{if } x_m^{(Diagnose)} \text{ is benign and } x_m^{(Location)} \text{ is UpperQ, then } x_m^{(Diagnose)} \text{ is malignant}$$

$$R_2: \text{if } x_m^{(Diagnose)} \text{ is malignant and } x_m^{(Location)} \text{ is LowerQ, then } x_m^{(Diagnose)} \text{ is benign}$$

Other anatomical regions of interest are the pectoralis muscle, the nipple, and the lobes.

This kind of rule is used in combination with the ordinary rules to make the diagnosis more reliable. It is also possible to combine information from two projections (see Figure 8) to reduce the number of false positive diagnoses.

4.4. Performing the refinement

Since the method chosen produces information in the form of images, it is suitable to use the π-function to make the transition from I to X where $X = \{x_m, x_c, x_p, x_s\}$. A fair assumption is that the system's performance would not decrease if we considered the different subsets of X as disjoint up to the final decision. This means that we might construct four parallel systems whose outputs are sent to a rule base for forming the final diagnosis.

With Figure 5 in mind, the various steps in the refinement function are described.

Digitalization. Mammograms are one of the most demanding of all images used in medicine [21]. According to the Swedish health board [29], a mammogram must have a resolution of at least 15 line pairs per millimeter, which corresponds to 25 μm^2 per pixel [1], to be used in a mammographic screening. This imposes great demands on the scanner when digitizing the mammogram. There are, however, examples of fully digital mammography (see, e.g., [5]).

Pre-processing. Since we have to detect extremely small objects, i.e., micro-calcifications, it is very important to be careful when removing noise. This together with the fact that high demands are imposed on mammographic equipment, makes us, in many cases, choose not to apply any noise reduction techniques to the mammogram.

Segmentation. In the segmentation step it is possible to use the fact that the different subsets of X are disjoint to divide the different segmentations needed. This means that there are four different phenomena that must be detected. These are the following.

Mass lesions. Mass lesions are the primary signs of breast cancer, although it may be hard to distinguish between benign and malign lesions. Some examples of research carried out in this field are found in [3], [10], [12].

Micro-calcifications. Most work in the field of computer-assisted mammography is concentrated on detecting micro-calcifications. This is due to two reasons. They may be hard to detect by the naked eye and it is relatively easy to define their appearance in a mammogram. Some examples of systems that detect calcifications may be found in [1], [2], [24], [33].

Parenchymal deformations. To be able to detect parenchymal deformations we must be able to model the general structure of the breast parenchyme. One example of this is found in [26], [27]. It is hard to detect a parenchymal deformation by observing a single mammogram, thus, when a radiologist judges mammograms, he/she always compares images from the left and the right breast.

Secondary signs. Secondary signs are skin thickening and nipple retraction. Just as parenchymal deformations, the secondary signs are relative measures that require comparison between breasts. We have not found any work done detecting these signs.

Representation. From the segmentation step, the different mass lesions and micro-calcifications, etc., are obtained and the characteristics of these objects must be captured with numerals.

4.5. The integrated system

The combination of all subsystems suggested in this section yields the results shown in Figure 10. Note that non-mammographic facts are used in the final diagnosis. This corresponds to having two methods, m_1 and m_2, each producing an I_i. To I_i, corresponding to the method of mammography, the π-function is applied. It is not necessary to apply any refinement to I_j that corresponds to the non-mammographic method since it is directly suitable for the final interpretation. The non-mammographic subsystem is mainly used to decide what calling frequency is to be used, that is, how often should a person be called to participate in the screening. There is also an anatomical subsystem for using the anatomical knowledge of where in the breast certain phenomena should be visible, i.e., to reduce the false positives in the final diagnoses.

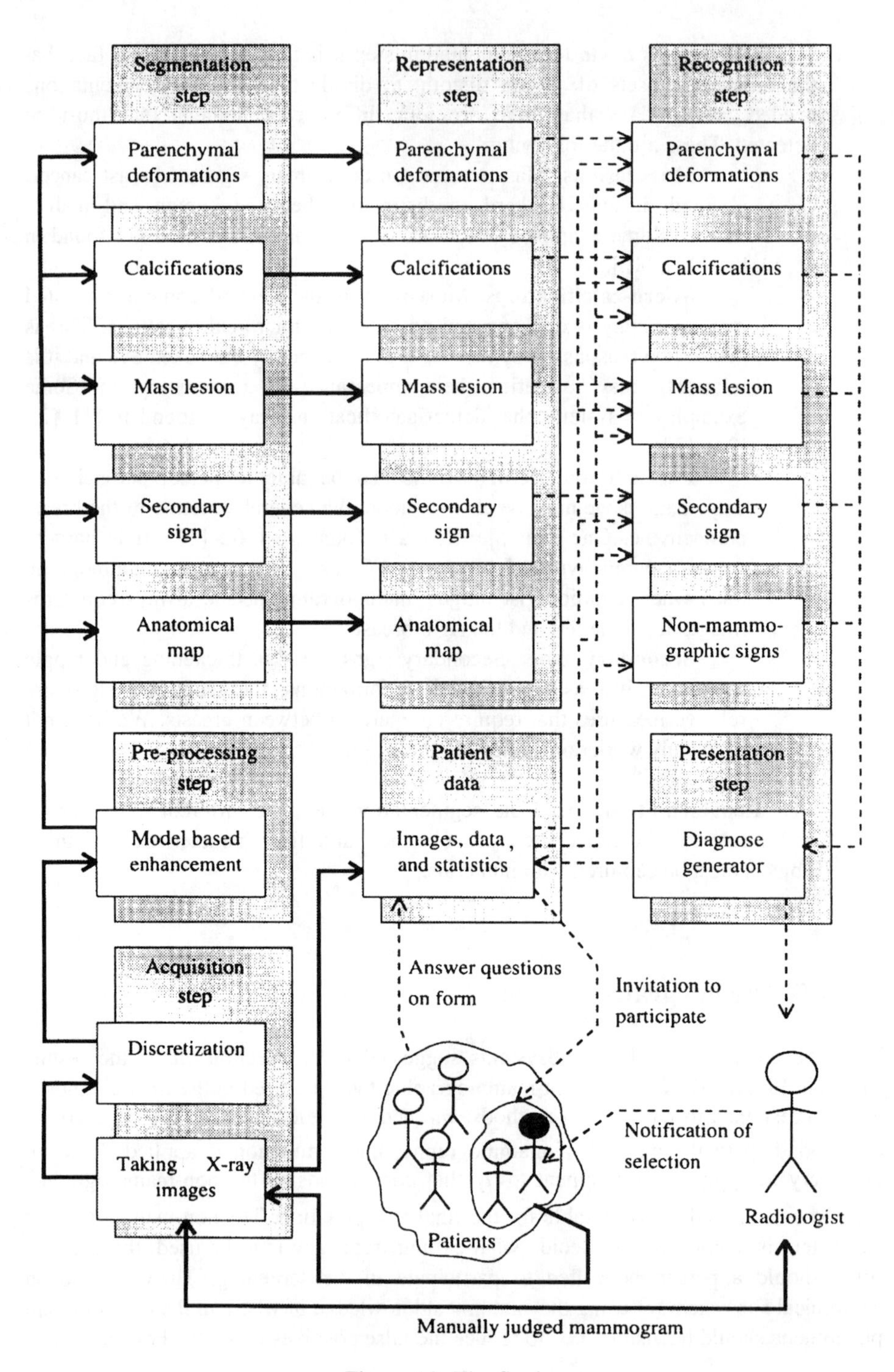

Figure 10: The final system

5. CONCLUSIONS AND FURTHER WORK

A formal framework for the medical screening problem was introduced in the first sections of the chapter. The capabilities of this formalism have been tested within a case study referring to mammographic screening. Difficulties, especially in the refinement function, must be dealt with and large data sets must be obtained for the different knowledge bases that are the core of a screening system with this formalism. The next step is to implement the system and subsystem and test them in a real screening environment.

Much work has been carried out by other researchers within the different sub-areas, but it is believed that no one has taken this top-down approach to the problems often associated with screening. When doing a literature search it is clear that in most cases it is the method used that is of importance and not how to fit the proposed system into an existing health care organization. The theory and the presented methodology may fit for a large number of practical problems in preventive medicine.

Acknowledgments. This work is part of the project "Computer Aided Mammographic Screening" and is funded by the County Council of Västerbotten.

REFERENCES

[1] H. Bårman, G. Granlund, and L. Haglund. Feature extraction for computer-aided analysis of mammograms. *International Journal of Pattern Recognition and Artificial Intelligence,* 7(6): 1339-1356, 1993.

[2] L. M. Bruce and R. R. Adhami. Wavelet-based feature extraction for mammographic lesion recognition. In K. M. Hanson, editor, *Medical Imaging 1997, Image Processing*, volume 3034 of SPIE, *Proceeding series*. SPIE, 1997.

[3] J. W. Byng, N. F. Boyd, E. Fishell, R. A. Jong, and M. J. Yaffe. Automated analysis of mammographic densities. *Physics in Medicine and Biology*, 41(5): 909 - 923, May 1996.

[4] S. Carlson. Personal communication, 1997.

[5] J. D. Cox, S. R. Sharma, and R. B. Schilling. Advanced digital mammography. In H. U. Lemke, M. W. Vannier, and K. Inamura, editors, *Computer Assisted Radiology and Surgery*, International Congress Series 1134. Excerpta Medica, 1997.

[6] I. Dinstein, R. M. Haralick, and K. Shanmugam. Textural features for image classification, *IEEE Transactions on Systems, Man and Cybernetics,* (3): 610-621, 1973.

[7] D. L. Dixon and L. M. Dugdale. An introduction to clinical imaging, Churchill Livingstone, 1988.

[8] F. Georgsson and P. Eklund. A metric framework for texture measurements. In O. Eriksson, editor, *Proceedings of SSAB '98*, 1998, 117-120.

[9] Gershon, M., and Cohen, A., *Atlas of Mammography*. Springer-Verlag, Berlin, 1970.

[10] M.L. Giger, R. M. Nishikawa, M. Kupinski, U. Bick, M. Zhang, R. A. Schmidt, D. E. Wolverton, C. E. Comstock, J. Papaioannou, S. A. Collins, A. M. Urbas, C. J. Vyborny, and K. Doi. Computerized detection of breast lesions in digitized mammograms and results with clinically implemented intelligent workstation. In H. U. Lemke, M. W. Vannier, and K. Inamura, (editors), *Computer Assisted Radiology and Surgery*, International Congress Series 1134. Excerpta Medica, 1997.

[11] R. C. Gonzales and R. E. Woods. *Digital Image Processing*, Addison Wesley, 1993.

[12] B. R. Groshong and W. P. Kegelmeyer, Jr. Detecting circumscribed lesions with the Hough transform. In M. H. Loew and K.M. Hanson (Editors), *Medical Imaging 1996, Image Processing*, volume 2710 of *SPIE Proceedings Series*, pages 59 - 70, 1996.

[13] S.-O. Hietala, B. Johansson, and L. Johansson. *Nuklearmedicin*. Umeå University, 1993.

[14] M.-K. Hu. Visual pattern recognition by moment invariants. *IRE Transactions on Information Theory*, 8:179 - 187, 1962.

[15] National Cancer Institute. Screening of cancer. Technical report 208/03092, 1996.

[16] J. Istas and A. Trubuil. Non-parametric segmentation for star-shaped objects. In *The 9th Scandinavian Conference on Image Analysis*, Uppsala, Sweden, June 1995.

[17] B. Jacobson. *Medicin och Teknik*. Studentlitteratur, 1995.

[18] A. K. Jain. *Fundamentals of Digital Image Processing*. Prentice-Hall International Series, 1989.

[19] R. A. Jones, J. Kvaerness, P. A. Rinck, and T. E. Southon. *Magnetic Resonance in Medicine*. Blackwell Scientific Publications, 1993.

[20] C. E. Kahn, Jr., L. M. Roberts, K. A. Shaffer, and P. Haddawy. Construction of a Bayesian Network for Mammographic Diagnosis of Breast Cancer. *Comput. Biol. Med.*, 27:19-29, 1997.

[21] A. Karellas. Digital X-ray imaging with emphasis on mammographic applications. Short Course at SPIE Medical Imaging 1997, University of Massachusetts, 1997.

[22] S. A. Karkanis. Statistical texture discrimination based on wavelet decomposition, Number 2762 in *SPIE Proceeding Series*, 1996.

[23] B. Kovalerchuk, E. Triantaphyllou, J. F. Ruiz, and J. Clayton. Fuzzy logic in computer-aided breast cancer diagnosis: analysis of lobulation. *Artificial Intelligence in Medicine,* (11):75-85, 1997.

[24] S.-C. B. Lo, H. Li, J.-S. Lin, A. Hasegawa, O. Tsujii, M. T. Freedman, and S. K. Mun. Detection of clustered micro-calcifications using fuzzy modeling and convolution neural network. In M. H. Loew and K. M. Hanson, (Editors), *Medical Imaging 1996, Image Processing,* volume 2710 of *SPIE Proceedings Series*, pages 8 - 15, 1996.

[25] HeartInfo Navigation. Coronary artery scanning and coronary disease risk assessment. Technical report, Center for Cardiovascular Education, 1996.

[26] T. C. Parr, C. J. Taylor, S. M. Astley, and C. R. M. Boggis. Statistical modeling of oriented line patterns in mammograms, In K. M. Hanson, (Editor), *Medical Imaging 1997, Image Processing*, volume 3034 of *SPIE, Proceedings Series*, pages 44 - 55, 1997.

[27] C. E. Priebe, J. L. Solka, R. A. Lorey, G. W. Rogers, W. L. Poston, M. Kallergi, W. Qian, L. P. Clarke, and R. A. Clark. The application of fractal analysis to mammographics tissue classification. *Cancer Letters,* 77(2-3): 183-189, March 1994.

[28] T. H. Reiss. *Recognizing Planar Objects Using Invariants Image Features.* Number 676 in Lecture Notes in Computer Science. Spinger-Verlag, 1993.

[29] *Mammografiscreening* (in Swedish). Allämna råd från socialstyrelsen. Liber, Stockholm, 1993.

[30] I. Soini. Risk factors and selective screening for breast cancer, *Acta Universitatis Tamperensis*, Ser. A, Vol 88, 1977.

[31] N. A. Stillings, S. E. Weisler, C. H. Chase, M. H. Feinstein, J. L. Garfield, and E. L. Rissland. *Cognitive Science. An Introduction*, The MIT Press, second edition, 1995.

[32] L. Tabar. Diagnosis and in-depth differential diagnosis of breast cancer. Uppsala School of Medicine, Uppsala, Sweden, 1991.

[33] H. Yoshida, R. N. Nishikawa, M. L. Geiger, K. Doi, and R. A. Schmidt. An improved CAD scheme using wavelet transform for detection of clustered micro-calcifications in digital mammograms. *Academic Radiology*, 3, 1996.

Chapter 7

A Fuzzy System For Dental Developmental Age Evaluation

Masao Ozaki

Establishing occlusal harmony and correct masticatory function demands an accurate assessment of dental developmental age. A dental expert system for the evaluation of the dental developmental age, which is based on fuzzy system models, is presented in this chapter. The preparation process and the optimization, based on a Genetic Algorithm (GA), of the fuzzy system are discussed. The results generated by the fuzzy system models are contrasted with the results obtained using direct visual inspection data adapted to practical human estimation techniques by the medical experts. The fuzzy system evaluations were compared with those carried out by pediatric dentists who had special expertise in dental age assessment. The clinical development of these methods may facilitate accurate assessment of dental age in addition to reducing patient exposure to X-rays.

1. INTRODUCTION

In diagnosis and therapy in dentistry for pediatric patients, it is necessary to consider the growth stage condition of the infant, namely, to evaluate the physiologic age. The concept of physiologic age involves the understanding that different tissues mature at different rates. For example, the values for skeletal age, morphological age, dental developmental age, etc. may differ in the same individual at a given chronological age [1]. The dental developmental age is used for planning the treatment and the evaluation of oral function in relation to maxillo-facial growth [2], [3], [4]. For example, it is important to be able to evaluate the significance of spaces created by the early loss of primary teeth. Dental radiographs are the most commonly used technique. X-ray images give dentists the most accurate assessment of a person's dental development [5], [6]. However, dental radiograph exposures can produce undesirable health risks and, thus, the number of exposures should be minimized, especially in young children. However, in the absence of information from dental radiograph

0-8493-9806-1/98/$0.00+$.50

images, our available data are limited to the knowledge of average ages of eruption and their standard deviations (SD) [7]. The large values for standard deviation of the premolars and canines suggest that modeling the data as a normal distribution may be inadequate and calls for a more powerful technique. It seems reasonable (from experience in other dental and medical applications) that the use of engineering methods could be helpful in refining our evaluation and decision-making system for dental developmental age.

In this study, we report on a computer system to evaluate the dental developmental age of a patient. The system is used in direct connection to the visual inspection data adapted to practical human estimation techniques [8]. The development of the fuzzy system incorporated medical expertise. In the second step, Principal Component Analysis (PCA) was used to reduce the complexity of the system. Finally, several optimization methods were applied, such as Genetic Algorithms (GA), to improve the system performance.

2. Technical Consideration

2.1. Basic conception of the teeth evaluation system

Our implementation of a teeth fuzzy evaluation system employs the method of Mamdani (Figure 1) [9]. In this implementation, the control is performed by three distinct software modules; namely, the fuzzification module, the rule evaluation module, and the defuzzification.

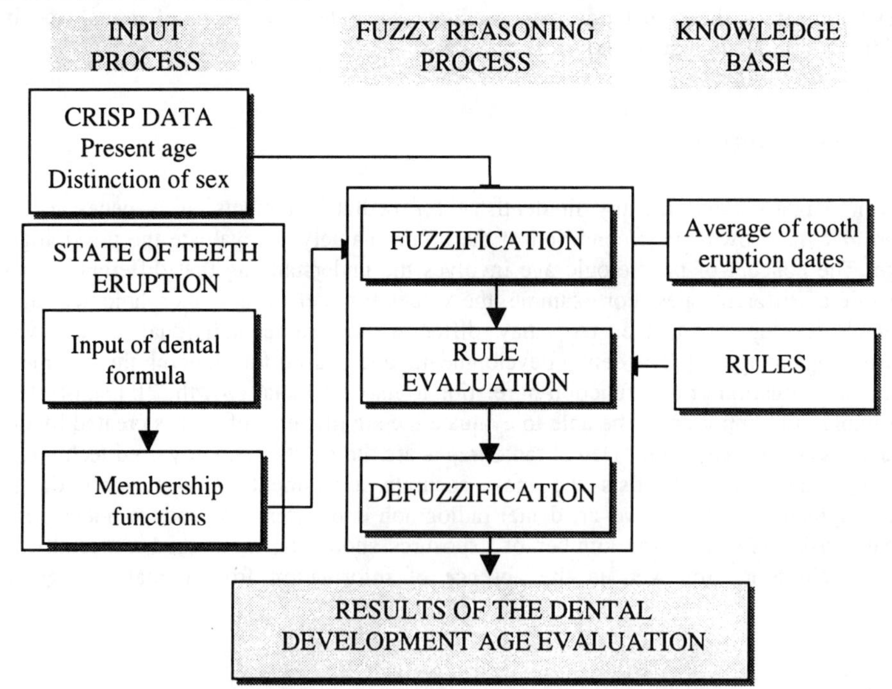

Figure 1: The basic concept of the fuzzy reasoning method in the system.

The first module performs the fuzzification for a tooth eruption state. Its operation is based on the relationship between general data on average time for tooth eruption, the number of teeth erupted, and other crisp data such as sex and present age. The method of evaluation of the stage of eruption (eruption state) and the use of fuzzy operations are illustrated in Figure 2.

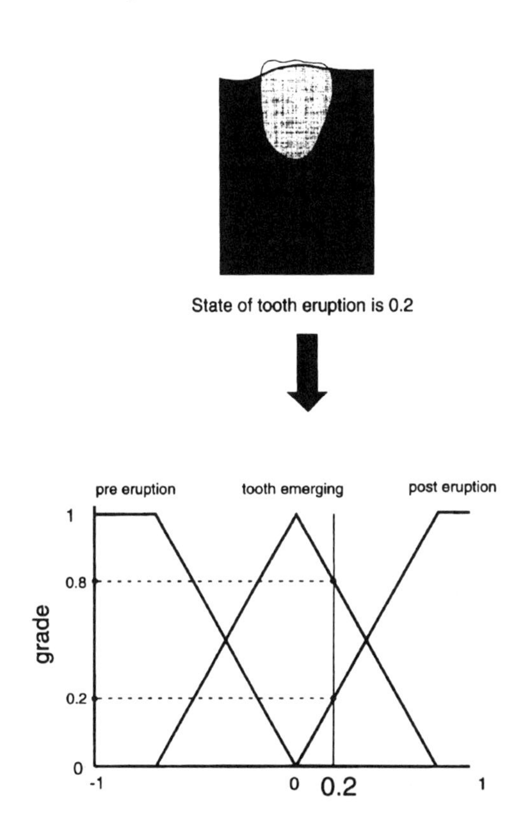

Figure 2. Fuzzification method for the tooth eruption state

Three linguistic degrees were chosen to represent, in conjunction with the corresponding three membership functions, the tooth eruption. These degrees are: "*pre-eruption*," "*tooth emergence*" and "*post-eruption.*" In Figure 2, the state of eruption of one tooth for one particular subject is obtained by representing a vertical line with an abscissa value of 0.2. Since this line does not intersect the "*pre-eruption*" membership function, the degree to which tooth eruption can be said to be "*pre-eruption*" for this subject is "*not*" or zero. Since the line intersects the "*tooth emergence*" function at 0.8, the degree to which tooth eruption is said to be beginning may be said to be true to a large degree. Since the line also intersects the "*post-eruption*" function at 0.2, it may also be said that, to a small degree, the tooth belongs to the post-eruption group. Similarly, we use a fuzzy representation for the relationship between individual present age and the general data of average and standard deviation from time of tooth eruption (Figure 3). For instance, if the target tooth is the lower permanent first molar of a male, we can know the average time of the tooth eruption in the center of the group, i.e., 6.58 years (see Figure 3).

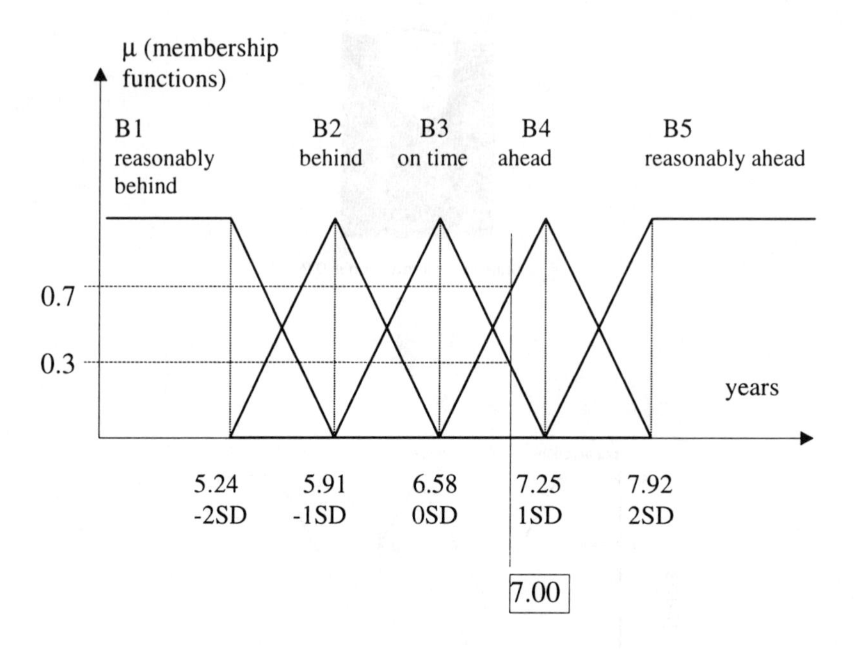

Figure 3. An example of fuzzification for the relationship between present age and the average time of tooth eruption (for the lower permanent first molar in male). Average time of tooth eruption is 6.58 years; standard deviation (S.D.) is 0.67. Actual (chronological) age: 7 year old.

Other groups were fixed in connection with standard deviation. Assuming that patient presenting age is 7.0 years old, the relationship between individual present age and the general data of average time of tooth eruption were represented by the vertical line. Since the line intersects the "*ahead*" membership function in a point corresponding to a value of 0.7, the degree to which relationship between presenting age and the average time of tooth eruption is said to be ahead may be said to be true to a large degree. Since the line also intersects the "*on time*" function at 0.3, it may also be said that to a small degree the presenting age belongs to the "*on time*" group. Other groups (*reasonably behind, behind,* and *reasonably ahead*) are "not" or zero degrees. All input values were fuzzified using appropriate membership functions.

2.2. Rule evaluation module

The second software module performs the Rule Evaluation using input patient data and a rule knowledge base (Table 1).

Table 1: Rule base for the evaluation of dental developmental age*

Rule 1 :	IF	STE is A1	and	DATTE is B1	THEN	EDDL is C5
Rule 2 :	IF	STE is A2	and	DATTE IS B1	THEN	EDDL is C6
Rule 3 :	IF	STE is A3	and	DATTE is B1	THEN	EDDL is c7
Rule 4 :	IF	STE is A1	and	DATTE is B2	THEN	EDDL is C4
Rule 5 :	IF	STE is A2	and	DATTE is B2	THEN	EDDL is C5
Rule 6 :	IF	STE is A3	and	DATTE is B2	THEN	EDDL is C6
Rule 7 :	IF	STE is A1	and	DATTE is B3	THEN	EDDL is C3
Rule 8 :	IF	STE is A2	and	DATTE is B3	THEN	EDDL is C4
Rule 9 :	IF	STE is A3	and	DATTE is B3	THEN	EDDL is C5
Rule 10 :	IF	STE is A1	and	DATTE is B4	THEN	EDDL is C2
Rule 11 :	IF	STE is A2	and	DATTE is B4	THEN	EDDL is C3
Rule 12 :	IF	STE is A3	and	DATTE is B4	THEN	EDDL is C4
Rule 13 :	IF	STE is A1	and	DATTE is B5	THEN	EDDL is C1
Rule 14 :	IF	STE is A2	and	DATTE is B5	THEN	EDDL is C2
Rule 15 :	IF	STE is A3	and	DATTE is B5	THEN	EDDL is C3

*) State of Tooth Eruption (STE): *A1 (pre-eruption), A2 (tooth emerging), A3 (post-eruption).* Deviation from the Average Time of Tooth Eruption (DATTE): *B1 (reasonably behind), B2 (behind), B3 (on time), B4 (ahead), B5 (reasonably ahead).* Evaluation results to Dental Developmental Level (EDDL): *C1 (very delayed), C2 (delayed), C3 (slightly delayed), C4 (standard), C5 (slightly premature), C6 (premature), C7 (very premature)*

The set of rules is called a "fuzzy IF-THEN rule base". The variables include *State of Tooth Eruption* (STE) with values from A_1 to A_3, *Deviation from the Average Time of Tooth Eruption* (DATTE) with values ranging from B_1 to B_5, and *Evaluation Results of Dental Developmental Age* (EDDA) with values from C_1 to C_7.

The following example illustrates the way the process operates (see Figure 4).

Assume we have the following two rules:

Rule 8 : IF STE is A2 and DATTE is B3 THEN EDDA is C4
Rule 9 : IF STE is A3 and DATTE is B3 THEN EDDA is C5

For example, suppose STE_0 and $DATTE_0$ are the input values for the linguistic variables *State of Tooth Eruption* (STE) and *Deviation from the Average Time of Tooth Eruption* (DATTE). Then, their truth values can be represented by $\mu_{A2}(STD_0)$ and $\mu_{B3}(DATTE_0)$ respectively for *Rule 8*, where μ_{Ai} and μ_{Bi} represent the membership functions for the fuzzy sets A_i, B_i (these set allow us to represent "divisional" linguistic values, i.e., use hedges, as "very," "rather," etc.). Similarly, for *Rule 9* we have $\mu_{A3}(STD_0)$ and $\mu_{B3}(DATTE_0)$ as the truth-values of the preconditions.

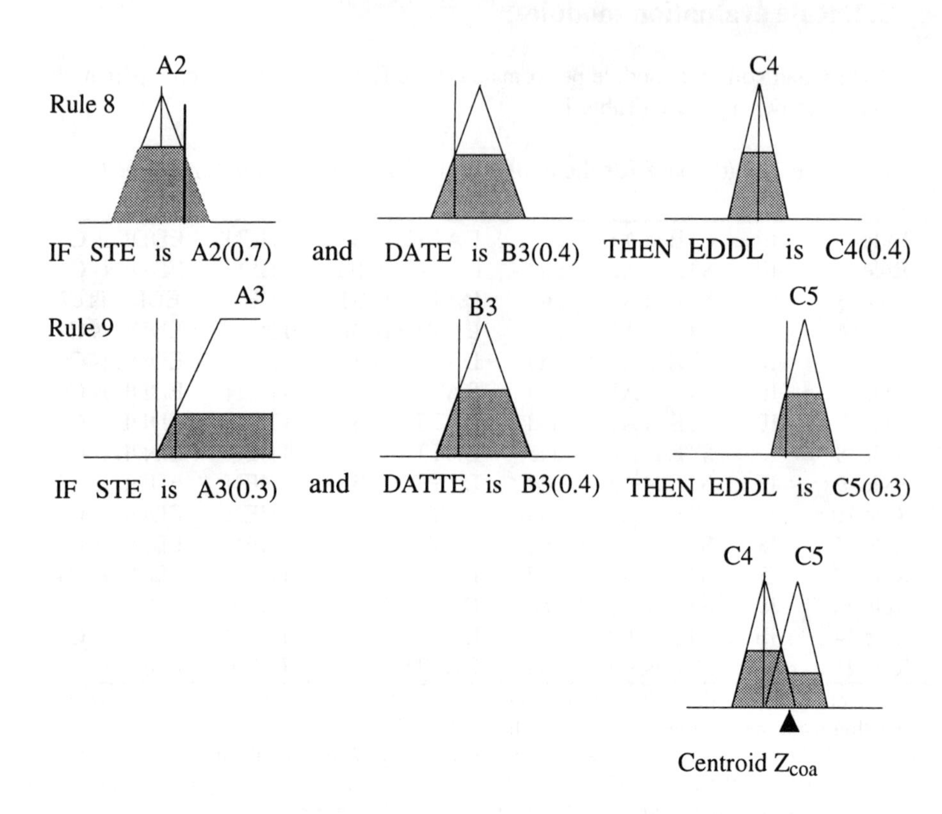

Figure 4. Defuzzification of the combined conclusion of rules described in the example.

The strengths of *Rule 8* and *Rule 9* can be calculated by

Rule 8 : $W_8 = \mu_{A2}(STD_0) \wedge \mu_{B3}(DATTE_0)$

Rule 9 : $W_9 = \mu_{A3}(STD_0) \wedge \mu_{B3}(DATTE_0)$

The control outputs of Rules 8 and 9 are calculated by applying the matching strength of their preconditions on their conclusions:

Rule 8 : $\mu_{C'4}(EDDA) = W_8 \wedge \mu_{C4}(EDDA)$

Rule 9 : $\mu_{C'5}(EDDA) = W_9 \wedge \mu_{C5}(EDDA)$

Here, EDDA ranges over the values of the conclusions. The range of values is specific to the application. This means that, as a result of reading values STE_0 and $DATTE_0$, *Rule 8* is recommending a control action with $\mu_{C'4}(EDDA)$ as its membership function and *Rule 9* is recommending a control action with $\mu_{C'5}(EDDA)$ as its membership function:

$$\mu_{C'}(EDDA) = \mu_{C4}(EDDA) \vee \mu_{C5}(EDDA) = \left[W_8 \wedge \mu_{C4}(EDDA)\right] \vee \left[W_9 \wedge \mu_{C5}(EDDA)\right]$$

Here, $\mu_{C'}(EDDA)$ is a pointwise membership function for the combined conclusion of Rule 8 and Rule 9. Finally, we can obtain the evaluation result of individual dental developmental age by defuzzification. Defuzzification assumes that a control action with a pointwise membership function μ_C has been produced; the *Center Of Area* (COA) method calculates the center of gravity of the distribution for the control action. Assuming a discrete universe of discourse, we have

$$Z_0 = \frac{\sum_{i=1}^{Q} Z_i \cdot \mu_C(Z_i)}{\sum_{i=1}^{Q} \mu_C(Z_i)}$$

Above, Q is the number of quantization levels of the output, Z_i is the amount of control output at the quantization level i, and $\mu_C(Z_i)$ represents its membership value of C.

Table 2: An example of experimental result. The lower permanent first molar in a male (average time of tooth eruption: 6.58 years, S.D. 0.67)* was evaluated for state of the tooth eruption and represented to deviation from the average time of tooth eruption.

State of eruption	66 months (5.5 years)	78 months (6.5 years)	90 months (7.6 years)
-1.0	0.5906	-0.8474	-2.1016
-0.8	0.8428	-0.5572	-1.8475
-0.6	1.0088	-0.4067	-1.6904
-0.4	1.1316	-0.2596	-1.5422
-0.2	1.3132	-0.0882	-1.4473
0.0	1.5906	0.1526	-1.3968
0.2	1.6348	0.4428	-1.1454
0.4	1.7080	0.5933	-0.9800
0.6	1.7653	0.7404	-0.8560
0.8	1.9391	0.9118	-0.6771
1.0	2.2460	1.1526	-0.3968

*According to the Japanese Society of Pedodontics Standard [7]

For example, applying W_8 to the conclusion of *Rule 8* generates the result in the shaded trapezoid figure shown in Figure 4 for C_4. Similarly, applying W_9 to the conclusion of *Rule 9* generates the result in the dashed trapezoid shown in Figure 4 for C_5. By superimposing the resulted memberships over each other, the membership function for the combined conclusion of these rules is found.

Furthermore, using the *Center Of Area* method, the defuzzified value for the conclusion is found. Table 2 shows an example of experimental results.

3. System Optimization By Using Clinical Data

The teeth evaluation system reported in this chapter applies the fundamental concepts of evaluation of the dental developmental age by means of fuzzy logic. However, a direct application of the basic methods of evaluation would require us to input a huge amount of data about all teeth, as in classic clinical evaluation by human experts. We used a simplified method of evaluation, based on data reduction. A system automatic design based on clinical data was needed to assess and validate the fuzzy system for dental age estimation. Expert pediatric dentists performed evaluations of the state of individual permanent dentition and the results were contrasted to the results provided by the system. The optimization of the fuzzy system using Principal Component Analysis and Genetic Algorithm was performed.

3.1. Material and method

a. Database

Although a normal adult has 28 teeth (except the third molar), seven in each quadrant, only the teeth in the left half side of the mouth were used in this study. In this way, the dimension of the problem is significantly reduced, without loss of precision. Six hundreds and thirty five patients (303 males, 332 females, age range 27 to 184 months) were examined. We divided the original data into nine different training and testing sets. Moreover, we collected another 50 cases for comparing human expert estimation with three data reduction methods.[1]

b. Valuation of Tooth Eruption State

The eruption state of each tooth is represented for a scale of pre-eruption to perfect eruption (-1.0 to 1.0). A value of -1 indicates that the tooth is below the gumline and it has not erupted. A tooth that has erupted is evaluated according to several linguistic degrees, from "*tooth emerging*" to "*fully erupted*" (Figure 5).

---

[1] The reader may find more technical details on these topics in the paper [19]. (Editors' note).

Although a normal adult has 28 teeth (excepting the third molar), seven in each quadrant, we reduce the dimensionality by considering only those in the left half of the mouth. Our data consists of eruption estimates of the 14 teeth in the left half of the mouth (14 teeth). This reduction is reasonable due to the symmetry of normal dentition.

3.2. Dimensionality Analysis by Principal Component Analysis

Fuzzy system models, like neural network models, are trainable nonlinear function approximators. However, conventional fuzzy systems use linear combinations of rules. This makes the design of high dimensional input systems difficult. In our case, it is difficult to design the fuzzy system, because of the number of input variables and values (14 teeth eruption state inputs). Imagine the number of rules needed when each input space was partitioned into three subsets (Figure 2). To solve this problem, we map the given input space to a new input space whose dimension is lower. We used Principal Component Analysis (PCA) to perform a dimensionality analysis [10]. The space constructed by PCA has the property that the dimensions are sorted, according to the amount of variance of the original data each captures. Keeping only the components that account for the most variance reduces the dimensionality of the space.

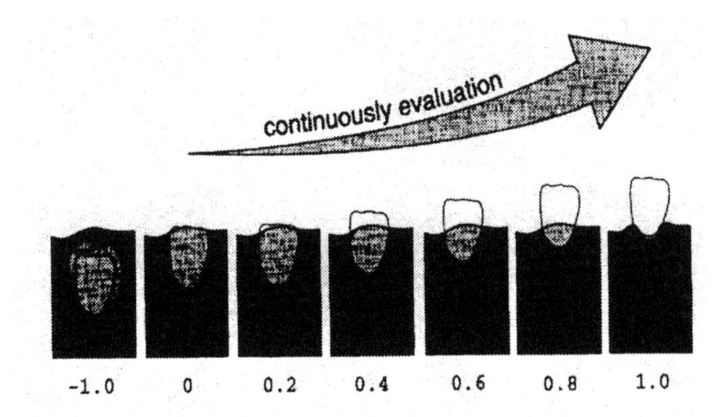

Figure 5: Schematics of the degree of tooth eruption state. Pre-eruption is evaluated as "-1," tooth emerging is "0," and complete eruption is "1". The total scale has 20 sub-steps (degrees).

In our experiments, we ran a PCA analysis on each of the training sets and obtained a transformation matrix for each training set. Using the PCA transformation of the data, it was found that, on average, the first four components account for about 78.2% of the variance of the original data set. Based on this result, we decided to use only the first four dimensions as inputs to the fuzzy system.

Table 3 shows an example of the relation between the variance and the number of principal components used. Here, the new input variables (principal components) are Z_i, as obtained from the given input variables, X_i $(1 \leq i \leq 14)$.

Table 3: The results of the principal component analysis.

Dimension	Variance	New Input Variables
1	48.2%	$Z1 = -0.32286\,X1 + 0.89189\,X2 - 0.06556\,X3 - 0.04366\,X4 - 0.00395\,X5 - 0.20257\,X6 - 0.01304\,X7 + 0.77998\,X8 - 0.25441\,X9 - 0.01536\,X10 - 0.06342\,X11 - 0.01370\,X12 - 0.16068\,X13 + 0.00944\,X14$
2	66.3%	$Z2 = -0.11514\,X1 - 0.14347\,X2 - 0.013116\,X3 - 0.01639\,X4 - 0.03888\,X5 + 1.36535\,X6 + 0.00962\,X7 - 0.12832\,X8 - 0.01189\,X9 - 0.00303\,X10 - 0.01164\,X11 - 0.02126\,X12 - 0.29866\,X13 - 0.01959\,X14$
3	73.5%	$Z3 = -0.00524\,X1 + 0.00857\,X2 - 0.00313\,X3 - 0.02214\,X4 - 0.08489\,X5 - 0.01738\,X6 - 0.29695\,X7 + 0.00911\,X8 + 0.00955\,X9 - 0.06057\,X10 + 0.00783\,X11 - 0.09956\,X12 - 0.00537\,X13 + 1.23122\,X14$
4	78.2%	$Z4 = -0.00495\,X1 - 0.00201\,X2 - 0.04693\,X3 - 0.06926\,X4 - 0.23196\,X5 - 0.02494\,X6 - 0.22947\,X7 - 0.00029\,X8 - 0.02810\,X9 - 0.04932\,X10 - 0.22230\,X11 + 1.45738\,X12 - 0.01415\,X13 - 0.13545\,X14$
5	81.7%	$Z5 = -0.0.01702\,X1 + 0.00304\,X2 - 0.30530\,X3 - 0.11537\,X4 - 0.24432\,X5 - 0.00411\,X6 - 0.10675\,X7 - 0.00163\,X8 - 0.10117\,X9 + 1.61662\,X10 - 0.21565\,X11 - 0.05682\,X12 - 0.00527\,X13 - 0.09351\,X14$
6	84.9%	$Z6 = 1.32811\,X1 - 0.24304\,X2 + 0.00554\,X3 - 0.00161\,X4 + 0.01071\,X5 - 0.11144\,X6 - 0.00564\,X7 - 0.22455\,X8 - 0.00900\,X9 - 0.00711\,X10 + 0.00320\,X11 + 0.00010\,X12 - 0.15069\,X13 - 0.00500\,X14$

Each new input variable has the accumulated contribution ratio as shown in the table. For example, when we use Z_1, Z_2, Z_3, and Z_4 instead of X_1 - X_{14}, the four components Z_i $(1 \leq i \leq 4)$ can "explain" 78.2% of the results given by the 14 variables X_i.

3.3. System Optimization by Using Genetic Algorithm

Genetic Algorithms (GAs) can help solve complex real-world problems because of their flexibility [11]. They are effective at finding optimal solutions to a variety of problems. A genetic algorithm technique was used to design a fuzzy system for training set of clinical data [12]. We use a four-input-one-output TSK fuzzy system in our experiments (Figure 6) [13]. The membership functions are triangular and, together, they realize a fuzzy partition of the space. All possible combinations of input fuzzy partitions were included in the rule base. Because of the reduced dimensionality, the number of possible rules is only 81, much lower than the number required by the initial input variables. The fuzzy system was trained using a genetic algorithm technique. The fuzzy system was allowed to evolve for 10,000 generations, during the training.

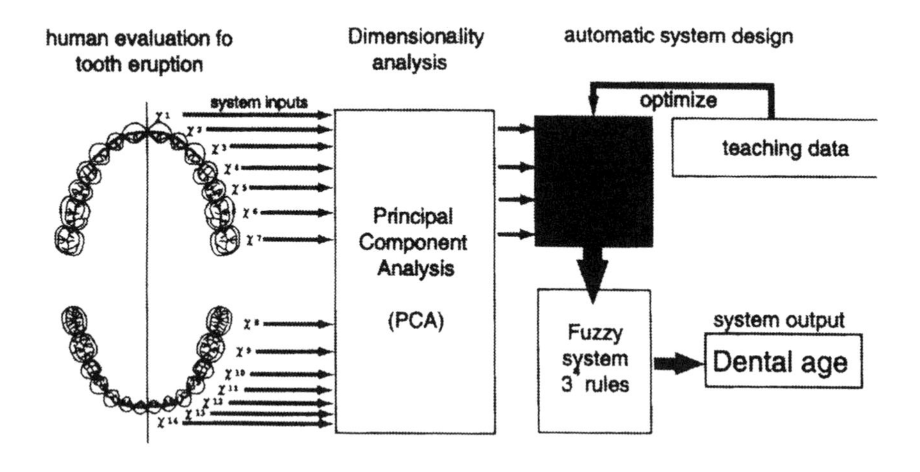

Figure 6: Fuzzy system model for dental age evaluation.

3.4. System Evaluation and Results

Two different error measures were used to assess the quality of prediction. Namely,

$$\varepsilon_1^2 = \frac{\sum \left(a_{predicted} - a_{actual}\right)^2}{n^2}$$

is the squared error, based on the predicted age $a_{predicted}$ and the actual age a_{actual}.

The second error measure was the relative (%) standard deviation of the squared (individual) error:

$$\varepsilon_2 = \frac{\sqrt{\sum \left(\frac{a_{predicted} - a_{actual}}{a_{actual}}\right)^2}}{n} \cdot 100\%$$

Figure 7 shows system evaluation results, based on standard deviation error at different ages. The results were satisfactory for all cases between 60 to 130 months old. The results for ages lower than 40 months are poor, showing the tool is not appropriate for these ages.

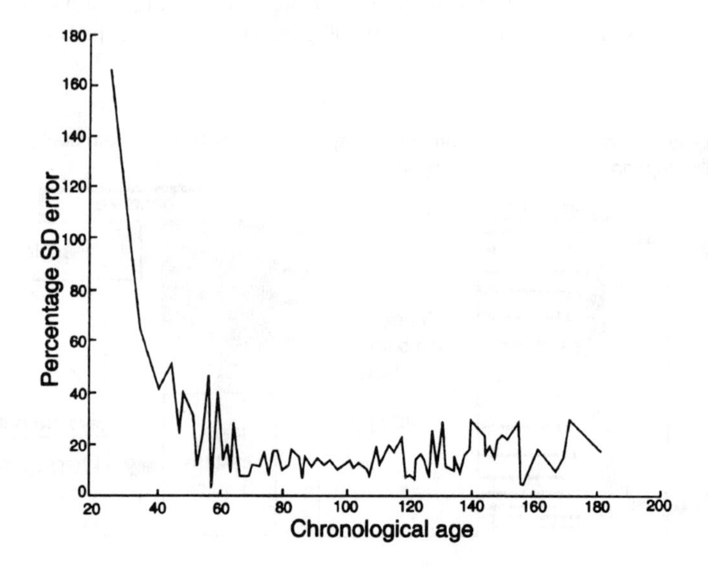

Figure 7. System evaluation results of percentage S.D. error vs. chronological age.

For system evaluation, the computed results were compared with those obtained by two pediatric dentists (A and B), who had special expertise in dental age assessment. They then performed evaluations on the group of 50 patients. The results of the average squared error for our fuzzy system models are given in Table 4. The models have about the same average squared error (about 15 to 18 months) made from testing data sets.

Table 4. Comparison of results by human experts and by the fuzzy system

	Doctor A	**Doctor B**
Human estimation	14.173	11.997
Fuzzy system	17.939	15.771

Figure 8 shows the results of the estimation of these cases. The clinical example was a 95-month old female, who is congenitally missing the lower left lateral incisor. The state of teeth eruption of lower left lateral incisor was evaluated " –1" degree in "Input A" (see arrow). This result of our fuzzy system output was only 3 months (91.21

months) different than the age of the patient. For the input example of "Input B" we used the lower right lateral incisor as the missing tooth. The result provided by the system in this case shows no difference compared to the estimation result of a physician, or to the age of the patient.

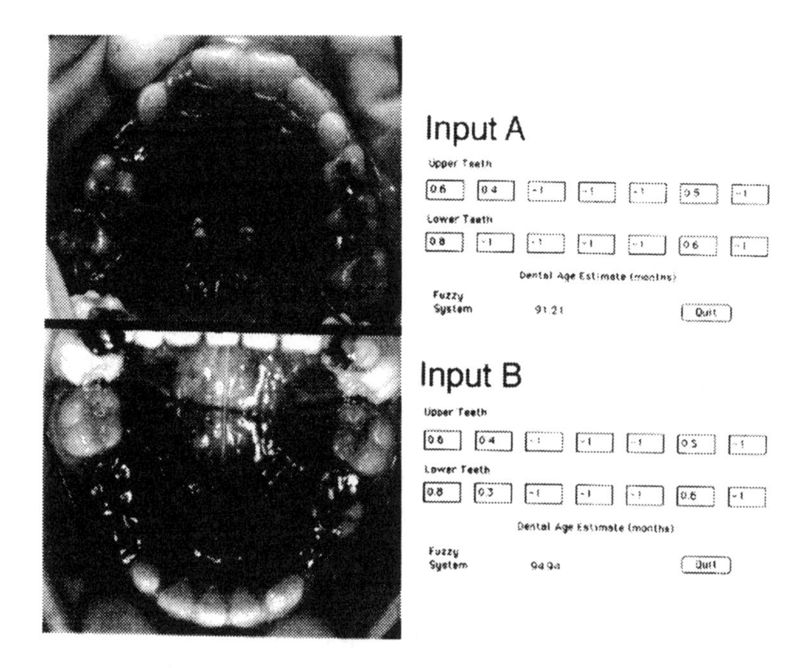

Figure 8. An example. This case was 95-month old female who had congenitally missing teeth.

Beyond these examples, our experience with the system shows that no significant errors were produced when clinical data were used. After numerous tests, we concluded that the results obtained with the described fuzzy system proved its flexibility and validity for medical practice.

4. Discussion and Conclusions

With recent progress in the field of personal computers, AI (Artificial Intelligence) has developed rapidly and its possibilities have become a center of attraction for the medical community. Some types of medical systems using AI have already been developed and manufactured [14], [15]. However, when we attempt to use these systems in real clinical situations, we find that they often yield outputs that contradict the diagnoses performed by expert clinicians.

Zadeh discussed the theory of fuzzy sets and established an important scientific foundation for future development by research workers in this field [16-18]. The present simulation uses fuzzy set theory for the mathematical analysis of tooth developmental age. Zadeh's theory is convenient for this purpose.

From a practical point of view, the two most important questions are how to use the fuzzy sets and how to make the rules of the diagnosis. We feel it is necessary for doctors to make clear their own views on these points. We approached the optimization technique based on GAs because it can produce automatic fuzzy system design. However, it is difficult to automatically design a fuzzy system, which takes into account 14 teeth eruption state inputs. Imagine the number of rules when each input space was partitioned into three subsets (this number is 3^{14} = 4,782,969!). With such huge spaces, time and computer memory limitations become a problem.

To cope with the dimensionality problem, the input space dimension was first reduced using PCA analysis. Using this technique, the input data space was compressed to only four dimensions that covered 78.2% of all input eruption state data [19]. Then, the GA technique was used for automatic fuzzy system design. Performance of the fuzzy system model was only slightly worse than experts in pediatric dentistry. The lower performance is probably due to the input space dimensional reduction. According to the procedure used, the fuzzy system input space is based on only four principal components and the learning algorithm is also based on them. This system is used as a current tool in evaluating the dentition age.

Imprecision has been a problem for medical or dental expert systems, but fuzzy logic provides ways to process such rules easily and efficiently, even with a low-end PC. We believe that the fuzzy logic reasoning method approach to the measurement of dental developmental age is a powerful tool in the context of the other measurement and inference methods. Scholars and other fuzzy expert system designers are almost of the same opinion. However, the clinical value and correctness of these approximate reasoning methods have yet to be adequately clarified. It may be concluded that the application of fuzzy logic is a valuable tool in medicine and a diagnosis system in pediatric dentistry based on fuzzy methods of inference is already available and proved to be fruitful in a clinical setting.

REFERENCES

[1] Proffit, W. R. (Ed.), Contemporary Orthodontics. Inc. Missouri, Mosby-Yearbook, 1993.

[2] Hellman, M., The face and occlusion of the teeth in man. *Int. J. Orthodont & Oral Surg.* 13, pp. 921-945, 1929.

[3] Hellman, M., The face in its developmental career. *Dental Cosmos*, 77, pp. 685-699, 1935.

[4] Hellman, M., Changes in the human face brought about by development. A. *J. Orthodontics*, 40, 475-516, 1954.

[5] Nolla, C. M., The Development of the permanent. *J. Dent. Child*, 27, 254-266, 1960.

[6] Moorrees, F.A.C., Fanning A.E., and Hunt, E.E., Age Variation of Formation Stages For the Permanent Teeth. *Journal of Dental Research*, 42, pp. 1490-1502, 1963.

[7] The Japanese Society of Pedodontics, The chronology of deciduous and permanent dentition (in Japanese children). *Jap. J. Ped. Dent.*, 26, pp. 1-18, 1988.

[8] Ozaki, M., Baumrind, S., Braham, R. L., and Motokawa, W., A new approach to the measurement of dental developmental age by using fuzzy logic reasoning. Part 1: Basic conception. *Pediatric Dental Journal*, 5, pp. 61-67, 1995.

[9] Mamdani, E.F., Application of fuzzy algorithms for control of simple dynamic plant. *Proc. IEE*, 121, 12; 1585-1588, 1974.

[10] Everitt, B.S. and Dumm, G., *Applied Multivariate Data Analysis*, Oxford University Press, New York, 1992.

[11] Kennedy, S. A., Five ways to a smarter genetic algorithm. *AI EXPERT*, pp. 35-38, December, 1993.

[12] Lee, M.A. and Takagi, H., Integrating design stages of fuzzy systems using genetic algorithms. *Proc. IEEE Int. Conf. on Fuzzy Systems (FUZZY-IEEE '93)*, San Francisco, CA, pp. 612-617, 1993.

[13] Takagi, T. and Sugeno, M., Fuzzy Identification of System and Its Applications and Control, *IEEE Transaction on Systems, Man and Cybernetics*, 5, pp. 116-132. 1985.

[14] Shortliffe, E.H., Computer Based Medical Consultation: MYCIN. American Elsevier, New York, 1976.

[15] Davis, R., Buchanan, B.G., and Shortliffe, E.H., Production rules as a representation of a knowledge-based consultation program. *Artificial Intelligence*, 8, pp. 15-45, 1977.

[16] Zadeh, L.A., Fuzzy sets. *Information and Control*. 8, pp. 338-353, 1965.

[17] Zadeh, L.A., A fuzzy-set-theoretic approach to fuzzy quantifiers in natural languages. *Computers and Mathematics*, 9, 149-184, 1983.

[18] Yager, R.R. and Zadeh L.A. (Eds.), *An introduction to fuzzy logic applications in intelligent systems*. Kluwer Academic Publishers, Boston, pp. 1-25, 1992

[19] Ozaki M., Lee, M.A., and Takagi, H., Neural Network, Fuzzy System, and Multiple Regression Models for Dental Age Prediction. *Proc. 3rd Int. Conference on Fuzzy Logic, Neural Networks and Soft Computing, IIZUKA'94*, Japan, Aug.1-7, 1994, pp. 313-314

Chapter 8

Fuzzy Expert System For Myocardial Ischemia Diagnosis

Sorina Zahan, Christian Michael, and Stephanos Nikolakeas

1. Introduction

Hippocrates characterized medicine as an art and, as such, it is very often found to be subjective, ambiguous, and not amenable to austere criteria for the definition between right and wrong or, in this case, between health and disease. In fact, in medicine, uncertainty is the rule, while certainty is the exception. Parameters that cannot be quantified and contribute to increased uncertainty include the physicians' experience and intuition, the underlying pathophysiologic mechanisms and the variability between individuals regarding the expression of clinical signs or their susceptibility to medications.

Nevertheless, in everyday clinical practice, doctors are facing inevitable questions posed by the patients like "how severe is my disease?" or "will I be cured?," questions demanding explicit answers. Very often, such questions cannot be answered accurately; thus, physicians are forced to apply linguistic expressions like *very severe*, or *not so severe*, or to probabilistic estimations about the patient's prognosis.

In medical diagnosis, especially, diagnostic criteria usually only increase or decrease the probability that a certain patient has a certain disease. This is one of the reasons that the Bayesian concept[*)] has been adopted with regard to the truth value of

---

[*)] The Bayesian approach is an alternative concept to calculate the true predictive value of a possible test result. It is based on Bayes' theorem which, in brief, defines the predictive value of a certain test result in detecting a certain disease depending on the test's sensitivity and specificity along with this certain disease's prevalence in the certain patient's population, and it is expressed by the following equations:

Probability of disease presence with a

0-8493-9806-1/98/$0.00+$.50

diagnostic methods [1]. There seems to exist no "gold standard" for the precise definition of the boundaries between not only different diagnoses, but also between the various degrees of severity of a certain disease. Actually, the physician has to select from a continuum rather than from a defined climax. That is why fuzzy set theory has been applied in medicine and has offered some promising results [2], [3], [4].

The advantage of fuzzy systems over crisp systems is their ability to incorporate and process data with a graded degree of abnormality [5]. Several such systems have been introduced, with a varying degree of success, especially in the field of medical diagnosis [6-9]. Fuzzy logic applications were presented, in a relatively high number of publications, in the field of cardiological diagnosis and cardiac imaging. A wide range of subjects have been addressed with fuzzy techniques, including cardiac arrhythmia's classification [9] and recognition [10], the estimation of ventricular function using radionuclide [11] or computed tomographic [12] cardiac imaging, hemodynamic control studies [13], or ischemic heart disease detection and evaluation [8], [14]. Exercise test electrocardiography is one of the areas most aided by the blossoming of fuzzy logic techniques in medical diagnostics. The inherent uncertainty of exercise test criteria provides an appropriate ground for the application of fuzzy set theory. Combinations of exercise test criteria for the evaluation of cardiac ischemia, using the Bayesian probability methods, use different estimations of the significance of each criterion [15], [16]. These approaches tend to ignore the "gray zone" between normal and abnormal and between the various degrees of abnormality. On the other hand, combining various exercise stress criteria using fuzzy cluster analysis seems to be a practical and advantageous approach [14]. With a proper selection of these diagnostic criteria, such multimethod techniques are expected to be of increased diagnostic value. Among all these fuzzy approaches, medical expert systems seem to be particularly successful.

2. FUZZY EXPERT SYSTEMS

The applicability of fuzzy set theory to expert systems has been studied since 1975 [17]. Real applications appeared during the 1980s, shortly after some early (non-fuzzy) expert systems as DENDRAL [18] had been presented. In 1981, Weiss [19] proposed EXPERT, a fuzzy expert system used for rheumatological and ophtalmological diagnosis. Shortly, (in 1982), one of the most famous fuzzy expert systems - CADIAG, [20], [21] was created.

Classical expert systems approach the uncertainty – as usual – on a probabilistic basis. Starting with MYCIN [22], [23], a milestone in the history of expert systems

---

$$\textit{positive test} = \frac{\textit{sensitivity} \times \textit{prevalence}}{(\textit{sensitivity} \times \textit{prevalence}) + [(1 - \textit{specificity}) \times (1 - \textit{prevalence})]}$$

Probability of disease presence with a

$$\textit{negative test} = \frac{(1 - \textit{sensitivity}) \times \textit{prevalence}}{[(1 - \textit{sensitivity}) \times \textit{prevalence}] + [\textit{specificity} \times (1 - \textit{prevalence})]}$$

(performing diagnosis and therapeutic consultation of infectious diseases), some factors expressing the statistical uncertainty, belief, or evidence were assigned to the consequents of the rules used in the inference. The generalization of these factors was the first fuzzy "wave". These factors can be involved in a large range of operations [24]. The second fuzzy wave – a much more consistent one – was and still is the involvement of fuzzy predicates within the expert system. Technically speaking, this is the most significant advantage offered by the fuzzy approaches. Thus, fuzzy logic and approximate reasoning are used within the inference mechanism. In addition, fuzzy sets are involved in the expression of indicators, symptoms, or production rules, while fuzzy classification methods are introduced within the diagnosis process. Further, in order to handle the uncertainty, fuzzy techniques can be mixed with other theories (probability theory, theory of evidence, etc.) [2].

Generally speaking, a good fuzzy or non-fuzzy expert system is (or should be) characterized by the following attributes [2], [25]:

1. There is an explicit separation between the knowledge and the methodology employed to derive conclusions from it. The expert system has to include two different concepts: the *knowledge base (KB)* and the *inference engine (IE).*
2. The expert system should reason in a manner similar to the human expert.
3. The knowledge base should be dynamic. This means that it can be easily expanded or modified. New knowledge modules should be added without difficulty.
4. To transfer the knowledge from the human expert to the knowledge base, an interactive transfer technique should be used. This technique should minimize the time required in this transfer.
5. The interaction between the expert system and the expert domain should be performed by means of a natural language. This would allow the user to think in terms that are specific to the current problem. The system should be adapted to the user and not vice versa. The user should be detached from the implementation details.
6. The main component of the knowledge transfer, the knowledge engineer, should be eliminated.
7. The control strategy should be simple and transparent to the user. The user should be able to understand and evaluate the effect of adding new knowledge in the KB. Also, the expert system should be powerful enough to solve complex problems.
8. There should be a general and cheap environment for the expert system development and testing.
9. The expert system should be able to reason under uncertainty and incomplete knowledge constraints. In addition, it should be able to perform probabilistic reasoning.
10. The expert system should be able to explain its own actions and reasoning.
11. The expert system should be able to learn from experience.

The general structure of an expert system is illustrated in Figure 1.

The acquisition module supports the knowledge engineer in creating the KB. It may consist of a friendly rule editor, or it may be based on an automatic acquisition system using some learning technique. The knowledge base contains the available expert knowledge, which can be represented in different ways. So far, the most common representation is by production rules. In the case of production systems, the knowledge base has three components: the data-base (containing data structures), the rule base (containing the production rules), and the control knowledge base (containing rules that control the inference process).

The inference mechanism uses the KB knowledge to construct the reasoning that leads to a conclusion. The explanation module supports the user with all the information that is required to understand the system's reasoning. The user interface is involved in explanation and also supports dialogue between the user and the system.

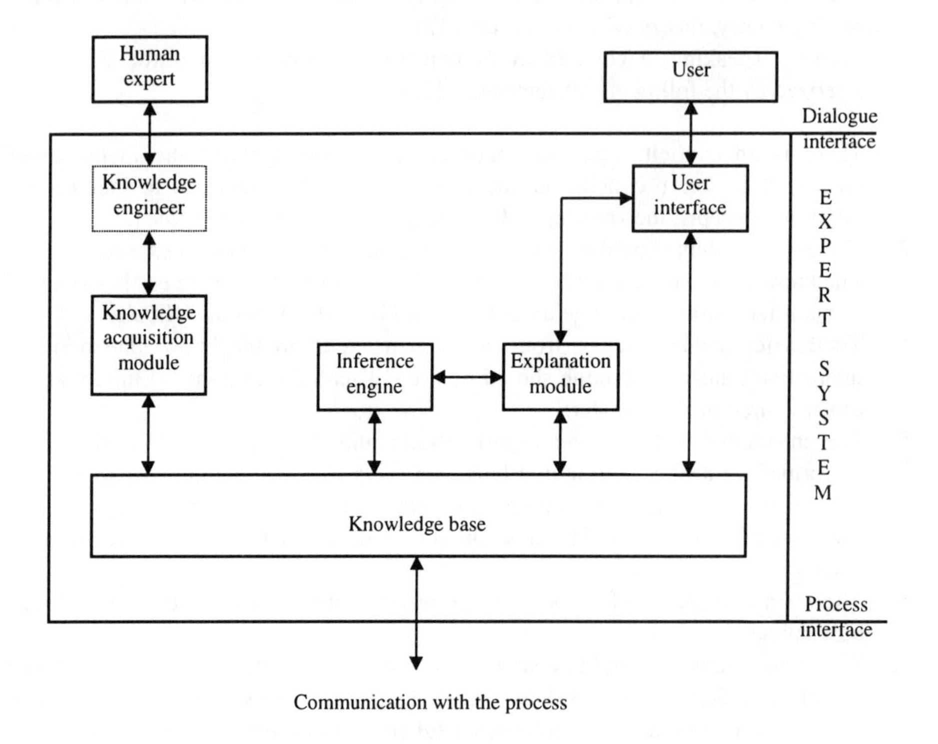

Figure 1. The structure of an expert system.

The theory of fuzzy sets has been largely involved in the development of production systems. This "fuzzification" involved the usage of fuzzy production rules [26] and, generally, the use of fuzzy predicates within the knowledge acquisition module [27] and for the knowledge representation [28], including the control knowledge [29]. Different semantic considerations can be taken into account in the representation of production rules and/or metarules. Their action is similar to the one of hedges, and different types of rules can be represented, modifying the membership function of the consequence. Dubois [24] proposed limitation of the membership degree, resulting in the possibility and certainty rules. He also proposed gradual rules that can be represented either by exclusion (the elements that have a small membership degree are excluded from the support of the consequence) or by maximization (the membership degrees that are larger than a specific threshold are maximized).

In the next section, we briefly present a fuzzy approach, called *diagnosis fuzzy system* (DIFUS) [30], [31]. We developed it as a general fuzzy diagnosis environment and used it to build a myocardial ischemia diagnosis tool. In Section 4, we discuss the principles of the multimethod diagnostic technique. It consists of five different diagnostic techniques that present different degrees of interdependence and rely on the information provided by the most widely used test in the evaluation of patients with coronary heart disease: the exercise stress test electrocardiography (EST-ECG). In Section 5, we present the implementation of score-based tests – a major component of the medical method we used. The next section is devoted to the presentation of the multimethod myocardial ischemia diagnosis system (MMIDS), including its structure, operation, and the main clinical results obtained so far. Finally, some conclusions will be presented.

3. DIFUS – HIERARCHICAL DIAGNOSIS FUZZY SYSTEM

The DIFUS system was designed as a hierarchical fuzzy approach to medical diagnosis [30]. Technically, it is used as a development tool for different diagnostic applications. It has a specific internal structure that is based upon a hierarchy of groups and reflects knowledge granularity. Structurally, it is based on some elementary diagnosis units (cells) that, depending upon the type of knowledge, are fuzzy systems, neural networks, neuro-fuzzy networks, and other types of hybrid systems or even simple crisp or fuzzy analytical expressions. This flexibility allows DIFUS to include and effectively handle different types of knowledge.

3.1. Characteristics

The main characteristic of DIFUS is that the universe of discourse of the output variable represents the possibility degree of a certain diagnosis, and not the diagnosis itself. When it is supposed to perform a diagnosis regarding the presence or absence of a specific disease, the output universe of discourse consists of fuzzy sets ranking from *impossible* to *very possible* [2] (*extremely possible*). In systems such as MILORD [32], such degrees represent linguistic labels assimilated to truth values within a multivalued logical system [29], whereas in DIFUS they form a *family of fuzzy sets* over the output universe of discourse. The latter approach has several advantages. Indeed, the similarity between the inferencial process of the expert system and human reasoning is higher (attribute no. 2 – refer to Section 2); moreover, it can be easily understood by the user (attribute no. 5) and, implicitly, the explicative module requires minimal additional processing. Most important, the expert knowledge needs minimal or no processing at all before being included into the knowledge base (attribute no. 6).

There are other characteristics that increase DIFUS performances in terms of flexibility, capacity to embed different types of knowledge, and of diagnosis accuracy. These structural and functional characteristics, although not so apparent, accentuate the difference between DIFUS and other fuzzy approaches like, for example, the well-known CADIAG-2 [6], [20]. They refer to the way DIFUS deals with different types of knowledge involved in a diagnosis; the uncertainty related to the observed symptoms, the relative importance of different signs, symptoms, syndromes, disease stages, and diseases, as well as the dependencies existing among them, and the accuracy of the knowledge embedded into the knowledge base. These features mainly result from the way knowledge is organized within DIFUS.

3.2. Knowledge Organization

The knowledge that is involved within the medical diagnosis process is heterogeneous, both in its meaning and in its form. It may refer to very different things such as signs or symptoms, patient's lifestyle, or heredity. It may be expressed in different ways such as crisply or fuzzily, numerically or linguistically. In order to build a sound diagnostic tool, the knowledge base should be dynamic (attribute no. 3), i.e., it should be easily controlled, improved, or extended. Therefore, the first thing to be done is to organize this heterogeneous knowledge.

In our approach, the knowledge is organized in two steps. The first, and the most important one, takes into account *the meaning of the knowledge*. The hierarchical structure of the diagnostic tool is the direct result of this first step. In the second stage, the knowledge is further organized according to its form, subsequently resulting in the implementation of the individual modules pertaining to the knowledge base. This second step has no influence upon the hierarchical structure of the system. In the following, we will focus on the crucial stage of the DIFUS development: the organization of the knowledge taking into account its meaning. Thus, inside DIFUS, the knowledge is classified into two categories: *basic knowledge* (BK) and *relational knowledge* (RK). The classification is done regardless of the way knowledge is expressed (mathematically, linguistically, mixed [33], [34]) and regardless of its source (experiments, tests, human experts, etc.). Within the *basic knowledge* (BK) class is included all the knowledge that is related to the basic elements of the diagnosis: signs, symptoms, results of various methods of clinical investigations, laboratory tests, the medical history of the patient, her/his life style and heredity, and clinical evolution. The *relational knowledge* (RK) class contains all the knowledge referring to any relation existing among different groups of basic elements or between an element and one or more groups of elements. Relational knowledge may also refer to relations existing between a basic element and a certain diagnostic, other than the indication of that diagnosis. The most common of these is the specificity of the element. We may note that the diagnosis may refer not only to a single disease, but also to its stage, and to a whole family of diseases.

In order to build the modules inside the knowledge base, a more refined organization within each class is necessary. First, all the BK regarding a certain basic element will form an *individual knowledge base*. This will be embedded in what we call a *diagnosis cell* (DC) or elementary cell [30]. Again, this classification is done regardless of the way in which knowledge is expressed. Such an individual knowledge base may contain linguistic as well as numerical (crisp and/or fuzzy) knowledge, experts' statements, and experimental data.

Similarly, *individual RK bases* will be formed according to the elements (groups of elements) the relational knowledge refers to. Thus, the elements among which there are some known relations are grouped together. The respective knowledge will form the individual RK base of the group. These RK bases may also include other types of knowledge, such as relations between a specific element and the diagnostic or knowledge that controls the inference. They represent the key points in the development of the system hierarchical structure.

The RK organization within the individual bases should be flexible. It is important to allow an element to belong simultaneously to different groups. This implies that the relational knowledge should be split into different RK bases, although they partly refer to the same elements. For example, blood tests may be involved in the diagnosis of different diseases and, implicitly, they will be related to elements organized in different RK bases. This flexibility is important for diagnosis accuracy, but the automation of the knowledge organization becomes considerably more difficult. Within the RK bases the knowledge can also be expressed in different ways and obtained from different sources (experts, statistics, tests, etc.).

Before concluding this section, we should note that, although relational knowledge can take various forms, practice has demonstrated that usually it is not very comprehensive and is simply and typically expressed. The most usual types of relational knowledge are [31]:

RK1. The relative importance or the relevance of an element pertaining to a certain group. This is generally stated either by linguistic labels as: *not too important, quite relevant,* or by numerical – crisp or fuzzy – scores.

RK2. The confidence degree of the result given by a diagnosis cell or group. The confidence is diminished by the inherently limited accuracy of the inputs (observations, laboratory test results, measurements, and clinical investigations) and/or of the BK base itself. The confidence degree is usually given by linguistic labels as *good accuracy, low precision, highly reliable, rough approximation*, etc. Sometimes it may be numerically expressed, as is the case of automatic measurements for which the equipment allowance is specified.

RK3. The relevance. It represents the degree in which the presence or absence of a certain element (symptom, sign, test, etc.) is relevant to a certain diagnostic (disease, stage of a disease, family of diseases). For example, the relevance of an element may be given by the frequency of its occurrence – which is equivalent to the measure of being typical for that diagnosis, or by its specificity which represents the measure that exclusively confirms that particular diagnosis. RK3 differs from RK1 by referring to absolute instead of relative measurements of the relevance. Some RK3 type results may be derived from RK1 types. For example, depending upon the particular application, symptoms that are more frequent might be considered as more important, or less important (as is the case of a diagnosis within a family of diseases having typical common symptoms). The relevance is commonly characterized by statistic indicators [20], [35]. These can be numerically or linguistically expressed (*always, often, sometimes*).

RK4. The dependencies among the elements or combinations of elements within the group. This type of relational knowledge is usually given by linguistic labels as *strongly related, independent*, and *always accompanied*.

In order to deal with these common types of relational knowledge, we propose a computation scheme that will be described in subsequent sections.

3.3. Structure

The aim of DIFUS is to offer a comprehensive and friendly environment for the development of concrete applications. All the applications that are developed with DIFUS share, in principle, the same hierarchical structure, but the structural details, i.e., the number of groups, and the number of elements/group, differ from one application to another and has to be carefully designed. This is done by taking into account

- the hierarchy of elements
- the individual relational knowledge bases.

In medical diagnosis, the hierarchy of elements generally presents small variations for a wide range of applications. Perhaps the most usual hierarchy is

(0) symptoms and signs→ *(1) group of symptoms*→
→ *(2) disease*→ *(3) family of diseases*

Differences may occur when multiple methods of diagnosis are simultaneously considered [34]. A *(diagnosis) method* layer, situated between layers *(1)* and *(2)* should be considered. In addition, a *pre-processing* layer might be used to prepare the inputs of a diagnosis cell (DC).

Within DIFUS, every layer is composed of modules. The internal structure of each layer, i.e., the modules that pertain to each layer, is defined by the individual relational knowledge bases.

There are three types of modules:

- **diagnosis cells**
- **output computation cells**
- **groups**

Each layer may contain all three types of modules. Diagnosis and output computation cells are contained by groups.

Diagnosis cells (DC's) are the elementary modules. A DC contains a unique individual BK base (BKB). It is in the DC where the inference is performed, according to the BKB. DCs may belong to any layer. Their inputs can be directly taken from the system's interface. They can also be outputs of some groups belonging to the lower layers, as well. The DCs that are related to each other, i.e., they are referred within the same RK base, are clustered into *groups.* A group contains at least two elements and has a unique RK base. Its elements cannot be only DCs but also groups belonging to a lower layer. The output of any group is given by an *output computation cell* (OCC). The OCCs embed and exploit all the relational knowledge available for the group they belong to.

The groups pertaining to the first layer of the hierarchy are called primary groups (PGs). They are composed exclusively of DCs and the corresponding OCC. At this level, the organization of relational knowledge – and implicitly the organization of PGs usually reflects natural criteria like heart symptoms, kidney symptoms, patient medical history, her/his clinical evolution, and heredity. Any PG output may be further related to other PG or DC outputs thus forming a secondary group (SG). The chain may continue with upper level groups (3G, 4G, etc.). Starting from the secondary layer, the elements of a group can be DCs and/or groups from any of the inferior layers. An example of a DIFUS hierarchy is depicted in Figure 2 (from [31]).

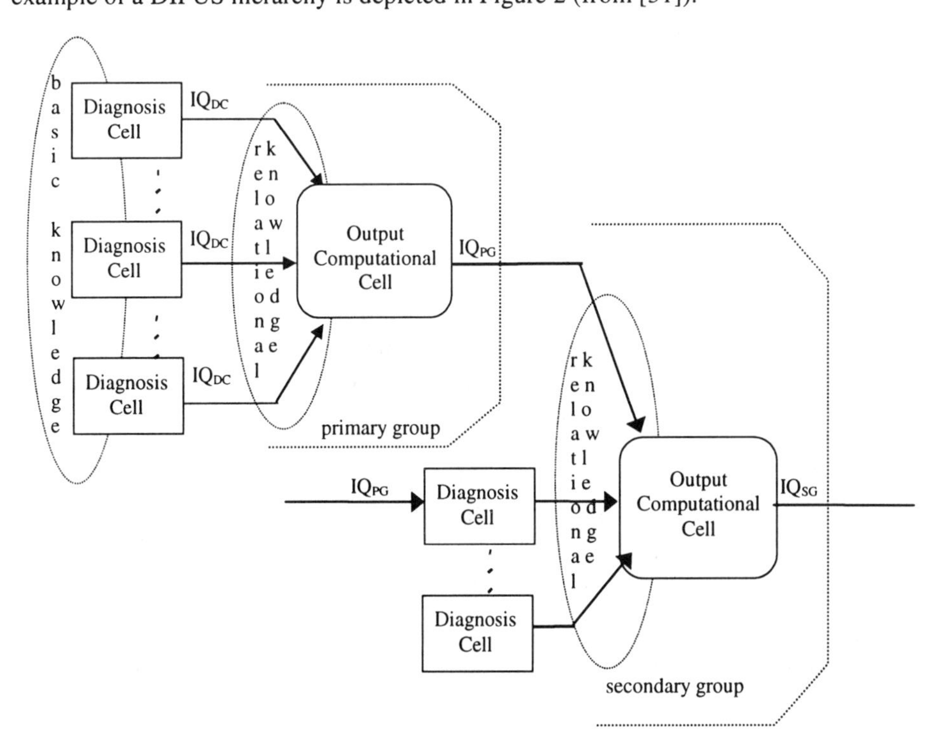

Figure 2. Example of a DIFUS structure. (From Ref. [31], ©IEEE 1997. With permission).

We should note that the internal structure of the groups may differ considerably, even when they belong to the same layer. For example, a disease (ternary group) could be diagnosed by means of a three-layer hierarchy (symptoms, group of symptoms, diagnosis methods), while another disease (another ternary group), belonging to the same family, might require only a two-layer hierarchy (symptoms, groups of symptoms). We point out again that, in order to set the hierarchical layers, one has to take into account only the meaning of the knowledge and not its source, nor the way it is expressed.

3.4. Operation

In all fuzzy approaches to medical diagnosis, the result of the diagnosis process is represented by the degree in which all the available knowledge indicates one particular diagnosis. In DIFUS, this degree is a *fuzzy set*. It is called the *illness quotient* (IQ). Because IQ usually results from a chain of fuzzy inferences, it is frequently a non-normal and, possibly, multi-modal fuzzy set. Each module from the DIFUS hierarchy gives its own indication regarding the diagnosis. Referring to the example that is depicted in Figure 2, IQ_{DC}s represent the degrees in which different symptoms presented by a patient indicate a particular disease. IQ_{PG} is the indication of the whole group of symptoms, while IQ_{SG} represents the degree in which the patient might have that disease, taking into consideration all the available knowledge.

From the implementation point of view, the three types of modules have different requirements. Groups, being merely organizational modules, are specified by the relations existing between the outputs of DCs, lower level groups and the inputs of the corresponding OCC. DCs and OCCs implementation is directly dependent upon the way in which the knowledge (basic and relational, respectively) is expressed. This is the point where the form of the knowledge is relevant, but not its meaning. Thus, if BK (RK) was stated in the terms of fuzzy IF-THEN rules, the DC (OCC) should be implemented by a fuzzy or some hybrid (neuro-fuzzy) system. In this case, the output of the cell (IQ_{DC}) is obtained by performing fuzzy inference upon the IF-THEN rules contained by the cell's knowledge base. The diagnosis cell delivers its output IQ_{DC} to the OCC cell, where it will be processed together with the other related IQs (as we will see in the following), in order to obtain the indication of the whole group. Then, the OCC will deliver the group output to the higher level OCC and so on. A DC or an OCC can be implemented not only a as fuzzy system, but also as a neural network or even simple fuzzy or crisp mathematical expressions.

The implementation of the output computational cells deserves a closer look. As we have already mentioned, in many practical cases the relational knowledge is given by some statistically derived indicators and/or linguistic labels. In these situations, neither fuzzy nor neural/hybrid systems are necessary. The OCC has only to deal with the relative importance (RK1), confidence degree (RK2), relevance (RK3), and/or dependencies among the elements (RK4). For these types of knowledge, we propose a computational scheme that is composed of three operations; namely,

- IQ weighting
- IQ expansion/compression
- IQ aggregation

The implementation of the OCC is very simple, as its action is limited to one or more of these operations. They will be described in some detail in the following.

- ***IQ weighting***

This rather typical operation corresponds to relational knowledge of types RK1 and RK2. It can be applied when the knowledge is numerically or linguistically expressed, fuzzy or crisp. The linguistic labels or the numerical data that describe the relative importance of the elements, and/or different confidence degrees, are applied to the corresponding IQ as weights. As we have previously stressed, IQs are usually obtained by fuzzy inference, and are thus frequently defined by irregular membership functions. The application of a fuzzy weight requires many computations, and is not justified by the slight increase of the diagnosis accuracy. For this reason, we prefer to assign to each linguistic label a crisp numeric value belonging to the unit interval. This assignment is done according to a predefined scale. On this scale, the modal values of all the fuzzy sets corresponding to linguistic labels (weights) are defined. The assigned value (crisp weight) represents the modal value of the corresponding fuzzy set (*important, less important*, etc.). This crisp value is used to weight the IQ by product or max operators. If no weights are available within the RK base, the default values, which are set to 1, are considered.

- ***IQ expansion/ compression***

This operation corresponds to the RK3 class of knowledge (relevance). By this operation each IQ may be expanded or compressed on its universe of discourse. We remark here that the output universe of discourse (on which all IQs are defined) contains the degrees in which a certain diagnosis is applicable to the patient. In order to explain this choice for the implementation of RK3 knowledge, we should first note that in this case we generally have to deal with a number of statistically derived indicators. DIFUS uses two such indicators: **H** and **I**. **H** represents the average degree in which the patients that were proven not to have the disease ('healthy patients') manifested a particular symptom at the diagnosis moment. **I** represents the average degree in which patients proven to have the disease ("ill patients") manifested that symptom.

For practical reasons, the patients were crisply classified as having or not having the disease or, more generally, manifesting or not the subject of the diagnosis (which might be, for example, a disease stage). Thus, **H** and **I** are computed as SCounts of all the IQs inferred for healthy/ill patients (defuzzified values), divided by the number of healthy/ill patients. Nevertheless, different choices can be made. For example, a third indicator, corresponding to the patients with inconclusive diagnosis, could be computed. On the other hand, **H** and **I** could be implemented as fuzzy numbers instead of crisp ones, but it is doubtful that the increase of the diagnosis accuracy justifies the increased computational load.

All of these statistical indicators have to be included into the application database. Also, the IQs corresponding to each patient who was diagnosed by the system should be included. All IQ values might be stored (generally after defuzzification), or only those that contribute to **H/I** computation. The database is automatically updated whenever a patient obtains a final (confirmed) diagnosis from the physician. After **H/I** updating, the IQ values of that patient can be removed from the database.

The initial values of the **H** and **I** indicators have to be derived during the knowledge acquisition process. Numerical or linguistic information regarding the frequency of symptoms' occurrence, as well as their specificity, is generally available. From the definition of the **I** indicator, we note that it represents a general form of the occurrence frequency of the associated symptom. If we take the crisp case, when a symptom is present (IQ = 1) or absent (IQ = 0), **I** equals the ratio between the number of the 'ill patients' that presented the symptom and the total number of the ill patients. So, if the "nuances" regarding the presence of a symptom were excluded, **I** equals its occurrence frequency.

The more a symptom discriminates between the persons that suffer from a certain disease and the ones that are free from the disease, the more it is specific to that particular disease. Introducing nuances, we can say that the more the degree to which a symptom is presented by 'ill patients' exceeds the degree to which it is presented by the 'healthy', the more it is specific to that disease. This means that the difference **I–H** gives a fuzzy measure of specificity.

From the available information regarding the frequency of symptom occurrence, the initial value of **I** is derived. When in the same group the specificity is also known, **I–H** takes this value and then **H**[*)] is computed. If no such knowledge exists, the initial values of **H** and **I** are set to 0 and 1, respectively. When this information is expressed linguistically, a numerical value is assigned to each linguistic label according to a predefined scale. The initial values of **H** and **I** are then updated as our database is developed. Thus, the system learns from experience (attribute no. 11).

Indicators such as **H** and **I** have to be included in the knowledge base of any medical diagnosis system. In our approach their meaning is enlarged, to allow the gradual presence of signs and symptoms. Another truly specific characteristic of the DIFUS system is the way in which these indicators are involved in the diagnosis process. *They act upon the corresponding IQ by compressing or expanding it on the output universe of discourse.* Compression/expansion is simply achieved by a variable changing, according to relation (1)

$$x' = \frac{x - H}{I - H} \tag{1}$$

[*)] The CADIAG-2 indicators (occurrence and confirmation) are equivalent to these initial values of **I** and **I–H**, respectively.

Although it seems similar to the action of aggravating factors [36] or to gradual rules [27], compression/expansion is different not only formally, but also in principle. Its aim is to adapt the indication of the module (IQ) to the particular case of that module. For example, a presence of a symptom in the degree 0.3 (IQ = 0.3) may signify a normal health state, while in the case of other symptoms, the same degree may indicate a possible presence of the disease. The traditional way to solve this problem would be to adapt each term set of the output variable of the modules (IQ). These term sets are composed of linguistic labels as *normal, quite possible, possible, very possible*, etc. The adaptation can be done by scaling the whole term set between **H** and **I** (instead of 0 and 1). For DIFUS this approach is not practical because, on the next hierarchical layer, IQs coming from different groups and/or DCs have to be aggregated. If each IQ was referred to its own scale, the aggregation would become a difficult task.

From the DIFUS operation point of view, a much more practical approach is to use, in all modules, the same term set as the output variable IQ. It is composed of fuzzy sets ranging from *normal* to *extremely possible (ill).* Different modal values of *normal* and *ill* can be chosen. They can be 0 and 1, suggesting a possibility degree, or 0 and 10, suggesting a mark (see Figure 7), or 0 and 100, suggesting a percent, etc. *The adaptation to the module-specific values of normal and abnormal is performed by expanding/compressing the inferred IQ and not the term set.*

For the example mentioned above, in the group where the **H** indicator is 0.3, an IQ = 0.3 (modal value) should indicate a normal state, in spite of the fact that 0.3 is the modal value of the fuzzy set *relatively possible*. Applying the compression/expansion operation relative to **H**, the IQs modal value will become 0, the value that corresponds to a *normal* indication, according to the IQ term set. Actually, the inferred IQ is not only translated on the universe of discourse, but, depending upon the difference between **I** and **H**, it is expanded or compressed. During the system operation we may have different situations:

- If 0< **I-H** <1 — IQ is expanded.
- If **I-H** >1 — IQ is compressed.
- If **I-H** = 1 — IQ is translated on its universe of discourse (for $H \neq 0$).
- If **H** = **I** — the symptom is not conclusive.
- If **H** > **I** — the knowledge base is unrealistic.

Possible errors within the knowledge base may be detected by checking the last two conditions.

• *IQs aggregation*

This operation corresponds to the RK4 type of knowledge (dependencies among elements). This type of knowledge is usually expressed by statements such as, "the method/symptom *a* is totally independent from method/symptom *b*, while one method/symptom (*c*) is relatively strongly related to *b*" (see also next section). This type of knowledge is taken into consideration *by taking various fuzzy unions of IQs.*

Thus, in case of strong dependencies a hard union is performed,

$$a \text{ or } b = \max(a, b) \tag{2}$$

For independent modules (symptoms, groups of symptoms, diagnostic methods), a soft union is applied

$$a \text{ or } b = \min(1, a + b) \tag{3}$$

Any of the union classes [37] can be used. For example, if we take the Yager's union class (4), larger values of the parameter w reflect increasing degrees of dependency:

$$a \text{ or } b = \min\,[1, (a^w + b^w)^{1/w}]\ ;\ w \in (0,\infty) \tag{4}$$

When no dependencies are explicitly contained in the knowledge base, the aggregation still has to be performed to obtain the group output. In this case, the hard union is the usual choice.

The control strategy within DIFUS is implemented according to the hierarchical layer of the module. For upper layers (third layer or higher), the control strategy is described by metarules. Each group has its own control knowledge base (CKB). The first step of the diagnosis is the inference in the highest layer group CKB. Its aim is to decide the order in which the inference within the modules that compose the group will be performed. The inferences in the CKBs pertaining to lower layer groups is then performed in the same sequence, determining the inference order within each group. Once the upper layer groups are organized, the inference of the first layer BKB's may start. The third layer (family of diseases) is the lowest layer having a CKB. In primary and secondary groups, forward reasoning is performed.

A diagnosis is possible even if the input data are not complete, i.e., some tests were not performed or measurements are missing. For this, the default values of all input variables should be set to produce no activation of the rules in which they are involved. The diagnosis is less thorough, but it is still correct in relation to the available knowledge (attribute no. 9).

After the computation of all the outputs belonging to a certain layer, the user may choose to display one or more of those IQs. The software provides the user with a graphical representation of the membership function of each output within the group (represented as a singleton if it is crisp). The defuzzified value of each output (by MOM, COA, or other methods) is also available. A linguistic label, provided by a linguistic approximator, can be attached to each output, too. Another option of the user interface is to forward the indication to the next layer. In forwarding, the overall outputs of all groups belonging to that layer are computed.

4. Multimethod Myocardial Ischemia Diagnosis

Although we are aware of numerous risk factors related to heart disease, such as smoking, high blood pressure, age, gender, hyperlipidemia, diabetes mellitus, and positive family history, ischemic heart disease remains the leading cause of death in U.S.A. and most European countries. It is known that almost twenty percent of patients surviving after a myocardial infarction previously had no symptoms. The various patterns of symptoms which symptomatic ischemic heart disease may manifest make the diagnosis difficult and, in some cases, highly uncertain.

One of the most important myocardial ischemia diagnostic tools is the stress exercise ECG. Since it represents a non-invasive procedure, is easy to perform, and requires equipment of moderate cost, the exercise stress test remains a highly valuable diagnostic tool for the evaluation of patients, with both suspected and proven ischemic cardiac disease. Although commonly performed, an accurate diagnosis based on the exercise stress test is often difficult and requires a great deal of expertise.

In order to increase the accuracy of the diagnosis, and obtain a more global perspective of diagnostic criteria for ischemia during the exercise stress test, a multimethod diagnostic technique has been developed. This technique takes into account different parameters such as symptoms, total time of the exercise test, percentage of target blood pressure, and the heart rate. It provides more complex evaluation of ECG signs that include ST segment depression, and QRS wave and ST/heart rate alterations.

In the case of multimethod myocardial ischemia diagnosis there is a factor that strongly favors a computer-assisted approach: the modern stress testing equipment continuously computes and displays status information including heart rate, blood pressure, test time, ST segment alterations, etc. during the test. All the parameters, including the ECGs, can be easily and automatically acquired by the expert software.

We have already mentioned in the first two sections of this chapter the advantages of a fuzzy approach to computer assisted medical diagnosis. Here we present an example of such an approach in order to create a multimethod myocardial ischemia diagnostic tool, based on the exercise stress test. We selected five exercise test diagnostic techniques that are well established in literature and clinical practice.

I. *Treadmill score (Kansal's score)*

This is a score-based test. In the classical (non-fuzzy) approach a score higher than 7 points is indicative of ischemia. The score is computed by taking into account the following parameters: a) age > 55 years = 3 points, b) exercise time < 14 min = 3 points, c) heart rate achieved < 80% of age-predicted = 9 points, and d) max. ST depression >1 mm = 6 points.

II. *ST/Heart rate recovery loop*

A clockwise loop is considered normal, whereas an anti-clockwise loop is indicative of ischemia. Ill-defined loops are considered as inconclusive tests.

III. *The multivariate method*

This is a more complex method, which takes into account the degree of ST depression (e.g., < 1 mm, 1-1.9 mm, ≥ 2 mm) in relation to various parameters, including exercise time, maximum heart rate, the double product (heart rate × blood pressure), and the persistence of ST depression in recovery.

IV. *ST/Heart rate adjustment*

This method analyzes the ratio

$$\frac{STmax - ST\ resting}{max\ Heart\ rate - Resting\ Heart\ rate} \tag{5}$$

The greater the value of (5), the stronger the indication of ischemic heart disease. Typically, a value over ≥ 0,013 mm/beats/min indicates a very possible presence of the disease.

V. *Athens QRS score*

QRS score accumulates the Q, R, S deflections in a VF and V_5 leads during rest and at maximum exercise. A score over +5 indicates a negative test while a score lower than –5 is strongly indicative of severe coronary disease.

These methods present various degrees of interdependence [16], [38-41], which has to be taken into account along with their varying dependence on the classic ischemia criterion, that is the ST segment depression. The Athens QRS score (method V) is totally independent from ST alteration, while the Treadmill score (method I) is strongly related to this criterion. The other three methods rely heavily on ST segment changes. The sensitivity of each method was accepted as the average value found in literature. We consider 81% for Treadmill score [16], [41], [42], 71% for ST/HR recovery loop [41], 70% for multivariate method [43], 75% for ST/HR adjustment [44], and 85% for Athens QRS score [38], [45].

5. Multimethod Myocardial Ischemia Diagnosis System

The multimethod myocardial ischemia diagnosis system (MMIDS) is implemented as a DIFUS application. Its structure contains two layers. The upper layer consists of a secondary group, whose output represents the degree in which the evaluated patient is likely to have myocardial ischemia. This group is composed of five modules. Every module corresponds to a diagnostic method. Because three of these methods rely on a score computation, we will start the presentation of MMIDS system with some considerations regarding the implementation of (fuzzy) score-based tests.

5.1. The Implementation of Fuzzy Score-Based Tests

The score-based tests (SBT) are rather common structural elements of the medical diagnosis processes. They are composed of a number of elements (subtests), to each of which a score is assigned. In classical diagnosis, the scores are represented by integers (usually positive). A set of integers (scores) can also be assigned to a subtest. It corresponds to a finite set of degrees of satisfaction of the subtest.

SBTs are fuzzy if they involve elements of the fuzzy set theory, such as fuzzy predicates or fuzzy scores (fuzzy numbers), and fuzzy inferences. We should note that in fuzzy diagnoses the scores can be fuzzy as well as crisp numbers. Generally, the scores are numerical. However, as we will see, there are situations in which some linguistic labels perform the same action as the numerical scores. During the diagnosis process, the appropriate score (*partial score*) is assigned to each subtest by means of some fuzzy processing (fuzzy inference, fuzzy arithmetic operations, etc.). Then the partial scores are added, resulting in the final score of the test (*total score*). The total score (TS) may itself represent the output of the fuzzy diagnosis, or – as is the case of MMIDS system, another indication may be taken into account for the systems output. In the latter case, the correspondence $TS \rightarrow$ *system output* has to be implemented as well (Figure 3).

There are several possible implementations of fuzzy SBTs. They can be compared according to various criteria such as the accuracy of the diagnosis, the transparency of the diagnosis process, or their computational requirements. In the following we will briefly overview a simple implementation choice introduced by Sanchez [27], and the straightforward generalization (according to the extension principle) of the classical implementation. We will also present the compact representation of SBTs, [46], [47], an implementation choice we derived by means of fuzzy graphs. For a more detailed discussion, see [46], [47].

5.1.1. Medical Patterns

One possibile fuzzy implementation of SBTs is to use simple *medical patterns* [27]. We will denote by $ST1, \ldots, STn$, the subtests that form the SBT. The fuzzy value VAL_i for which a maximum score is given (for which the subtest ST_i is considered completed) represents the characteristic value of the subtest. The output of the system will be given by:

$$DIAG = VAL_1(w_1) \text{ or } VAL_2(w_2) \text{ or } \ldots \text{ or } VAL_n(w_n) \quad (6)$$

where $VAL_i(w_i)$ denotes the (fuzzy) characteristic value of ST_i weighted by w_i.

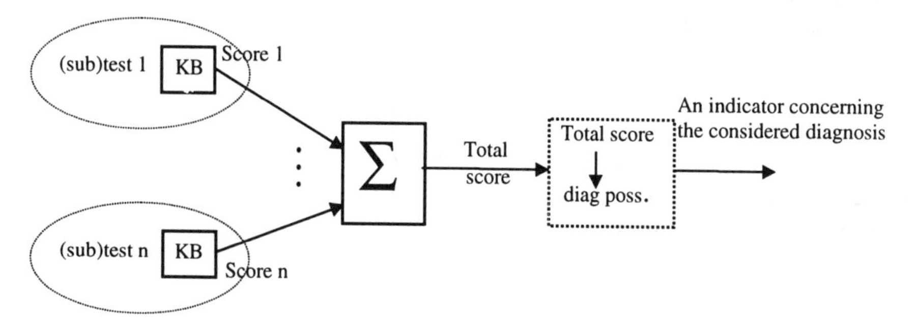

Figure 3. Diagnosis by means of score-based test.

The weights $w_i = \dfrac{Partial\ score_i}{\sum_n Partial\ score_i}$ reflect the relative importance of the subtests.

For a given input fuzzy vectors: SYMPT = [$SYMPT_1$ ..... $SYMPT_n$], the output of the system (DIAG) is given by:

$$DIAG = (SYMPT_1 \cap VAL_1(w_1))or \ \ .... \ \ or(SYMPT_n \cap VAL_n(w_n)). \qquad (7)$$

In this implementation, we do not actually have a summation of scores. Still, if a union operator like $T_\infty = \min(1, a+b)$ is chosen, the summation of the weighted (scored) degrees of subtest completion is obtained. Different methods of weighting can be used to compute $VAL_i(w_i)$.

The main advantage of this implementation is that it requires minimum computational time. The price paid for this advantage is the poor accuracy of the diagnosis. Therefore, its applicability is limited to rough (primary) diagnosis like simple telemedical assistance.

5.1.2. Sequential Processing

Another possible approach is to implement fuzzy SBTs sequential, as in the classical case (Figure 3). This means that partial scores have to be computed and then added. Nevertheless, each subtest has to be implemented as a fuzzy system (ST_i) that performs inference upon its inputs. The fuzzy output of each subtest (*Score$_i$*) is delivered to a (fuzzy) "adder". The resulting score (*Total score*) can be further processed in order to obtain the desired diagnostic indicator, as depicted in Figure 4.

If the scores are crisp, ST_i are Takagi–Sugeno systems of zero order [48]. The outputs *Score$_i$* will be crisp and their summation is a trivial task (no extension principle being involved, of course). This implementation is particularly suitable for medium accuracy diagnosis, when the different diagnostic possibilities are well separated. If we refer to hierarchical structure of DIFUS such implementation is recommended for ternary groups and above.

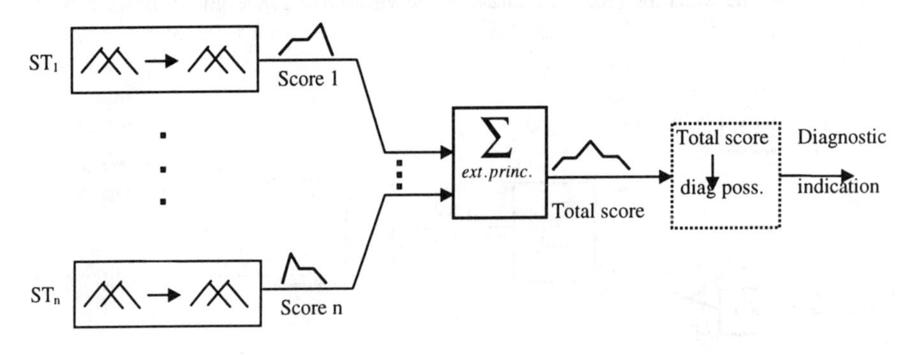

Figure 4. Fuzzy SBT with sequential implementation.

When the scores are fuzzy, the ST_i systems are 0-order Takagi–Sugeno systems with fuzzy consequent rules (FC-TS systems). Briefly, FC-TS systems [33], [34] are Takagi–Sugeno type systems that operate with fuzzy IF-THEN rules, whose consequents are *fuzzy analytical expressions*. That is, these expressions (usually functions) involve at least one fuzzy element (parameter, argument, etc.). We have discussed the inference mechanisms, rule weighting procedures, and some practical aspects in another paper [34]. Generally, an SBT using sequential implementation provides remarkable diagnostic performance. It certainly is the best choice for a large range of applications. The diagnostic process is also very transparent. We call an SBT *fully transparent* if all the partial scores are "visible" for the user. It is easy to observe that in this implementation, the SBTs are fully transparent.

The main disadvantage of this approach is the presence of the fuzzy "adder". In the case of fuzzy scores, it has to perform – according to extension principle – the summation of n fuzzy sets. Because the outputs $Score_i$ result from fuzzy inferences, their membership function may have very irregular shapes and, consequently, their summation is an extremely time-consuming task. The most attractive possibility for solving this problem is to approximate the inferred scores by triangular or trapezoidal fuzzy numbers (TFN/TrFN). The approximation itself has modest computational requirements and the summation becomes much simpler. The cost of this solution is a certain loss of diagnostic accuracy. However, different approximation techniques that minimize this loss can be taken into consideration.

There are, however, applications where disputable decisions have to be taken, i.e., the separation between diagnostic possibilities is very tight. This does not necessarily happen on the whole universe of discourse, but rather in certain critical regions. When the system inputs lie within these regions, the approximation of partial scores by TFN/TrFN may lead to costly incorrect decisions. For these cases, it is necessary to have an implementation that avoids the *on-line* summation of the partial scores (which is prohibitively time consuming) without using their approximation. Subsequently, we will derive this implementation by means of fuzzy graphs.

5.1.3. Compact Representation of Fuzzy Score-Based Tests

Zadeh [49] showed that any rule base containing n fuzzy rules "IF A_i THEN B_i" can be equivalently represented by the fuzzy graph [49], [50]

$$f^* = A_1 \times B_1 + A_2 \times B_2 + \ldots + A_n \times B_n = \sum_{i=1}^{n} A_i \times B_i \qquad (8)$$

We denoted by $A_i \times B_i$ the Cartesian product of A_i and B_i and by + the union operation. If X and Y are linguistic variables in **X** and **Y**, respectively, then for any $x \in Y$ and $y \in Y$ the membership degree of (x, y) to the graph f^* is

$$\mu_{f^*}(x, y) = \vee_i (\mu_{A_i}(x) \wedge \mu_{B_i}(y)) \qquad (9)$$

where $\vee$, $\wedge$ denote a t-norm and a t-conorm, respectively.

The main idea underlying the compact representation is to remove the tedious fuzzy summations from outside the subtests – where they have to be done on line – and to place them inside the subtests' knowledge base – where they are performed off line.

Zadeh [49] showed that two functions represented as fuzzy graphs $f(X)$ and $g(Y)$ can be combined through a binary operation (*) such as multiplication or addition. They will result in the function

$$h(X,Y) = f(X) * g(Y) = \sum_i (A_i \times B_i) * \sum_j (C_j \times D_j) \tag{10}$$

In its turn, the function $h(X, Y)$ can be represented by the fuzzy graph [49]

$$h(X,Y) = \sum_{i,j} (A_i \times C_j) \times (B_i * D_j) \tag{11}$$

On this basis we can now compute the summation of the partial scores $Score_i$. If we represent the knowledge bases of the subtests ST_i as fuzzy graphs $f_i(X_i)$, we obtain the function

$$h(X_1,\ldots,X_n) = \sum_{i=1}^{n} f_i(X_i) = \sum_{i_1} (A_{i_1} \times B_{i_1}) + \ldots + \sum_{i_n} (A_{i_n} \times B_{i_n}) \tag{12}$$

From (11) it results that

$$h(X_1,\ldots,X_n) = \sum_{i_1,\ldots,i_n} (A_{i_1} \times \ldots \times A_{i_n}) \times (B_{i_1} + \ldots + B_{i_n}) \tag{13}$$

or

$$h(X_1,\ldots,X_n) = \sum_k G_k \times H_k \tag{14}$$

We thus obtain a fuzzy graph equivalent to the following rule base:

h : IF $(X_1,\ldots,X_n)$ IS $A_{11} \times \ldots \times A_{1n}$ THEN *Total score* IS $(B_{11} + \ldots + B_{1n})$

....

IF $(X_1,\ldots,X_n)$ IS $A_{i1} \times \ldots \times A_{in}$ THEN *Total score* IS $(B_{i1} + \ldots + B_{in})$

....

The resulting diagnosis system is depicted in Figure 5. In this implementation the diagnosis consists merely of an inference performed by the fuzzy system $h(X_1,\ldots,X_n)$ upon the actual inputs $(X_1^*,\ldots,X_n^*)$.

The main advantage of this approach is that it manipulates the initial approximate knowledge without any further approximation, and within a reasonable computational time. The number of rules from the compact fuzzy system $h(X_1,\ldots,X_n)$ is $n = n_1 \times \ldots \times n_n$, where n_i is the number of values from the term set of the input variable X_i. This means that for large term sets and, possibly, for many subtests, the computational time can exceed that of the previous approach. However, it still remains well under the time required for the external summation without approximation of partial scores. One drawback of this approach is the loss of transparency, as the partial scores are hidden within the fuzzy system. The SBT will not be fully transparent, and the influence of each input to the total score cannot be directly identified by the user.

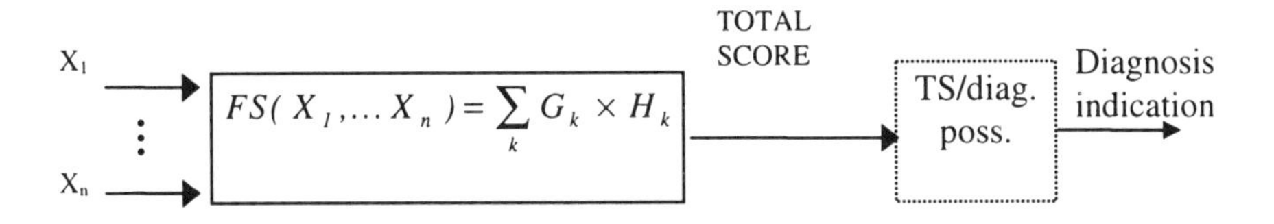

Figure 5. Fuzzy SBT represented by a unique fuzzy graph.

In order to obtain the compact structure, knowledge bases of the subtests have to be processed. This processing takes place off-line. It mainly implies the summation of the appropriate scores from the term set of the partial scores. Being generally represented as TFN/TrFN, their summation is not a difficult task.

5.2. MMIDS Structure and Operation

The MMIDS structure is depicted in Figure 6. The degree to which a patient is likely to have myocardial ischemia is obtained as the output of a secondary group (IQ_{SG}). The degree to which a certain method indicates a positive diagnosis is given by the first layer outputs.

5.2.1. MMIDS Secondary Group

The MMIDS secondary group contains five modules, corresponding to the five selected diagnosis methods. Among these, there are two primary groups and three diagnosis cells. The relational knowledge in the secondary group is given by

a. **I** and **H** indicators
b. Dependencies among elements

According to the medical knowledge described in Section 4, we have

I. *Treadmill score (TS)*
- type: primary group (fuzzy score-based test)
- implementation: sequential processing
- relational knowledge: - **I** = 0.81; **H** = 0.19 (initial values)
  - medium dependency upon multivariate method

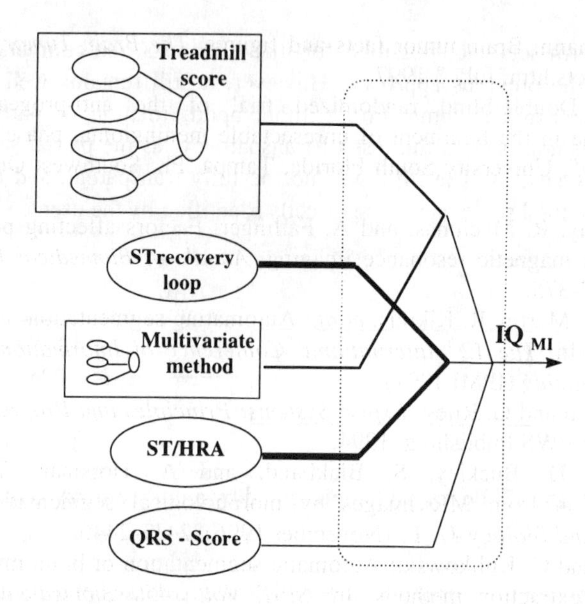

Figure 6. MMIDS structure. (From Ref. [31], ©IEEE 1997, with permission.)

II. *ST/ Heart rate recovery loop (ST)*
- type: diagnosis cell
- implementation: fuzzy system
- relational knowledge: - **I** = around 0.7; **H** = around 0.2 (initial values)
  - strong dependency upon ST/HRA method

III. *Multivariate method (MM)*
- type: primary group (fuzzy score-based test)
- implementation: compact representation
- relational knowledge: - **I** = 0.7; **H** = 0.24 (initial values)
  - medium dependency upon TS method

IV. *ST/Heart rate adjustment (ST/HRA)*
- type: diagnosis cell
- implementation: fuzzy system
- relational knowledge: - **I** = 0.75; **H** = 0.25 (initial values)
  - strong dependency upon ST/HRA method

V. *Athens QRS Score (QRS-S)*
- type: diagnosis cell
- implementation: fuzzy system
- relational knowledge: - **I** = 0.85; **H** = 0.28 (initial values)
  - independent upon the other methods

Taking into account the types of available relational knowledge, the computation scheme of the group output includes IQ compression/expansion, according to **I** and **H** indicators, and IQ aggregation (by Yager union class), according to the different strengths of dependencies among elements (illustrated by lines of different widths in Figure 5).

5.2.2. MMIDS Primary Groups

MMIDS includes two primary groups, namely TS and MM.

a) *TS (Treadmill Score)*

TS represents a score based test (SBT) which takes into account four criteria. In the DIFUS hierarchy it is implemented as a primary group consisting of four diagnosis cells, each corresponding to one criterion. These diagnosis cells are:

1. *Age (A)*
    - type: diagnosis cell
    - implementation: fuzzy system
    - number of rules: 5
    - relational knowledge: linguistic weight (*quite important*)
2. *Exercise time (ET)*
    - type: diagnosis cell
    - implementation: adaptive fuzzy system
    - number of rules: 5
    - relational knowledge: linguistic weight (*less important*)
3. *Heart rate achieved (HRA)*
    - type: diagnosis cell
    - implementation: adaptive fuzzy system
    - number of rules: 5
    - relational knowledge: linguistic weight (*very important*)
4. *Maximum ST depression (STD)*
    - type: diagnosis cell
    - implementation: fuzzy system
    - number of rules: 3
    - relational knowledge: linguistic weight (*important*)

The basic knowledge corresponding to each diagnosis cell is linguistically expressed (by IF-THEN rules). The cells *A* and STD are implemented by two fuzzy inference systems (FIS) having 5 and 3 rules, respectively. ET and HRA are implemented as adaptive fuzzy systems. The term set of ET, which is the input variable of the ET cell, is adapted by a fuzzy consequent Takagi–Sugeno (FC–TS) system [33], [34], which generates the ET_{MAX} value. Similarly, the term set of HRA, the input variable of the HRA diagnosis cell, is adapted by a 0 order FC–TS system having 7 rules. The output variable term sets of the two adapting systems ($GENET_{MAX}$ and $GENHRA_{MAX}$), as well as of the adapted HRA system, are different for male and female patients, respectively.

Since the relational knowledge in the TS group is given only by numerical weights, the output computation scheme includes IQ weighting and aggregation by summation.

b) MM (Multivariate Method)

The structure of the primary group MM is rather special. The multivariate method can be seen as a score-based test made up of three different (sub)tests: T1, T2, T3. There are some interesting characteristics of MM as SBT. The scores that are assigned to the subtests are *linguistically expressed*, and the computation of the total score is based on certain *linguistic rules*. Further, in order to obtain the diagnostic indication (IQ_{MM}), the total score is also processed according to some linguistic rules, which have an apparent arithmetic significance. Thus, we can say that IQ_{MM} is computed according to an arithmetic of words.

The subtests have the following input variables:

T1: ST depression (STD) and the product between the heart rate and the blood pressure (HR x BP).
T2: ST depression and exercise time (ET).
T3: ST depression persistence (STD persistence).

The appropriate limit values (thresholds) of the tests are: 1 mm and 9 mm (STD, low and high limit, respectively); about 2300 mmHg × beats/min (HR × BP, high limit); about 12 minutes (ET, high limit), about 3 mm, (STD persistence, low limit). From the medical knowledge, it is also known that:

a) If a single test is completed, then there is only a *faint possibility* (FP) of a coronary disease.
b) If one test is completed and another one is about 90% completed, then the evaluation given by the method is *inconclusive* (I).
c) If two tests are completed, it is *possible* (P) for a coronary disease to be present.
d) If all of them are completed, it is *very possible* (VP) for the coronary disease to be present.

From the above statements, we can observe that all subtests are equally important. No numerical scores are assigned, but each subtest introduces, at completion, a faint possibility of disease (FP). Thus, FP may be seen as a linguistic score of each subtest.

The knowledge base of each subtest could be represented by the fuzzy graph

$$T_i(X_i) = normal_i \times normal + almost\ limit_i \times almost\ FP + limit_i \times FP =$$
$$= \sum_j A_j \times B_j \qquad (15)$$

Here, X_i is the corresponding input vector. The fuzzy sets $normal_i$, $almost\ limit_i$ and $limit_i$ form the term set of the input variable corresponding to each subtest T_i. For a given input, the inference of the scores within the T_i systems is performed. These (partial) scores should then be added to obtain the total score.

Because of the critical decision region between an inconclusive diagnosis and a possible presence of the disease, we used the compact implementation for MM. From (13) we get

$$h(X_1, X_2, X_3) = \sum_{i=1}^{3} T_i(X_i) = \sum_{j,k,l} (A_j \times A_k \times A_l) \times (B_j + B_k + B_l) = \sum_k (G_k \times H_k) \qquad (16)$$

The resulting fuzzy system h that implements this SBT will have four inputs (STD, HR x BP, ET, STD persistence) and a single output (*Score*). The rule base contains 27 rules. The term set of the output variable *Score* contains fuzzy sets (H_k) that range from 3×*normal* to 3×*faintly possible*.

In order to obtain the indication of the group (IQ_{MM}), the correspondence between the inferred output *Score** and the diagnostic indication has to be determined. This is done according to the following computational rules:

faintly possible = *normal*
almost faintly possible + *faintly possible* = *inconclusive*
faintly possible + *faintly possible* = 2 × *faintly possible* = *possible*
3 × *faintly possible* = *very possible* (17)

It is worthwhile to notice that the rules (17) actually represent an arithmetic of words. The neutral element for the linguistic summation defined by (17) is the fuzzy set *normal* (corresponding to *Score* variable, not to IQ variable). This summation is functionally equivalent to fuzzy inference in fuzzy systems. Each computing rule from (17) is represented by a fuzzy relation defined on the Cartesian product of the input and output universe, respectively. The left side of the last three rules is computed according to the rules of fuzzy number addition (multiplication).

5.3. Experimental Results

The MMIDS system has been tested using data from 10 different cases. After the first exercise test, each of these 10 patients was evaluated angiographically. All patients with coronary stenosis received a percutaneous transluminal coronary angioplasty (PTCA). Within two months from PTCA an exercise stress test was performed by each patient, resulting in a total of 20 angiographically controlled exercise stress tests.

For each case, the complete physicians' diagnosis was available. Eight cases had a positive clinical conclusion; nine diagnoses were negative and three inconclusive. For each case, the physicians' diagnoses corresponding to each of the five selected methods were also available. These were expressed fuzzily (see Table 1) to assist in a comparison with MMIDS' output.

MMIDS testing had two main objectives.

- To compare MMIDS' diagnosis with the physicians' one – at secondary as well as at primary group (method) level.
- To observe the behavior of the system when incomplete knowledge is available.

All diagnosis cells and primary groups had an identical term set of the output variable IQ (Figure 7). The IQ variable is defined on the closed interval [0,10]. Here, "0" represents the modal value of the fuzzy set *normal* while "10" is the modal value of *very possible*. The term sets corresponding to the input variables were set according to experts' opinions. For variables other than IQ, the composing fuzzy sets were distributed much less uniformly over the corresponding universe of discourse. The partition was rather coarse in the case of the STD variable (three fuzzy sets), but refined in the case of the *Age* variable (9 fuzzy sets). Most of the variables, including IQ, included 5 fuzzy sets in their term sets.

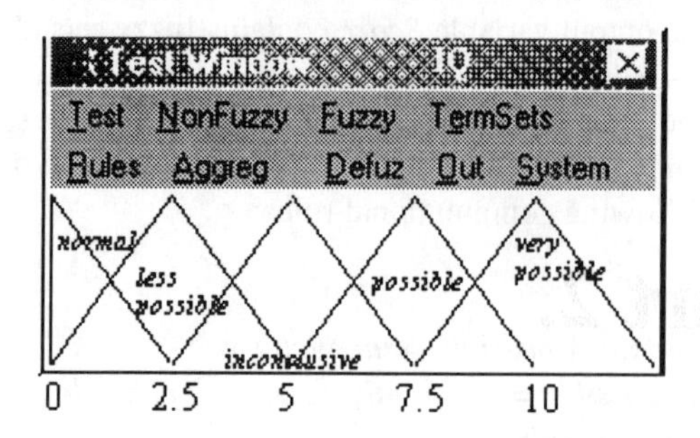

Figure 7. Term set of the output variable IQ.

Table 1
Clinical data corresponding to cases no. 9 and no. 10

PATIENT No.	PATIENT Clinical Conclusion	TEST Name	TEST Value	METHODS No.	METHODS Characteristic Values	METHODS Results
9	Negative	Age	78	1	8 (>7)	Less possible
		Ex. Time	7'37"	2		Inconclusive
		max. HR	197	3		Normal
		max. BP	210/85	4	<0.013	Not indicative
		max. ST	< –1 mm	5	– 1	Possible
		A-QRS	– 1			
10	Inconclusive	Age	75	1	8 (>7)	Less possible
		Ex. Time	8'	2	(max. ST: 0)	Inconclusive
		max. HR	116 (80%)	3		Normal
		max. BP	180/95	4	≈ 0	Not indicative
		max. ST	0	5	–2	Possible
		A-QRS	– 2			

During the first testing stage, all the input values corresponding to a patient were presented to the system. An early conclusion was that, in many cases, a bimodal overall IQ was inferred, rendering the centroid defuzzification methods useless. A maximum defuzzification is preferable, but only when accompanied by the undefuzzified values of IQ. The users unanimously preferred these undefuzzified representations, available at each level. Even the linguistic approximator proved to be, in a large extent, redundant.

From the statistics of this first testing stage we note that all 17 "clear" (8 positive, 9 negative) cases were correctly diagnosed by MMIDS, with the defuzzified IQ (maximum method) being larger than 6.25 and less than 3.75, respectively. In two of the three cases that were diagnosed by the physicians as "inconclusive," the defuzzified IQ was around 5 (corresponding to *inconclusive* in the term set), but with a negative nuance that was apparent in the graphical representation of IQ. The third case (case no. 10) was diagnosed by the system as *inconclusive/possible*. The clinical data corresponding to this case is presented in Table 1. The table also contains data corresponding to case no. 9 (the same patient before PTCA), which is a clearly negative diagnosis. The MMIDS output corresponding to both cases is presented in Figure 8, in which the defuzzified IQ value is obtained by COA. As we can see (especially from 8a) this value is not conclusive, a maximum defuzzification method being more appropriate (IQ_{MOM} =1.62). In case no. 10 (Figure 8b), a positive diagnosis was given after studying the MMIDS output. It was these kind of cases where such a system really proves its utility.

During the second testing stage, incomplete clinical data was presented to the system. It consisted of at least three of the five selected diagnostic methods, corresponding to 16 "clear" cases. Most of the time, the ST recovery loop and some multivariate method inputs were not presented to the system. All the cases had the input data for at least three elements of the Treadmill score (for some cases exercise time was missing). This time, MMIDS' output fits the physicians' diagnosis in 14 of the 16 cases. In the two 'wrong' cases, a *possible* diagnosis was inferred, while a negative one should have been correct. One of these was case no. 9. Following the MMIDS inference, it was observed that it was the age input that triggered the IQ to that diagnostic (78 and 75, respectively). This suggests that the importance of the age criteria within the treadmill score should be diminished.

Significant overall diagnostic errors occurred only when an entire treadmill score or multivariate method was omitted. As expected, this happened mainly for the clinical cases in which the other methods were inconclusive or had divergent conclusions. The conclusion of this stage of experiments was that the system displayed a reasonable behavior in the presence of incomplete input data.

6. CONCLUSIONS

Methods combining various exercise test criteria, including electrocardiographic and clinical criteria, have been traditionally based on a Bayesian probability approach. However, very often, in real conditions of everyday clinical practice, there is no clearly defined borderline between normal, mildly abnormal, or severely abnormal findings. Furthermore, this is the situation for almost every diagnostic approach in medicine. Fuzzy set theory has been introduced in medicine as an alternative to the conventional methods of reaching a medical diagnosis with the use of diagnostic techniques, which presents a graded degree of uncertainty.

Attempts to apply fuzzy sets in medical diagnosis have resulted in the creation and introduction of several interesting models with varying clinical value, which are presented throughout this book. The newly introduced DIFUS system, as described, represents a potentially advantageous approach, which takes its place in the increasing array of expert systems used in medical diagnosis. Its specified application in the evaluation of multivariate exercise stress test analysis for the diagnosis of coronary artery disease has proved its clinical value, as confirmed by the angiographic findings of the studied patients. It possesses, to a large degree, the attributes of a good expert system. The system ensures explicit separation between knowledge and inference mechanisms, and ensures the dynamism of the knowledge base and the interaction with the user in a natural language. It has the potential to reason in uncertain conditions using incomplete knowledge. It has the ability to motivate its own actions. By using fuzzy inference mechanisms and allowing the implementation of certain operations with words, its reasoning is quite close to human reasoning. It includes adaptation facilities. Still, in order to develop applications with more than three layers, future work is needed to improve the control strategy, the acquisition module, and, possibly, to have an automatic generation of the hierarchy.

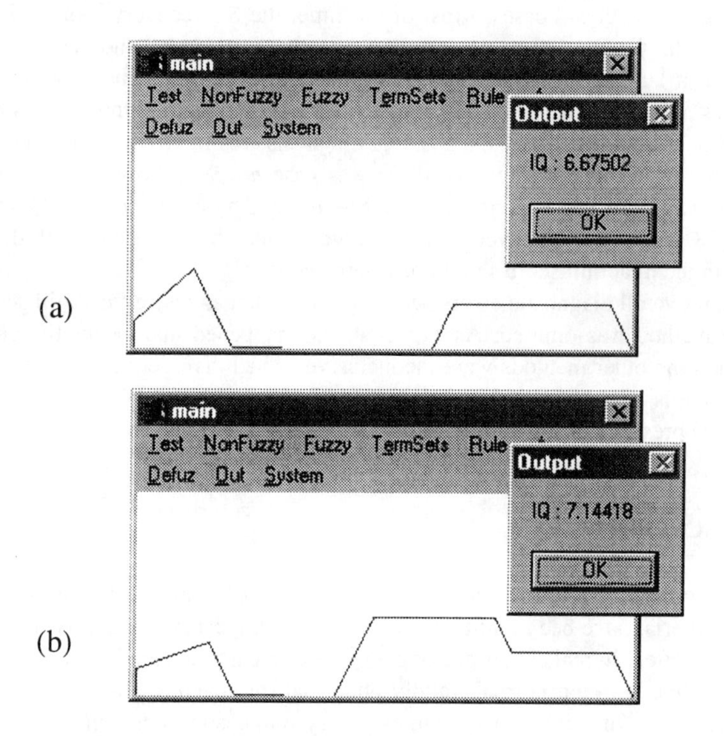

Figure 8. MMIDS output for cases no. 9 and no.10, respectively.

Experiments using the multimethod myocardial ischemia diagnosis system (MMIDS) have revealed a good level of diagnostic accuracy compared both to physicians' and classical automated diagnosis – based on bivalent logic and on classical score computing. Due to the flexibility of DIFUS, the application was developed in a short time, in spite of its heterogeneity. MMIDS exploits different types of knowledge – from statistical indicators to linguistic scores – which are expressed in different ways – from fuzzy production rules to operations with words. Different types of fuzzy systems have to be used for the implementation of the diagnosis cells, such as adaptive and non-adaptive, Mamdani, Takagi–Sugeno, and Takagi–Sugeno with fuzzy consequence.

The physicians' opinion seems to be that fuzzy approaches to these multimethod techniques are a diagnostic modality that is easy to use, more comprehensive, and, most important, of increased diagnostic value in everyday clinical practice. This encourages us to consider an expansion of the system. Thus, we take into account the inclusion of new symptoms into the knowledge base, correlation to other non-invasive diagnostic techniques (e.g., thallium scintiography) and, possibly, an ECG analysis module (image processor) as a sixth module (diagnostic method).

Meanwhile, the experiments continue both in order to realistically evaluate MMIDS performances – especially for incomplete set of inputs – and to increase the knowledge accuracy through a proper adjustment of the statistical indicators, term sets, rules, weights, etc. Automatic design and tuning techniques, such as the one presented in the previous chapter, may prove to be extremely helpful. Unfortunately, in this respect of "fine tuning," medicine succeeds to be even fuzzier. Many parameters used for diagnosis are continuously revised and adapted. Even normal values are not defined on the basis of the majority of values found in a certain healthy, homogeneous population. Also, and more important, the future clinical significance of such values on people's well-being is not defined. From a medical point of view, such an approach necessitates the conduction of long-term prospective studies, which sometimes have to be revised, thus making the problem even more complicated. Therefore, the fine tuning never ends. To deal with this, not only is a very manageable knowledge base a prerequisite, but benevolent physicians and at least one determined operator are necessary as well. Unfortunately, it is not very easy to satisfy some of these requirements!

REFERENCES

[1] Rifkin R., Hood W., Bayesian analysis of electrocardiographic exercise stress testing, *N. Engl. J. Med.*, 1977, 297, 681-685.

[2] Zimmerman H., *Fuzzy Set Theory and Its Application*. Kluwer Acad. Publ., Boston, 1991, 178-190.

[3] Rau G., Becker K., Kaufmann R., Zimmermann H., Fuzzy logic and control: principal approach and potential applications in medicine. *Artif. Organs*, 1995 Jan., 19(1), 105-12.

[4] Esogbue A., Elder R., Measurement and validation of a fuzzy mathematical model for medical diagnosis, *Fuzzy Sets and Systems* 1983, 10, 223-242.

[5] Klir G. H., Fuzzy logic: unearthing its meaning and significance, *IEEE Potentials*, 1995 Oct.-Nov., 10-15.

[6] Adlassnig K. P., Fuzzy set theory in medical diagnosis, *IEEE Trans. on Sys. Man and Cyb.*, SMC-16, 1986, 260-265.

[7] Holzmann C. A., San Martin M., Medical expert system on fuzzy analog ganglionar lattices, *Med. Prog. Technol.* 1996-97, 21(4), 195-203.

[8] Standera L. M., Goodenday L. S., Gos K. J., A neuro-fuzzy algorithm for diagnosis of coronary artery stenosis, *Comput-Biol-Med,* 1996 Mar., 26(2), 97-111.

[9] Ham F.M., Han S., Classification of cardiac arrhythmias using fuzzy ARTMAP, *IEEE. Trans. Biom. Eng* , 1996 Apr., 43(4) : 425-30.

[10] Ruttkay-Nedecky I., Riecansky I., Dipolar electrocardiotopographic evaluation of ventricular activation in various degrees of Coronary Artery Disease, *J. of Electrocardiol.*, 1994, 27:2, 149-155.

[11] Boudraa A. E., Arzi M., Sau J., Champier J., Automated detection of the left ventricular region in gated nuclear cardiac imaging, *IEEE Trans. Biom.Eng.*, 1996 Apr.: 43(4), 430-7.

[12] Dove E. L., Philip K., Gotteiner N. L., Vonesh M. J., A method for automatic edge detection and volume computation of the left ventricle from ultrafast computed tomographic images, *Invest-Radiol*, 1994 Nov. 29(11) : 945-54.

[13] Held C. M., Roy R. J., Multiple drug hemodynamic control by means of a supervisory-fuzzy rule-based adaptive control system, *IEEE Trans. Biom. Eng.*, 1995, Apr 42(4), 371-85.

[14] Peters R. M., Shanies S. A., Peters J. C., Fuzzy cluster analysis of positive stress tests, a new method of combining exercise test variables to predict extent of coronary artery disease, *Am. J. Cardiol.*, 1995, Oct 1, 76(10), 648-51.

[15] Froelicher V., Myers J., Follansbee W., Labovitz A., *Exercise and the Heart*, Mosby, 1993, 136-146.

[16] Ribisl M. P., Liu J., Mousa I., Comparison of computer ST criteria for diagnosis of severe coronary artery disease, *Am. J. Cardiol.*, 1993, 71, 546-551.

[17] Negoita, C.V., *Expert Systems and Fuzzy Systems*, Benjamin/Cummings, Menlo Park, 1985.

[18] Lindsay, R. K., Buchanan, B. G., Feigenbaum, E. A., Lederberg, J., Applications of Artificial Intelligence for Organic Chemistry, The DENDRAL Project, New York, 1980.

[19] Weiss, S. M., Kulikowski, C.A., Expert consultation systems: the EXPERT and CASNET projects, in *Machine Intelligence*, Infotech State of the Art Report, 9, no. 3, 1981.

[20] Adlassnig, K. P., Kolarz, G., CADIAG-2. Computer assisted medical diagnosis using fuzzy subsets, in: M.M. Gupta, E. Sanchez, Eds., *Approximate Reasoning in Decision Analysis*, North-Holland, New York, 1982, 141-148.

[21] Adlassnig, K. P., Kolarz, G., Scheihauer, W., Present state of the medical expert system CADIAG-2, *Medical Information* 24, 1985, 13-20.

[22] Shortliffe, E. H., *Computer Based Medical Consultations: MYCIN*, American Elsevier, 1976.

[23] Davis, R., Buchanan, B. G., Shortliffe, E. H., Production rules as a representation of a knowledge-based consultation program, *Artificial Intelligence* 8, 1977, 15-45.

[24] Dubois, D., Prade, H., Fuzzy rules in knowledge-based systems, in R.R. Yager, L.A. Zadeh, Eds., *An Introduction to Fuzzy Logic Applications in Intelligent Systems*, Kluwer, 1992, 45-69.

[25] Konopasek, M., Jayaraman, S., Expert systems for personal computers, *Byte*, 1984, 137-154.

[26] Yager, R.R., Zadeh, L.A., Eds., *An Introduction to Fuzzy Logic Applications in Intelligent Systems*, Kluwer 1992

[27] Sanchez, E., Fuzzy logic knowledge systems and artificial neural networks in medicine and biology, in R.R. Yager, L.A. Zadeh, Eds., *An Introduction to Fuzzy Logic Applications in Intelligent Systems*, Kluwer 1992, 235-253.

[28] Yager, R. R., Expert systems using fuzzy logic, in R.R. Yager, L.A. Zadeh, Eds., *An Introduction to Fuzzy Logic Applications in Intelligent Systems*, Kluwer, 1992, 27-44.

[29] de Mantaras, R. L., Sierra, C., Agusti, J., The representation and use of uncertainty and metaknowledge in MILORD, in R.R. Yager, L.A. Zadeh, Eds., *An Introduction to Fuzzy Logic Applications in Intelligent Systems*, Kluwer, 1992, 253-263.

[30] Zahan, S., Michael, C., Nikolakeas, S., A fuzzy hierarchical approach to medical diagnosis. An application to the multimethod myocardial ischemia diagnosis, *Proc. of EUFIT'96*, Aachen, 1996, 2081-2085.

[31] Zahan, S., Michael, C., Nikolakeas, S., A fuzzy hierarchical approach to medical diagnosis, *Proc. of 6th IEEE Int. Conf. on Fuzzy Systems*, Barcelona, 1997, 319-324.

[32] Godo, L., de Mantaras, R.L., Sierra, C., Verdaguer, A., MILORD, the architecture and management of linguistically expressed uncertainty, *Int. J. of Intelligent Systems* 4, 1989, 471-501.

[33] Zahan, S., Miron, C., An approach to generation of Takagi-Sugeno rules based on linguistically expressed knowledge, *Proc. of EUFIT'95*, Aachen, 1995, 534-538.

[34] Zahan, S., Michael, C., Fuzzy inference systems based on Takagi-Sugeno fuzzy consequent rules, *Proc. of EUFIT'96*, Aachen, 1996, 563-567.

[35] Nguyen, H. P., Approach to combining negative and positive evidence in CADIAG-2, *Proc. of EUFIT'95*, Aachen, 1995, 1653-1658.

[36] (a) H.N. Teodorescu et al., Fuzzy expert system for hearing loss risk assessing. *Second Int. Conf. of BUFSA*, Trabzon, 1992 (Turkey), pp. 127-128.
(b) O. Paduraru, H.N. Teodorescu, M. Costin, A. Ciobanu, S. Gradinaru: Fuzzy Expert System Shell for Diagnosis and Prediction Applications. *Proc. Fifth International Conference on Information Processing and Management of Uncertainty in Knowledge Based Systems*, Paris July 4-8, 1994. Vol. 2, pp. 759-764.
(c) Paduraru, O., Teodorescu, H.N., Costin, M., Gradinaru, S., Ciobanu, A., Knowledge-based generator for a fuzzy expert system, in H.N. Teodorescu, I. Bogdan, R. Strungaru, Eds., *Fuzzy Systems and Neural Networks*, Politechnic University of Iasi, 164-176, 1992.

[37] Klir, G. J., Folger, T., *Fuzzy Sets, Uncertainty and Information*, Prentice Hall, New York, 1988.

[38] Michailidis A., Triposkiadis P., Boudoulas H., New coronary artery disease index based in exercise induced QRS changes, *Am. Heart J.*, 1990, 120, 292-302.

[39] Lachterman B., Lehmann K., Detrano R., Neutel J., Froelicher V., Comparison of ST segment/heart rate index to standard ST criteria for analysis of exercise electrocardiogram, *Circulation*, 1990, 82, 44-50.

[40] Okin P., Bergman G., Kligfield P., Effect of ST segment measurement point on performance of standard and heart-rate adjusted ST segment criteria for the identification of coronary artery disease, *Circulation* , 1991, 84, 57-66.

[41] Parrens E., Douard H., The exercise recovery loop and exercise slope of ST segment changes/heart rate in the diagnosis of coronary disease and restenosis after angioplasty, *Arch. Mal. Coeur.*, 1994, 87(10), 1283-8.

[42] Hollenberg M., Zoltick J.M., Mateo G., Comparison of a quantitative treadmill exercise score with standard electrocardiographic criteria in screening asymptomatic men for coronary disease, *N. Engl. J. Med.*, 1985, 10, 600-606.

[43] Greenberg P.S., Cangiano B., Leamy L., Ellestad M.H., Use of the multivariate approach to enhance the diagnostic accuracy of the treadmill stress test, *J. Electrocardiol.*, 1980, 13, 227.

[44] Kligfield P., Ameisen O., Okin P, Heart rate adjustment of ST segment depression of improved detection of coronary artery disease, *Circulation* , 1989, 79, 245-255.

[45] Van Campen, C. M. C., Visser, F. C., Visser C. A., The QRS score: a promising new exercise score for detecting coronary artery disease based on changes of Q, R and S waves, *Eur. Heart J.*, 1996, 17, 699-708.

[46] Zahan, S., Fuzzy diagnosis–some implementation issues, *Proc. of 7th IFSA World Congress*, Prague, 1997, 427-432.

[47] Zahan, S., Implementation of score-based tests in fuzzy diagnosis, *Fuzzy Systems & Artificial Intelligence*, vol. 5, no. 1-3, 1996, 77-86.

[48] Takagi, T., Sugeno, M., Derivation of fuzzy control rules from human operator's control action, *Proc. IFAC Symp. on Fuzzy Inform., Knowledge Representation and Decision Analysis*, 1983, 55-60.

[49] Zadeh, L.A., Fuzzy logic = computing with words, *IEEE Trans. on Fuzzy Systems,* 2, 1996, 103-112.

[50] Zadeh, L. A., Fuzzy logic, neural networks and soft computing, *Comm. of the ACM*, vol. 37, no. 3, 1994, pp. 77-84.

Chapter 9

Design and Tuning of Fuzzy Rule-Based Systems for Medical Diagnosis

Alexander Rotshtein

The aim of this chapter is to present and exemplify a method to design and tune the fuzzy rules in a fuzzy system for medical diagnosis. This approach includes two fuzzy-rules-tuning stages: rough and fine. The rough stage consists of fuzzy rule generation by a medical expert, and the selection of the membership functions of the fuzzy terms by the method of paired comparisons. The fine stage consists of finding the parameters of the membership functions and the weights of the rules by using training data and solving the corresponding optimization problems, for the case of either continuous or discrete consequents. The effectiveness of the tuning models is demonstrated by examples. An important feature of the proposed method is that, for the design and tuning of fuzzy rules, only one simple form of the expert knowledge matrix is needed, both for the continuous and discrete consequent. A fuzzy expert system for differential diagnosis of ischemia heart disease based on this approach is discussed.

1. Introduction

Fuzzy logic is a useful tool for building expert systems for decision making in the field of medical diagnosis [1], [2], [3]. The quality of fuzzy expert systems strongly depends on the "quality" of fuzzy rules and the "quality" of membership functions used for fuzzy terms. The better the fuzzy if-then rules and the membership functions, the higher the quality of diagnostic decision making will be. However, no one can guarantee that the result of the fuzzy logic inference is the true diagnosis. That is why the problem of creating relevant fuzzy if-then rules and membership functions is important for the design of expert systems for medical diagnosis.

In this chapter, we propose an approach to increase the quality of an expert system by tuning the fuzzy logic inference model. This approach is based upon finding the

0-8493-9806-1/98/$0.00+$.50

weights of fuzzy if-then rules, and shapes of the membership functions, that minimize the difference between real (experimental) and inferred (theoretical) types of decisions. The problem of fuzzy model tuning is stated as a classical mathematical optimization problem.

This chapter is organized as follows: Section 2 introduces a problem statement and general methodology of design and tuning of fuzzy rules in the rough and fine stages. Section 3 presents the basic models necessary for the fuzzy logic inference and rough tuning of fuzzy rules by the method of pairwise comparison. Section 4 and Section 5 present fine tuning as a problem of optimization, mathematical models for fuzzy inference quality evaluation, and some results of computer experiments for the fuzzy models with continuous and discrete consequents, respectively. Section 6 is devoted to the application of the proposed approach in building a fuzzy expert system for the differential diagnosis of ischemia heart disease. Section 7 presents conclusions.

Parts of the concepts in this chapter are based on the monograph [4] and on the papers [5-9].

2. Problem Statement and General Methodology

From the formal point of view, the problem of fuzzy model creation for medical diagnosis can be considered as a problem of nonlinear object identification with multiple inputs (x_1, x_2, ..., x_n), and a single output (y). The inputs correspond to parameters of the patient's state which are relevant for decision making, and the single output corresponds to the type of diagnosis. For practical needs, it is interesting to consider the output variable y is either continuous or discrete (Figure 1).

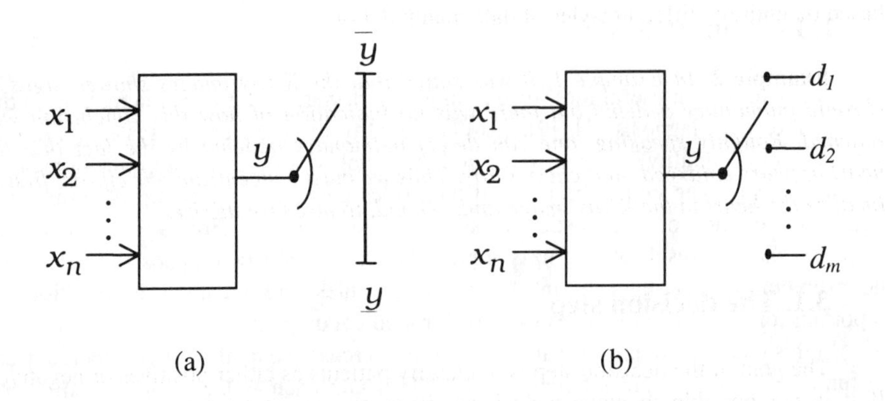

Fig. 1 Object with many inputs and a single output:
(a) continuous output, (b) discrete output.

We deal with continuous output (Figure 1a) when the result of fuzzy logic inference is any qualitative parameter $y \in \left[\underline{y}, \overline{y}\right]$, used for future decision making about the patient's diagnosis, where $\underline{y}\,(\overline{y})$ is lower (upper) value of y. For example, the blood volume lost during childbirth bleeding can be considered as a continuous output variable necessary to predict at the early stages of pregnancy. The approach considered by Zahan et al. [3], in which the universe of the output variable represents the possibility degree of a certain diagnosis, and not the diagnosis itself, can also be formalized in terms of identification with continuous output (Figure 1a).

We deal with discrete output (Figure 1b) when the result of fuzzy logic inference has to be obtained as a variable of $y = d_j \in \{d_1, d_2, ..., d_m\}$, where d_j is the j-th class of diagnosis. In the model with discrete output, we assume that classification of medical decisions is known *a priori*. This is usually possible in the tasks of the differential diagnosis [7].

Let us assume that the following data are known.

- Definitions of some parameters of a patient's state as input variables $x_i \in \left[\underline{x_i}, \overline{x_i}\right]$, where $\underline{x_i}\,(\overline{x_i})$ is lower (upper) value of x_i, $i = 1, 2, ..., n$. In the case of qualitative parameter x_i, the interval $\left[\underline{x_i}, \overline{x_i}\right]$ can be defined by some artificial (*min-max*) scale.

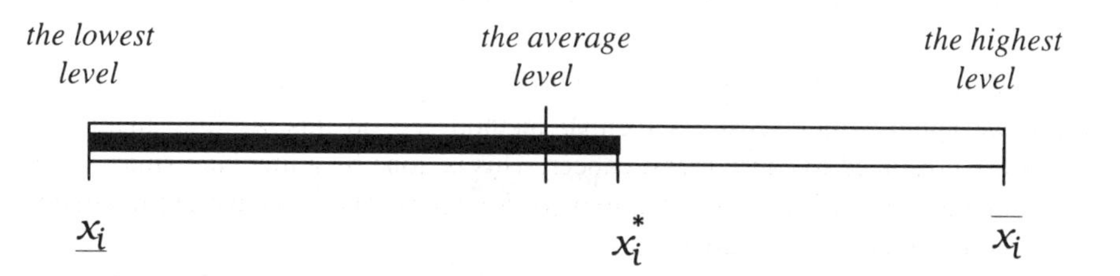

In the *min-max* scale, a qualitative value of parameter x_i, is indicated by the mark x_i^*. This mark on the (*min-max*) scale corresponds to an arbitrary linguistical term used for parameter x_i estimation. In the case above, the mark x_i^* can be interpreted as *little bit more than the average.*

- Definition of diagnoses as output variable

$$y \in \begin{cases} \left[\underline{y}, \overline{y}\right] & \text{for continuous output} \\ \{d_1, d_2, ..., d_n\} & \text{for discrete output} \end{cases}$$

- Fuzzy if-then rules – obtained from a medical expert – which connect diagnoses (output) with parameters of the patient's state.
- Training data «input-output» used for tuning the fuzzy rules.

- Testing data «input-output» used for evaluation of diagnostic decision making.

The problem is to build a model of $y = f(x_1, x_2, \ldots, x_n)$, both for continuous and discrete outputs, for which the results of the fuzzy logic inference is close to the relevant (real-life) medical decision.

The proposed methodology of this problem solving is shown in Figure 2. According to this methodology the process of diagnostic model design consists of two tuning stages: *rough* (1) and *fine* (2).

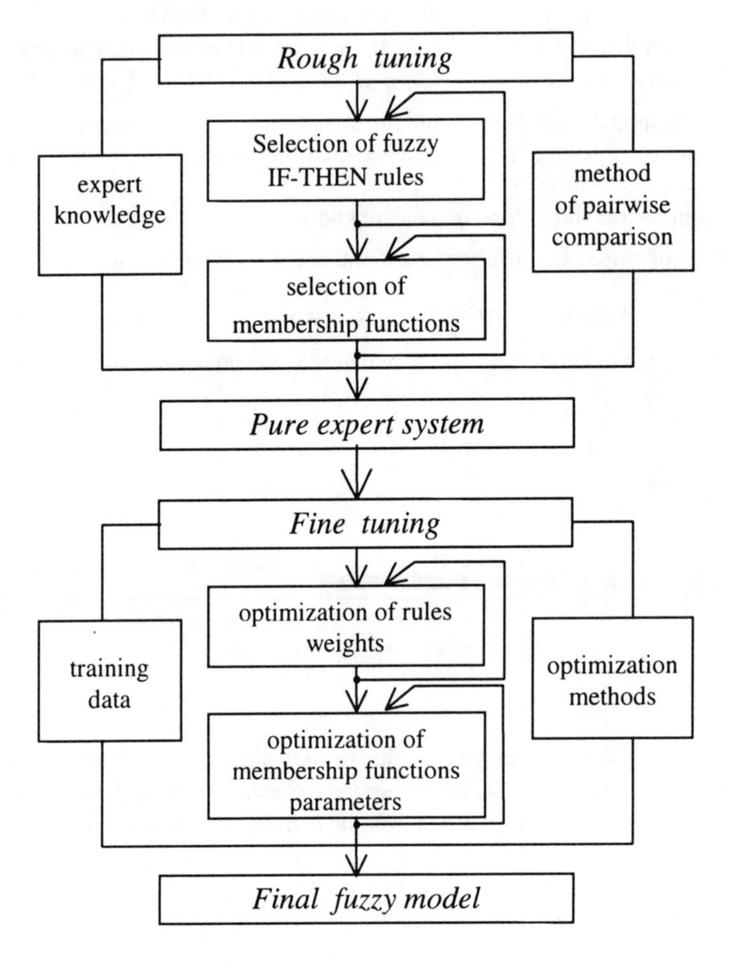

Fig. 2. General methodology of diagnostic model.

In rough model tuning, we use only expert information for the selection of fuzzy if-then rules and fuzzy terms' membership functions. That is why the expert system built at this stage can be called the *pure expert system*. For the rough tuning of fuzzy if-then rule weights and shapes of membership functions, it is convenient to apply Saaty's method of pairwise comparison [10], because this method demands only one expert. The higher the professional skill of this expert, the higher the chance to build a high quality pure expert system at the rough tuning stage. However, as we mentioned above, no one can guarantee the coincidence of fuzzy logic inference and real decisions. That is why we need the stage of fine fuzzy model tuning.

The aim of the fine tuning stage is to improve the pure fuzzy expert system by using training data. The procedure of pure fuzzy model improvement is formulated as a problem of optimization, in which the control variables are weights of if-then rules together with parameters of fuzzy terms membership functions, and the criterion of minimization is a distance between inferred and desired output.

3. DESIGN AND ROUGH TUNING OF FUZZY RULES

This section presents the basic models used in the process of pure expert system design, both for continuous and discrete outputs.

3.1. MATRIX OF KNOWLEDGE

For the extraction of fuzzy if-then rules from a medical expert, it is convenient to use the matrix of knowledge shown in Table 1. This matrix defines the plan of a virtual experiment, in which an expert must produce several rows for each type of diagnosis: k_1 rows for diagnosis of d_1, k_2 for d_2, . . . , k_m for d_m. It connects the linguistical variables of inputs (x_i) and output (y) by the following fuzzy rules:

$$\begin{aligned}
&IF \quad \left[\left(x_1 = a_1^{j1}\right) AND \left(x_2 = a_2^{j1}\right) \; AND ... \left(x_n = a_n^{j1}\right)\right] \quad \left(with\,weight\; w_{j1}\right)\\
&OR \quad \left[\left(x_1 = a_1^{j2}\right) AND \left(x_2 = a_2^{j2}\right) \; AND ... \left(x_n = a_n^{j2}\right)\right] \quad \left(with\,weight\; w_{j2}\right)\\
&...\\
&OR \quad \left[\left(x_1 = a_1^{jk_j}\right) AND \left(x_2 = a_2^{jk_j}\right) AND .. \left(x_n = a_n^{jk_j}\right)\right] \quad \left(with\; weight\; w_{jk_j}\right)\\
&THEN \quad y = d_j, \qquad\qquad j = \overline{1,m},
\end{aligned} \tag{1}$$

where

$a_i^{jk_j}$ is a fuzzy term for the evaluation of variable x_i in the j^{k_j} -th row,

w_{jk_j} is the real number in the interval [0,1], which characterizes a subjective degree of the expert's confidence in the j^{k_j} -th row.

Table 1. Matrix of knowledge

Rule number	IF (parameters of patient's state)			Weight of rule	THEN (type of diagnosis y)
	x_1	... x_i ...	x_n		
11	a_1^{11}	... a_i^{11} ...	a_n^{11}	w_{11}	
12	a_1^{12}	... a_i^{12} ...	a_n^{12}	w_{12}	d_1
...	...	...	...	...	
$1k_1$	$a_1^{1k_1}$	$a_i^{1k_1}$	$a_n^{1k_1}$	w_{1k_1}	
...	...	...	...	...	...
$j1$	a_1^{j1}	... a_i^{j1} ...	a_n^{j1}	w_{j1}	d_j
...	...	...	...	...	
jk_j	$a_1^{jk_j}$	$a_i^{jk_j}$	$a_n^{jk_j}$	w_{jk_j}	
...	...	...	...	...	...
$m1$	a_1^{m1}	a_i^{m1}	a_n^{m1}	w_{m1}	
$m2$	a_1^{m2}	a_i^{m2}	a_n^{m2}	w_{m2}	d_m
...	...	...	...	...	
mk_m	$a_1^{mk_m}$	$a_i^{mk_m}$	$a_n^{mk_m}$	w_{mk_m}	

Note that the knowledge matrix (Table 1) corresponds to a discrete output. To transform the continuous output $y \in \left[\underline{y}, \overline{y}\right]$ into a discrete one, the interval $\left[\underline{y}, \overline{y}\right]$ must be divided into the m sub-intervals as follows:

$$\left[\underline{y}, \overline{y}\right] = \underbrace{\left[\underline{y}, y_1\right)}_{d_1} \cup \underbrace{\left[y_1, y_2\right)}_{d_2} \cup \ldots \cup \underbrace{\left[y_{j-1}, y_j\right)}_{d_j} \cup \ldots \cup \underbrace{\left[y_{m-1}, \overline{y}\right]}_{d_m} \tag{2}$$

If the number of input variables (x_i) is higher than the "magic number 7±2" [11], it is difficult for an expert to compile all the *if-then* rules into a single matrix of knowledge. In such a case, it is desirable to use a hierarchical presentation of an input-output model, and to compile if-then rules for all classes of the hierarchy. According to [2] the most usual hierarchy in medical diagnosis is: (0) symptoms and signs → (1) group of symptoms → (2) disease → (3) family of diseases. An example of hierarchical presentation of some expert knowledge is shown in Section 6.

3.2. Fuzzy Model with Discrete Output

Let $\mu^p(x_i)$ be a membership function of input variable x_i to for the fuzzy term a_i^p, $p = \overline{1, k_j}$, $i = \overline{1, n}$, $j = \overline{1, m}$, and

$$a_i^p = \int_{\underline{x}_i}^{\overline{x}_i} \mu^p(x_i)/x_i ,$$

Here, $\mu^{d_j}(y)$ denotes one of the membership functions of output y for the decision set $d_j \in \{d_1, d_2, ..., d_m\}$.

Let us define these membership functions by fuzzy logic equations as follows:

$$\begin{aligned} \mu^{d_j}(y) = w_{j1} & \left[\mu^{j1}(x_1) \wedge \mu^{j1}(x_2) \wedge ... \wedge \mu^{j1}(x_n)\right] \vee \\ w_{j2} & \left[\mu^{j2}(x_1) \wedge \mu^{j2}(x_2) \wedge ... \wedge \mu^{j2}(x_n)\right] \vee ... \\ ...w_{jk_j} & \left[\mu^{jk_j}(x_1) \wedge \mu^{jk_j}(x_2) \wedge ... \wedge \mu^{jk_j}(x_n)\right], \quad j = \overline{1,m}. \end{aligned} \tag{3}$$

These equations are obtained from the if-then rules (1) by substitution of the variables x_i for their membership functions, and substitution of operations AND and OR for operations $\wedge$ and $\vee$. Because the logic operations $\wedge$ and $\vee$ in classic fuzzy set theory [13] correspond to operations *min* and *max*, the membership function of output y is defined by

$$\mu^{d_j}(y) = \max_{p=\overline{1,k_j}} \left\{ w_{jp} \min_{i=\overline{1,n}} \left[\mu^{jp}(x_i)\right]\right\}, \; j = \overline{1,m}. \tag{4}$$

For calculations using formula (4) it is necessary to define a membership function of linguistic terms used in the fuzzy rules (1). Generally, one uses parametrized fuzzy sets, i.e., fuzzy sets whose "shape" is determined by a set of parameters [13]. Triangular fuzzy sets, trapezoidal fuzzy sets, L-R type fuzzy sets, Gaussian fuzzy sets, etc., belong to this class. In this chapter, we use a simple membership function model of variable x to arbitrary term T in the form

$$\mu^T(x) = \frac{1}{1 + \left(\frac{x-b}{c}\right)^2} \tag{5}$$

Here, b and c are parameters that can be tuned. Note that

b is a coordinate of function maximum, $\mu^T(b) = 1$,

c is a coefficient defining the "concentration-dilation" of the function (Figure 3).

Consider that a patient is characterized by the vector $(x_1^*, x_2^*, ..., x_n^*)$ of state parameters. If $d_j^* \in \{d_1, d_2, ..., d_m\}$ is a type of diagnostic decision which corresponds to the patient with vector $(x_1^*, x_2^*, ..., x_n^*)$, then d_j^* can be defined from (4) as follows:

$$\mu^{d_j^*}(y) = \max_{j=\overline{1,m}} \mu^{d_j}(y).$$

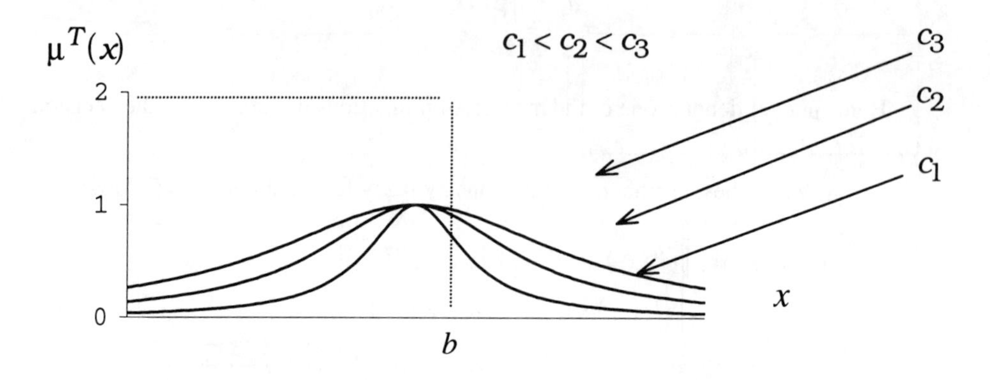

Fig. 3. Membership function.

3.3. FUZZY MODEL WITH CONTINUOUS OUTPUT

The result of applying the formula (4) is the fuzzy set for the output variable y

$$\tilde{y} = \left\{ \frac{\mu^{d_1}(y)}{\left[\underline{y}, y_1\right)}, \frac{\mu^{d_2}(y)}{\left[y_1, y_2\right)}, \ldots, \frac{\mu^{d_m}(y)}{\left[y_{m-1}, \overline{y}\right]} \right\} \tag{6}$$

Let us define the fuzzy set (6) defuzzification as the following. The result of the defuzzification is a number y

$$y = \frac{\underline{y}\mu^{d_1}(y) + y_1\mu^{d_2}(y) + \ldots + y_{m-1}\mu^{d_m}(y)}{\mu^{d_1}(y) + \mu^{d_2}(y) + \ldots + \mu^{d_m}(y)} \tag{7}$$

In the case where membership degrees have a probabilistic interpretation, the formula (7) can be considered as the mathematical expectation of a discrete random value. If the interval $\left[\underline{y}, \overline{y}\right]$ is divided into the m equal parts such as $y_1 = \underline{y} + \Delta$, $y_2 = \underline{y} + 2\Delta$, ..., $y_{m-1} = \overline{y} - \Delta$, $\Delta = \dfrac{\overline{y} - \underline{y}}{m-1}$, then formula (7) can be simplified as follows:

$$y = \frac{\sum_{j=1}^{m} \left[\underline{y} + (j-1)\Delta\right]\mu^{d_j}(y)}{\sum_{j=1}^{m} \mu^{d_j}(y)} \tag{8}$$

3.4. Rough Tuning of Fuzzy Rules

The idea of rough fuzzy rule tuning consists of finding the membership functions shapes and fuzzy rule weights based upon Saaty's method of pairwise comparison [10]. In this study we use the modified Saaty's method [8], which, in contrast to the usual method [10], needs no solution of the characteristic equation of the pairwise comparison matrix.

3.4.1. Rough tuning of membership functions

Let S be some property viewed as a linguistic term. The fuzzy set used to formalize the term S is represented by some population of pairs

$$\tilde{S} = \left\{ \frac{\mu_s(u_1)}{u_1}, \frac{\mu_s(u_2)}{u_2}, ..., \frac{\mu_s(u_n)}{u_n} \right\} .$$

where

$\{u_1, u_2, ..., u_n\} = U$ is a universal set necessary to describe the set $\tilde{S} \subset U$;

$\mu_S(u_i)$ is the degree of element $u_i \in U$ membership in fuzzy set $\tilde{S}$.

The problem is to define the value of $\mu_S(u_i)$ for all $i = \overline{1, n}$. The population of these values will account for the unknown membership function. The method proposed for solving the formulated problem is based on the idea of assigning membership degrees of the universal set elements according to their ranks. The same idea was used in the theory of structural analysis of systems [14], [15], where various definitions of an element's rank are treated. The technique of reliable assignment of system elements according to their ranks is stated in [14]. In our case, by *element rank* of $u_i \in U$ we mean some number $r_S(u_i)$, which characterizes some significance (or weight) of this element in establishing the property described by the fuzzy term $\tilde{S}$.

Let us assume that the following rule is satisfied:

The bigger the element rank, the higher is its degree of membership.

We use the notation

$$r_S(u_i) = r_i; \quad \mu_S(u_i) = \mu_i; \quad i = 1, ..., n .$$

The rule of membership degrees assignment can be given as

$$\frac{\mu_1}{r_1} = \frac{\mu_2}{r_2} = ... = \frac{\mu_n}{r_n} , \tag{9}$$

to which the condition of normalization is added

$$\mu_1+\mu_2+...+\mu_n=1 \tag{10}$$

Using equation (9), it is easy to define the membership degrees of all the elements of the universal set through the membership degree of the reference element.

If element $u_1 \in U$ is the reference element with membership μ_1, then

$$\mu_2 = \frac{r_2}{r_1}\mu_1, \quad \mu_2 = \frac{r_3}{r_1}\mu_1, \quad \cdots, \quad \mu_n = \frac{r_n}{r_1}\mu_1 \tag{11}$$

For the reference element $u_2 \in U$ with membership function μ_2, we get

$$\mu_1 = \frac{r_1}{r_2}\mu_2, \quad \mu_3 = \frac{r_3}{r_2}\mu_2, \quad \cdots, \quad \mu_n = \frac{r_n}{r_2}\mu_2 \tag{12}$$

Finally, for the reference element $u_n \in U$ with membership μ_n we have

$$\mu_1 = \frac{r_1}{r_n}\mu_n, \quad \mu_2 = \frac{r_2}{r_n}\mu_n, \quad \cdots, \quad \mu_{n-1} = \frac{r_{n-1}}{r_n}\mu_n \tag{13}$$

Taking into account the condition of normalization (10) and using equations (11) (13), we find

$$\left.\begin{array}{l} \mu_1 = \left(1+\frac{r_2}{r_1}+\frac{r_3}{r_1}...+\frac{r_n}{r_1}\right)^{-1} \\ \mu_2 = \left(\frac{r_1}{r_2}+1+\frac{r_3}{r_2}...+\frac{r_n}{r_2}\right)^{-1} \\ \cdots\cdots\cdots\cdots\cdots \\ \mu_n = \left(\frac{r_1}{r_n}+\frac{r_2}{r_n}+\frac{r_3}{r_n}...+1\right)^{-1} \end{array}\right\} \tag{14}$$

The formulas (14) allow calculation of the membership degrees $\mu_S(u_i)$ of the elements $u_i \in U$ for the fuzzy term $\tilde{S}$ in two independent ways.

1) Using absolute assessments of levels r_i, i = 1,2, ..., n, which are defined using methods suggested in the theory of system structural analysis [14], [15]. A nine mark scale (1 the lowest rank; 9 the highest rank), or *min-max* scale can be used for making expert assessments of ranks.

2) Using relative assessments of ranks $r_i / r_j = a_{ij}$, $i, j = \overline{1,n}$, and building the matrix

$$A = \begin{bmatrix} 1 & \frac{r_2}{r_1} & \frac{r_3}{r_1} & \dots & \frac{r_n}{r_1} \\ \frac{r_1}{r_2} & 1 & \frac{r_3}{r_2} & \dots & \frac{r_n}{r_2} \\ \dots & \dots & \dots & \dots & \dots \\ \frac{r_1}{r_n} & \frac{r_2}{r_n} & \frac{r_3}{r_n} & \dots & 1 \end{bmatrix} \quad (15)$$

This matrix has the following properties:

- It is diagonal, that is $a_{ii} = 1$, $i = \overline{1,n}$,
- The elements, which are symmetric relative to the main diagonal are tied up by the dependency $a_{ij} = 1/a_{ji}$,
- It is transitive, that is $a_{ik}a_{kj} = a_{ij}$, because $\frac{r_i}{r_k} \cdot \frac{r_k}{r_j} = \frac{r_i}{r_j}$.

These properties allow us to find the elements of all other rows using the known elements of one row of the matrix A. If the r-th row is known, that is elements $a_{kj}, k, j = \overline{1,n}$, then any arbitrary element a_{ij} will be found in this way:

$$a_{ij} = a_{kj} / a_{ki}, i, j, k = \overline{1,n}$$

Because the matrix (15) can be interpreted as the matrix of the pairwise comparison ranks, one can use the Saaty mark scale [10] for expert assessment of the elements of this matrix. In our case, this scale is formed in this way:

$$a_{ij} = r_i / r_j = \begin{cases} 1, & \textit{in case of no superiority of } r_i \textit{ over } r_j \\ 3, & \textit{in case of weak superiority of } r_i \textit{ over } r_j \\ 5, & \textit{in case of essential superiority of } r_i \textit{ over } r_j \\ 7, & \textit{in case of explicit superiority of } r_i \textit{ over } r_j \\ 9, & \textit{in case of absolute superiority of } r_i \textit{ over } r_j \\ 2, 4, 6, 8, & \textit{intermediate comparison assessments.} \end{cases}$$

Thus, using the resulting formulae (14), expert knowledge about ranks of elements or their pairwise comparison is transformed into the membership function of the fuzzy term.

To implement the suggested method, one can use the following algorithm:

Step 1. Set linguistic variable x,

Step 2. Define the universal set on which the variable x is based.

Step 3. Set the population of fuzzy terms $\{S_1, S_2, ..., S_l\}$, which are used for assessing the variable x.

Step 4. Form matrix (15) for each term $S_j, j = \overline{1,l}$.

Step 5. Calculate the membership functions of the elements for each term using formulae (14). Normalization of the functions thus found is accomplished by way of dividing into the highest degree of membership.

Example 1. Let us treat the linguistic variable of PATIENT'S_AGE, which is defined within the interval of [31], [57] years. The discrete interval set for this variable is defined as follows:

$$U = \{u_1 = 31, u_2 = 37, u_3 = 43, u_4 = 49, u_5 = 57\}.$$

The level of the patient's age will be assessed using such fuzzy terms as L-low, A-average, H-high. Matrix (15) for each term is formed in this way.

$$A_L = \begin{array}{c} \\ u_1 \\ u_2 \\ u_3 \\ u_4 \\ u_5 \end{array} \begin{array}{c} \begin{array}{ccccc} u_1 & u_2 & u_3 & u_4 & u_5 \end{array} \\ \left[\begin{array}{ccccc} 1 & \frac{7}{9} & \frac{5}{9} & \frac{3}{9} & \frac{1}{9} \\ \frac{9}{7} & 1 & \frac{5}{7} & \frac{3}{7} & \frac{1}{7} \\ \frac{9}{5} & \frac{7}{5} & 1 & \frac{3}{5} & \frac{1}{5} \\ \frac{9}{3} & \frac{7}{3} & \frac{5}{3} & 1 & \frac{1}{3} \\ 9 & 7 & 5 & 3 & 1 \end{array}\right] \end{array} \qquad A_A = \begin{array}{c} \\ u_1 \\ u_2 \\ u_3 \\ u_4 \\ u_5 \end{array} \begin{array}{c} \begin{array}{ccccc} u_1 & u_2 & u_3 & u_4 & u_5 \end{array} \\ \left[\begin{array}{ccccc} 1 & \frac{7}{5} & \frac{9}{5} & \frac{7}{5} & 1 \\ \frac{5}{7} & 1 & \frac{5}{7} & 1 & \frac{5}{7} \\ \frac{5}{9} & \frac{7}{9} & 1 & \frac{7}{9} & \frac{5}{9} \\ \frac{5}{7} & 1 & \frac{9}{7} & 1 & \frac{5}{7} \\ 1 & \frac{7}{5} & \frac{9}{5} & \frac{7}{5} & 1 \end{array}\right] \end{array}$$

$$A_H = \begin{array}{c} \\ u_1 \\ u_2 \\ u_3 \\ u_4 \\ u_5 \end{array} \begin{array}{c} \begin{array}{ccccc} u_1 & u_2 & u_3 & u_4 & u_5 \end{array} \\ \left[\begin{array}{ccccc} 1 & 3 & 5 & 7 & 9 \\ \frac{1}{3} & 1 & \frac{5}{3} & \frac{7}{3} & \frac{9}{3} \\ \frac{1}{5} & \frac{3}{5} & 1 & \frac{7}{5} & \frac{9}{5} \\ \frac{1}{7} & \frac{3}{7} & \frac{5}{7} & 1 & \frac{9}{7} \\ \frac{1}{9} & \frac{3}{9} & \frac{5}{9} & \frac{7}{9} & 1 \end{array}\right] \end{array}$$

After processing these matrices according to the formula (14) we obtain the membership functions shown in Figure 4.

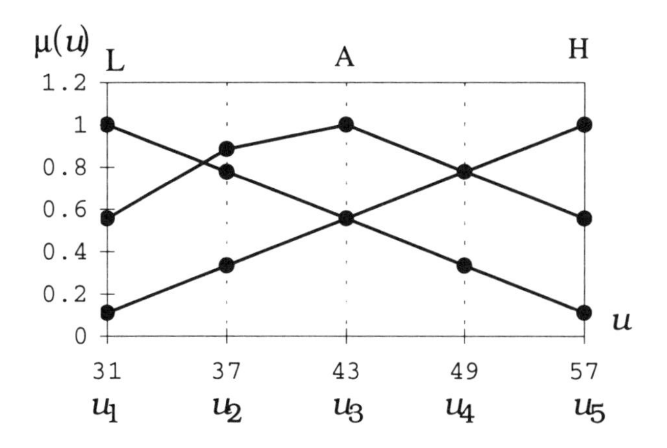

Figure 4. Membership functions of the fuzzy sets.

3.4.2. Rough tuning of rules weights

The same method can be used for the rough tuning of the rules weights in the knowledge matrix (Table 1). To apply this method, it is necessary

1) to form the matrix of pairwise comparison, according to (15), for each type of diagnosis, and
2) to find the weights of rules using models (14).

4. Fine Tuning of the Fuzzy Rules with Continuous Output

This section presents the optimization problem whose solution is equivalent to fine fuzzy model tuning with continuous output. Two examples of computer simulation for the demonstration of the tuning process are presented.

4.1. Tuning as a Problem of Optimization

The correlations (4), (5) and (7), (8) define the generalized model of a nonlinear object with continuous output (Figure 1a) which corresponds to the rules in (1). Let us denote this generalized model as

$$y = F(X, W, B, C)\,, \tag{16}$$

where

$X=(x_1,x_2,...,x_n)$ is a vector of the input variables,
$W=(w_1,w_2,...,w_N)$ is a vector of the rule weights in (1),
$B=(b_1,b_2,...,b_q)$ and $C=(c_1,c_2,...,c_q)$ are membership function parameter vectors, according to (5),
N is the total number of lines in the matrix of knowledge (Table 1),

$$N=k_1+k_2+...+k_m,$$

q is the total number of fuzzy terms in the matrix of knowledge (Table 1) and
F is an input-output connection operator corresponding to the correlations (4), (5) and (7), (8).

Let us define the training data as a set of M pairs as

$$\left\{\left(X_p,\hat{y}_p\right)\right\},\quad p=\overline{1,M}, \tag{17}$$

where

$X_p=\left(x_1^p,x_2^p,...,x_n^p\right)$ and $\hat{y}_p$ are the input vector and its corresponding output for the p-th input-output pair.

The vector of unknown parameters (W, B, C) that minimize the difference between theory (16) and experiment (17) can be found by the least squares method. That is why the problem of fine tuning of the fuzzy model (16) can be formalized by finding such a vector (W, B, C) that satisfies the restrictions

$$w_i\in\left[\underline{w}_i,\overline{w}_i\right],\ i=\overline{1,N},\ b_j\in\left[\underline{b}_j,\overline{b}_j\right],\ c_j\in\left[\underline{c}_j,\overline{c}_j\right],\ j=\overline{1,q}$$

and minimizes

$$\sum_{p=1}^{M}\left[F\left(X_p,W,B,C\right)-\hat{y}_p\right]^2 \tag{18}$$

To solve this nonlinear optimization problem, we apply a combination of genetic [16] and gradient descent [17] algorithms. Using genetic algorithms we can quickly find a permissible rough decision, and after that improve the rough decision using gradient descent.

The string needed in the genetic algorithm may be represented as

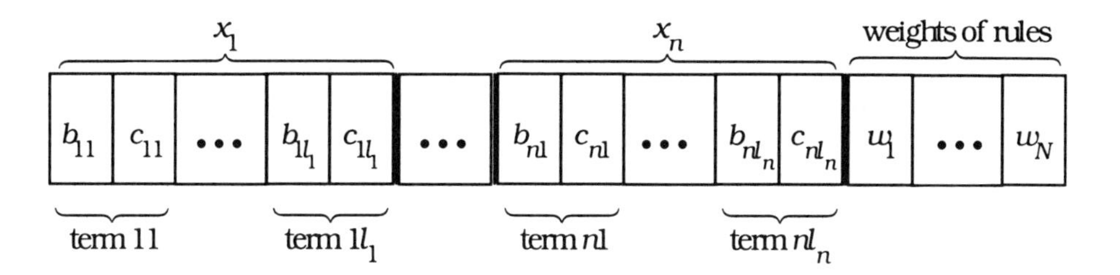

where $l_1, l_2, \dots, l_n$ are indexes of the fuzzy terms used for variables $x_1, x_2, \dots, x_n$ evaluation, and $l_1 + l_2 + \dots + l_n = q$.

For the crossover operation, we use $(n+1)$ random cut-points for the two parent strings

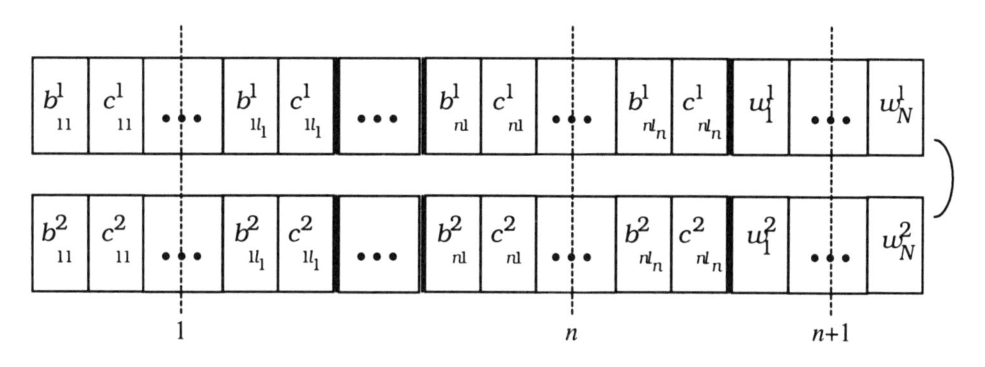

As a result, we obtain two new child strings as

b^1_{11}	c^1_{11}	...	$b^2_{1l_1}$	$c^2_{1l_1}$	...	b^1_{n1}	c^1_{n1}	...	$b^2_{nl_n}$	$c^2_{nl_n}$	w^1_1	...	w^2_N
b^2_{11}	c^2_{11}	...	$b^1_{1l_1}$	$c^1_{1l_1}$	...	b^2_{n1}	c^2_{n1}	...	$b^1_{nl_n}$	$c^1_{nl_n}$	w^2_1	...	w^1_N

4.2. Quality Evaluation of Fuzzy Inference

Let $y_F(X,M)$ be the fuzzy model (16) after tuning by *M* pairs of training data. To evaluate the quality of the fuzzy inference, the following criterion can be used:

$$R(M)=\frac{1}{\left|\{X_i\}\right|}\sqrt{\sum_{\{X_i\}}\left[y_F(X_i,M)-\hat{y}_i\right]^2}, \qquad (19)$$

where

$y_F(X_i,M)$ and $\hat{y}_i$ are the inferred and experimental outputs in a point:
$X_i=(x_1^i,x_2^i,...,x_n^i)\in[\underline{x_1},\overline{x_1}]\times[\underline{x_2},\overline{x_2}]\times...\times[\underline{x_n},\overline{x_n}]$,
$\{X_i\}$ is a set of elements of type X_i,
$\left|\{X_i\}\right|$ is the power of set of $\{X_i\}$.

The proposed criterion (19) is similar to a mean-square deviation between the inferred and experimental outputs corresponding to one element of the input space. The dependence of the *R(M)*– criterion (19) on the number *M* of training data pairs can be used to observe the dynamics of fuzzy model learning.

4.3. Computer Simulation

The aim of the computer simulation was to demonstrate the methodology of fuzzy models tuning for two nonlinear objects: 1) single input–single output, 2) two inputs-single output. The target systems were several analytical models. These models were used to generate training data. The gradient descent method was used to solve the optimization problem (18). The dynamics of fuzzy model learning was studied by criterion (19).

4.3.1. Experiment 1

Let us consider the object with single input $x\in[0,1]$ and single output $y\in[0.05,0.417]$. The analytical model as a target system is given by the formula

$$y=f(x)=\frac{(5x-1.1)(4x-2.9)(3x-2.1)(11x-11)(3x-0.05)+10}{40}.$$

We use this formula to analyze the ability of the fuzzy rule-based representation to fit the fifth-order model shown by the dotted line in Figure 5a.

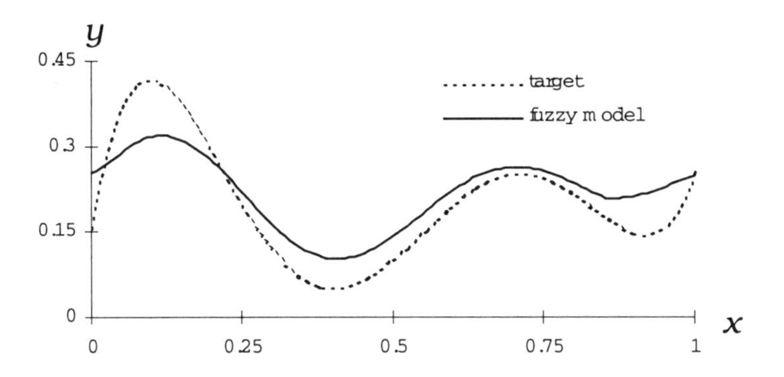

(a) comparison with target

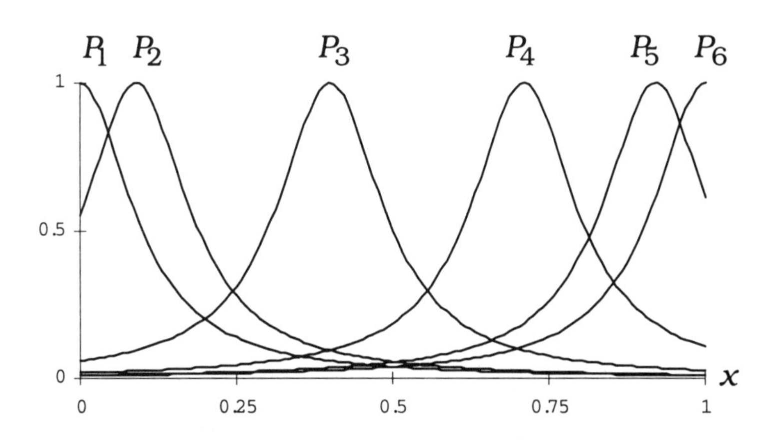

(b) membership functions of the fuzzy terms

Figure 5. Fuzzy model for *f(x)*, before tuning.

The rough fuzzy rules were formed by observing the behavior of *f(x)* as follows:

IF $x = P_1$, THEN $y \in [0.14, 0.23]$ (with weight w_1),
IF $x = P_2$, THEN $y \in [0.32, 0.42]$ (with weight w_2),
IF $x = P_3$, THEN $y \in [0.05, 0.14]$ (with weight w_3),
IF $x = P_4$, THEN $y \in [0.14, 0.23]$ (with weight w_4),
IF $x = P_5$, THEN $y \in [0.05, 0.14]$ (with weight w_5),
IF $x = P_6$, THEN $y \in [0.23, 0.32]$ (with weight w_6),

where P_1 = about 0, P_2 = about 0.09, P_3 = about 0.4, P_4 = about 0.71, P_5 = about 0.92, P_6 = about 1 are fuzzy terms, whose membership functions are shown in Figure 5b.

Before the tuning, all weights of the fuzzy rules were the same, that is w_i=1, $i=\overline{1,6}$. After tuning, we obtain the improved fuzzy model shown in Figure 6.

The parameters of the membership functions of the fuzzy terms, and the weights of the rules in the fuzzy model $y = f(x)$ before and after tuning, are shown in Table 2 and Table 3, respectively. The dynamics of fuzzy rule learning for the model $y = f(x)$ is shown in Figure 7.

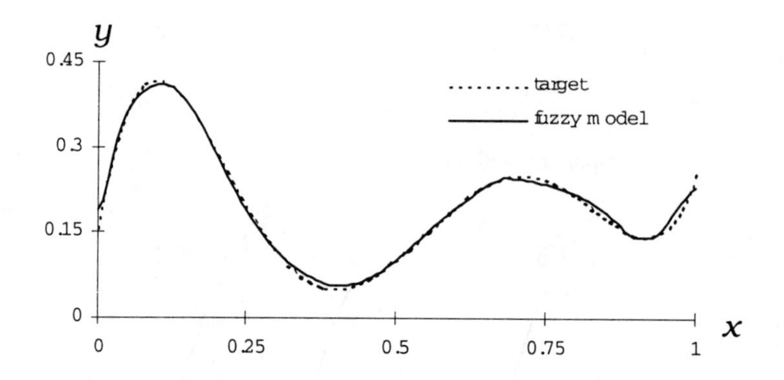

(a) comparison with target

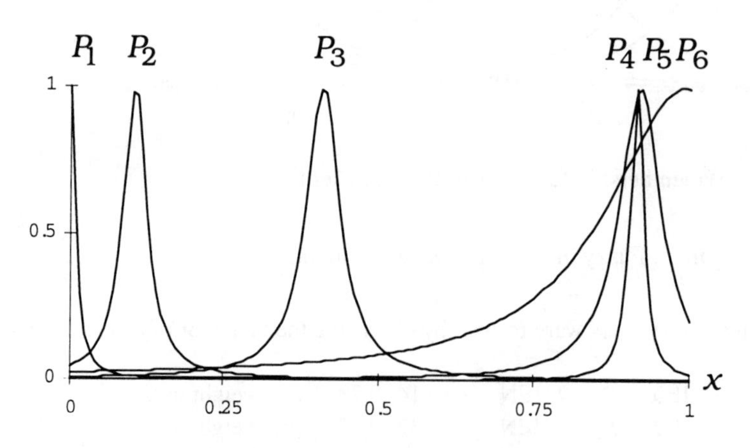

(b) membership functions of fuzzy terms

Figure 6. The fuzzy model for $f(x)$, after tuning.

Table 2. Parameters of membership functions in the fuzzy model $y = f(x)$

Fuzzy Term	Parameters			
	before tuning		after tuning	
	b	c	b	c
D_1	0.00	0.1	0.0000	0.0100
D_2	0.09	0.1	0.1050	0.0229
D_3	0.40	0.1	0.4094	0.0327
D_4	0.71	0.1	0.9148	0.0127
D_5	0.92	0.1	0.9196	0.0406
D_6	1.00	0.1	0.9904	0.1504

Table 3. Weights of rules in the fuzzy model $y = f(x)$

Weights of rules	w_1	w_2	w_3	w_4	w_5	w_6
before tuning	1	1	1	1	1	1
after tuning	1.000	0.501	0.401	0.300	0.296	0.800

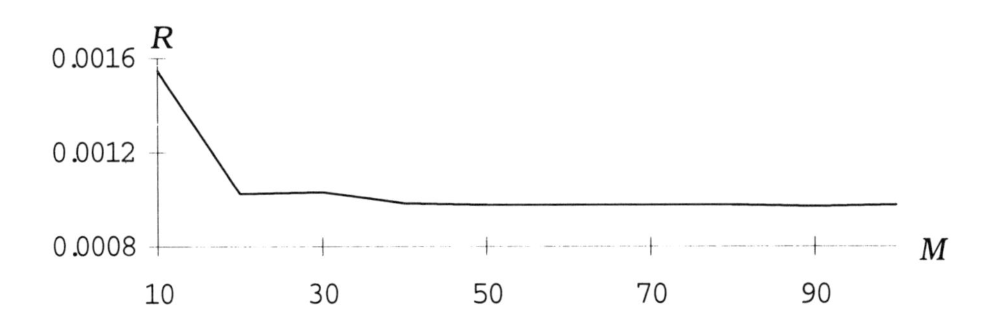

Figure 7. The dynamics of fuzzy model $f(x)$ learning.

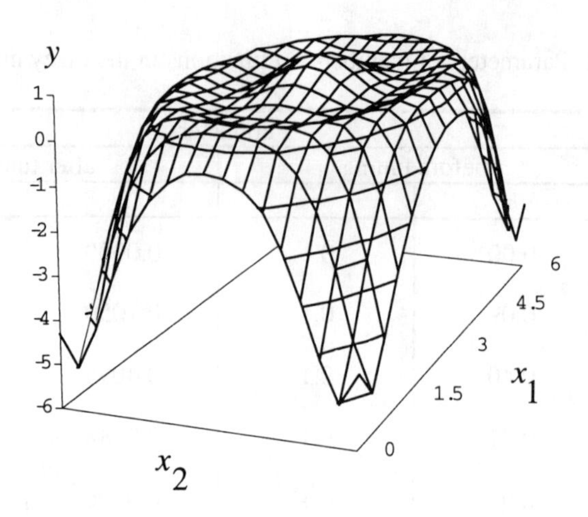

Figure 8. Behavior of $f(x_1, x_2)$.

4.3.2. Experiment 2

Let us consider the object with two inputs: $x_1 \in [0,6]$, $x_2 \in [0,6]$, and a single output $y \in [-5.08, 0.855]$. The target model (Figure 8) is given by the formula

$$y = f(x_1, x_2) = \frac{1}{40}(2z - 0.9)\,(7z - 1)\,(17z - 19)\,(15z - 2),$$

where

$$z = \frac{(x_1 - 3)^2 + (x_2 - 3)^2}{18}.$$

The rough fuzzy knowledge base was formed by considering the behavior of $f(x_1, x_2)$ resulting in a set of 49 rules as

$$\text{IF } x_1 = P_i \text{ AND } x_2 = Q_i \text{, THEN } y = B_j \text{, } i = \overline{1,7} \text{, } j = \overline{1,5}.$$

These rules are represented together in the 7×7 matrix

	P_1	P_2	P_3	P_4	P_5	P_6	P_7
Q_1	B_2	B_1	B_3	B_4	B_3	B_1	B_2
Q_2	B_1	B_3	B_4	B_5	B_4	B_3	B_1
Q_3	B_3	B_4	B_4	B_4	B_4	B_4	B_3
Q_4	B_4	B_5	B_4	B_5	B_4	B_5	B_4
Q_5	B_3	B_4	B_4	B_4	B_4	B_4	B_3
Q_6	B_1	B_3	B_4	B_5	B_4	B_3	B_1
Q_7	B_2	B_1	B_3	B_4	B_3	B_1	B_2

where

$P_1 = Q_1 =$ about 0, $P_2 = Q_2 =$ about 0.5, $P_3 = Q_3 =$ about 1.5,

$P_4 = Q_4 =$ about 3, $P_5 = Q_5 =$ about 4.5, $P_6 = Q_6 =$ about 5.5,

$P_7 = Q_7 =$ about 6 are fuzzy terms with membership functions shown in Figure 9a, $B_1 = [-5.08, -4.5)$, $B_2 = [-4.5, -3.0)$, $B_3 = [-3.0, -0.5)$, $B_4 = [-0.5, 0)$, $B_5 = [0, 0.855]$.

The weights of all fuzzy rules before tuning were the same and equal to one, resulting in the rough fuzzy model shown in Figure 9. The improved fuzzy model after tuning and the dynamics of fuzzy model learning are shown in Figure 10 and Figure 11, respectively. The modification of the dynamics of the weights of the fuzzy rules is represented in the matrix **WR** as:

WR =

	P_1	P_2	P_3	P_4	P_5	P_6	P_7
Q_1	1/0.01	1/0.42	1/0.03	1/0.64	1/0.05	1/0.42	1/0.03
Q_2	1/0.61	1/0.03	1/0.27	1/0.26	1/0.27	1/0.05	1/0.61
Q_3	1/0.05	1/0.44	1/0.71	1/0.03	1/0.71	1/0.44	1/0.09
Q_4	1/0.87	1/0.34	1/0.01	1/0.90	1/0.02	1/0.34	1/0.87
Q_5	1/0.03	1/0.43	1/0.70	1/0.05	1/0.70	1/0.43	1/0.01
Q_6	1/0.60	1/0.03	1/0.30	1/0.29	1/0.30	1/0.05	1/0.60
Q_7	1/0.03	1/0.48	1/0.05	1/0.83	1/0.09	1/0.48	1/0.01

Here, the numerators and denominators are the rules weights before and after tuning, respectively. It is easy to see from the **WR**-matrix that many rules with small weights can be eliminated from the fuzzy rule-based model of $f(x_1, x_2)$.

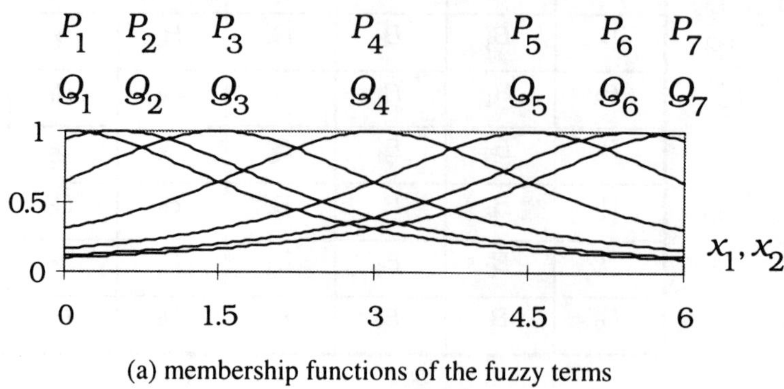

(a) membership functions of the fuzzy terms

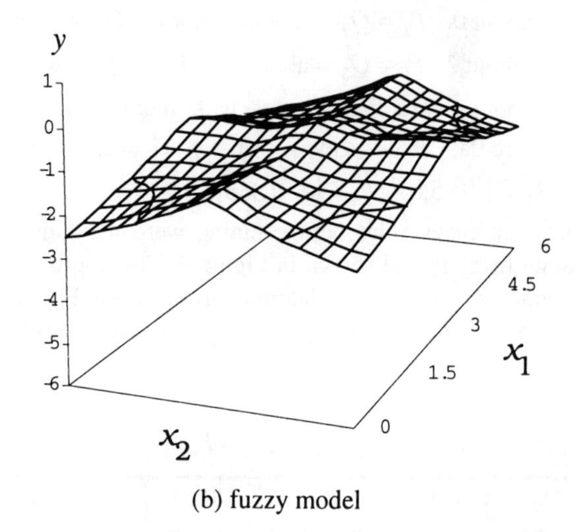

(b) fuzzy model

Figure 9. Fuzzy model $f(x_1, x_2)$ before tuning.

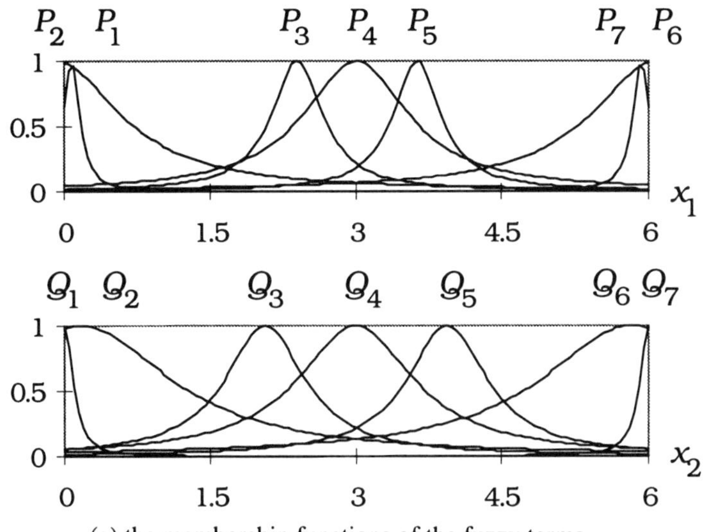

(a) the membership functions of the fuzzy terms.

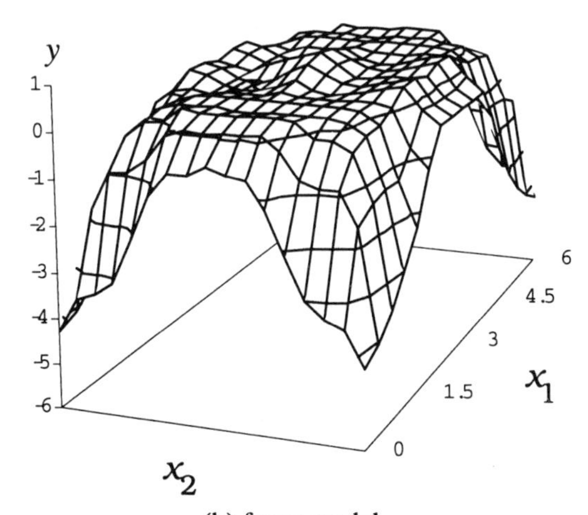

(b) fuzzy model

Figure 10. Fuzzy model $f(x_1, x_2)$, after tuning

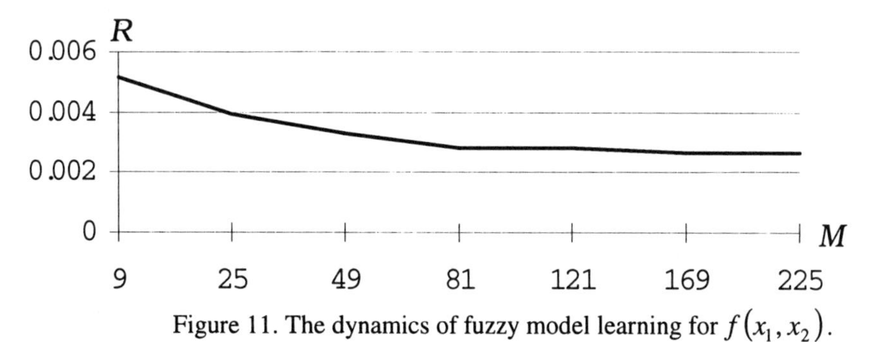

Figure 11. The dynamics of fuzzy model learning for $f(x_1, x_2)$.

5. Fine Tuning Of The Fuzzy Rules With Discrete Output

This section presents the optimization problem we need to solve for tuning the fuzzy model in the case of discrete output (Figure 1b). An example of computer simulation demonstrates the tuning process.

5.1. Tuning as a Problem of Optimization

The correlations (4) and (5) allow us to calculate the vector of inferred membership functions for all types of decisions (diagnoses), as

$$\left(\mu^{d_j}(X, W, B, C)\,,\ j = \overline{1, m}\right) \tag{20}$$

where *X, W, B* and *C* are vectors which have been defined in Section 4.1.

Let us define the desirable vector of membership degrees as

$$\left.\begin{array}{l} (1, 0, \ldots, 0)\text{ for the decision } d_1 \\ (0, 1, \ldots, 0)\text{ for the decision } d_2 \\ \ldots \\ (0, 0, \ldots, 1)\text{ for the decision } d_m \end{array}\right\} \tag{21}$$

Let us define the training data as a set of *L* pairs as:

$$\left\{\left(X_l, \hat{d}_l\right)\right\},\ l = \overline{1, L}\,, \tag{22}$$

where

$X_l = \left(x_1^l, x_2^l, \ldots, x_n^l\right)$ and $\hat{d}_l$ are the input vector and corresponding output for the *l*-th pair, $\hat{d}_l \in \{d_1, d_2, \ldots, d_m\}$.

The vector of unknown parameters (*W, B, C*) which minimizes the difference between theory (20) and experiment (22) can be found by least square method. That is why the problem of fuzzy model (20) fine tuning can be formalized by finding such a vector (*W, B, C*) which satisfies the restrictions

$$w_i \in \left[\underline{w}_i, \overline{w}_i\right],\ i = \overline{1, N}\,,\ b_j \in \left[\underline{b}_j, \overline{b}_j\right],\ c_j \in \left[\underline{c}_j, \overline{c}_j\right],\ j = \overline{1, q}$$

and minimizes the sum with respect to the *W, B, C* variables

$$\sum_{l=1}^{L}\left[\sum_{j=1}^{m}\left[\mu^{d_j}(X_l, W, B, C) - \mu^{d_j}(X_l)\right]^2\right] \tag{23}$$

where, according to (21)

$$\mu^{d_j}(X_l)=\begin{cases}1, \ if \ \hat{d}_j = d_j \\ 0, \ if \ \hat{d}_j \neq d_j\end{cases}, \quad j=\overline{1,m} .$$

To solve this optimization problem we use a combination of genetic algorithm and the gradient descent method.

5.2. Quality Evaluation of Fuzzy Inference

Let Q be the total number of situations (or patients) used for testing the fuzzy expert system. To evaluate the quality of fuzzy inference in our case of discrete output $y \in \{d_1, d_2, ..., d_m\}$, it is necessary to make a distribution of Q situations according to the tree shown in Figure 12, where

Q_j is the number of situations demanding the decision d_j, that is

$$Q = Q_1 + Q_2 + ... + Q_m \ ,$$

Q_{ji} is the number of situations demanding the decision d_j, but recognized by fuzzy inference as decision d_i , that is

$$Q_j = Q_{j1} + Q_{j2} + ... + Q_{jm} \ , \ j = \overline{1,m}$$

According to Figure 12, we can evaluate the quality of the fuzzy inference by

$$\hat{P}_j = \frac{Q_{jj}}{Q_j} \quad , \quad \hat{P}_{ji} = \frac{Q_{ji}}{Q_j} \ , \quad \hat{P} = \frac{1}{Q}\sum_{j=1}^{m} Q_{jj} \ , \tag{24}$$

where

$\hat{P}_j$ is the probability of correct inference of decision d_j,

$\hat{P}_{ji}$ is the probability of incorrect decision d_i when decision d_j was correct,

$\hat{P}$ is the average probability of correct decision inference.

By observing the dependence of the probabilities (24) from the number L of training data, we can study the dynamics of fuzzy expert system learning.

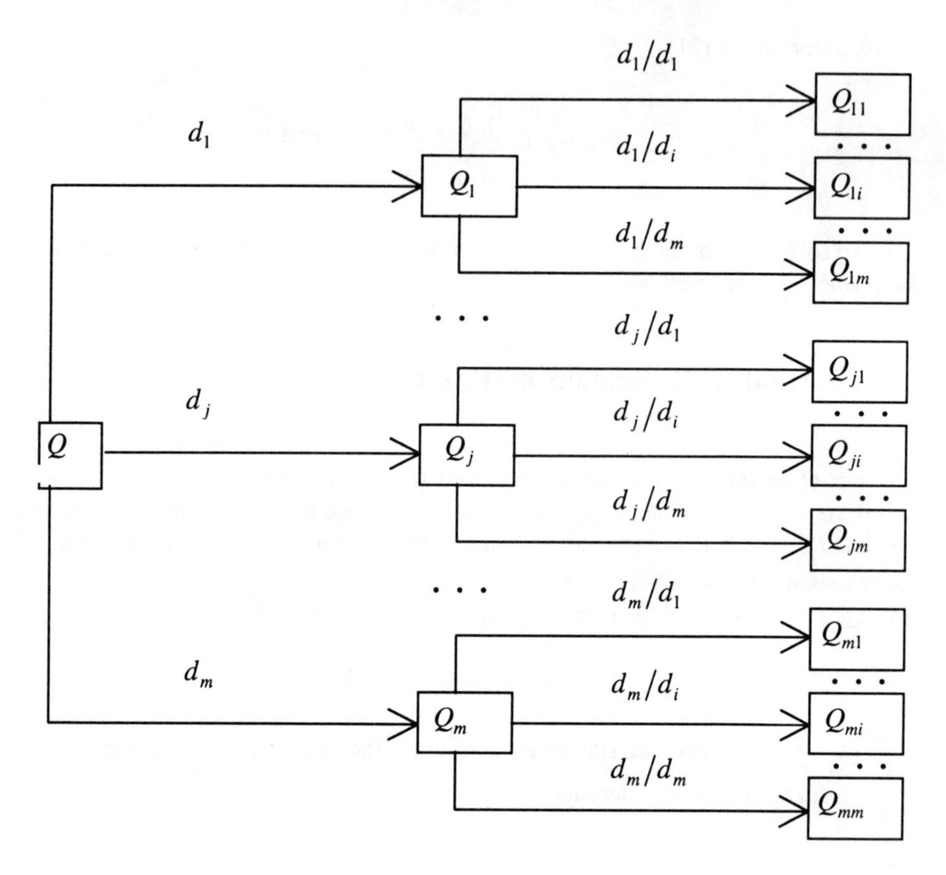

Figure 12. Tree of situations distribution.

5.3. Computer Simulation

The aim of this computer simulation was to demonstrate the effectiveness of the proposed method by comparing the inferred (theoretical) and the relevant (desired) decisions. To produce fuzzy if-then rules, and training and testing data, we used a two-dimensional space (x_1, x_2) divided into five areas of decision, as shown in Figure 13, where

①: $x_1 \in [0.5, 0.625]$, $x_2 = 32(x_1 - 0.5)^2 + 0.5$,

②: $x_1 \in [0, 0.25]$, $x_2 = -16(x_1 - 0.25)^4 + 0.8125$,

③: $x_1 \in [0.25, 0.5]$, $x_2 = -80(x_1 - 0.25)^4 + 0.8125$,

④: $x_2 \in [0, 0.5]$, $x_1 = (0.5 - x_2)(2 - x_2)(-0.1 - x_2) + 0.5$,

⑤: $x_2 \in [0.25, 0.5]$, $x_1 = \sqrt{(0.5 - x_2)} + 0.5$,

⑥: $x_1 \in [0.75, 0.1]$, $x_2 = -80(x_1 - 1)^4 + 0.75$.

Figure 13 corresponds to the nonlinear object $y = f(x_1, x_2)$ with discrete output $y \in \{d_1, d_2, d_3, d_4, d_5\}$. To obtain fuzzy if-then rules, we divided the input space (x_1, x_2) into 5×5=25 areas according to the fuzzy terms: *L* –low, *lA* – lower than average, *A* – average, *hA* – higher than average, *H* – high, as shown in Figure 14. In this figure, it is easy to see that the nonlinear object $y = f(x_1, x_2)$ can be described by the fuzzy if-then rules shown in the matrix of knowledge in Table 4.

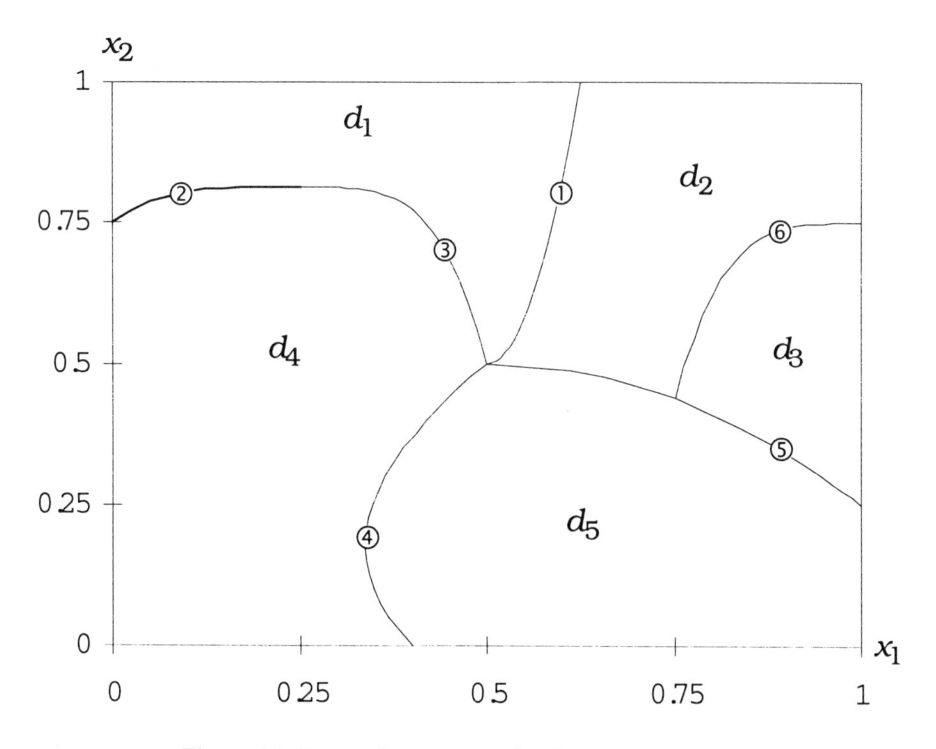

Figure 13. Space of parameters for five types of decisions.

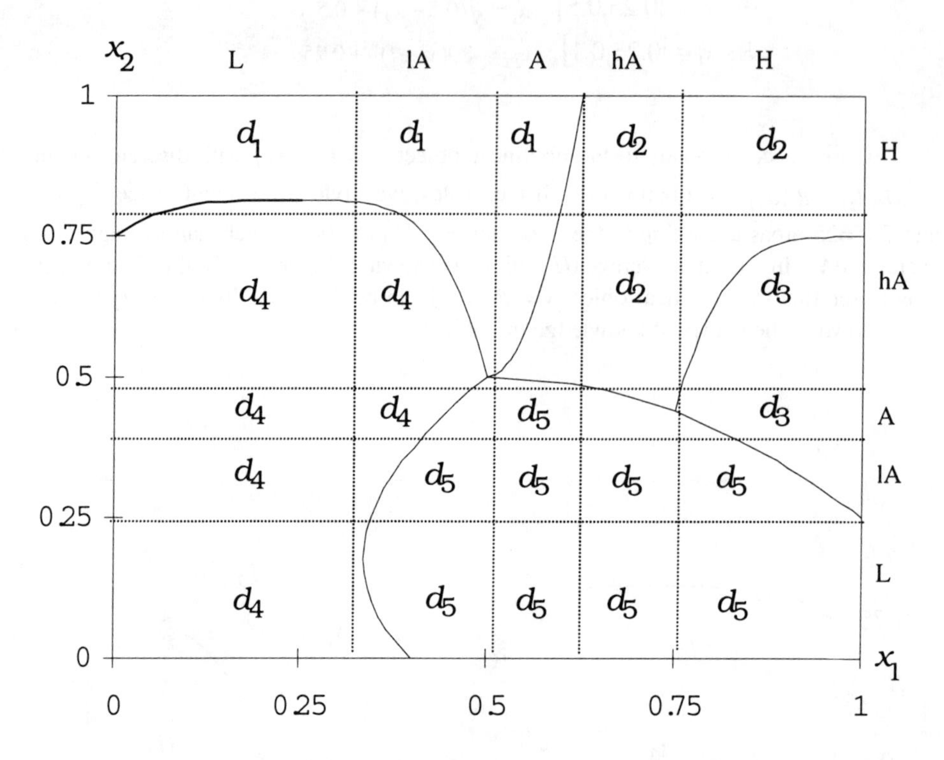

Figure 14. Division of input space (x_1, x_2) into the 25 areas.

In Figure 15b, some decisions, belonging to one class were inferred to be in some other class – for instance, on the boundary between classes d_4 and d_5. After solving the tuning problem (23), the membership functions (Figure 16a) and the weights of rules (Table 5) were changed and the results of the fuzzy inference were improved, as shown in Figure 16b.

The dynamics of the learning process of the fuzzy model, as obtained by (24), is shown in Figure 17.

Table 4

Matrix of knowledge for Figure 14

N	x_1	x_2	y
11	L	H	d_1
12	lA	H	
13	A	H	
21	hA	hA	d_2
22	hA	H	
23	H	H	
31	H	hA	d_3
32	H	A	
41	L	hA	d_4
42	L	A	
43	L	lA	
44	L	L	
45	lA	hA	
46	lA	A	
51	lA	L	d_5
52	A	A	
53	A	lA	
54	A	L	
55	hA	lA	
56	hA	L	
57	H	L	

Table 5

Weights of rules before (w_b) and after (w_a) tuning

N	w_b	w_a
11	1.0	0.943
12	1.0	0.987
13	1.0	0.935
21	1.0	1.000
22	1.0	1.000
23	1.0	1.000
31	1.0	0.909
32	1.0	0.984
41	1.0	1.000
42	1.0	0.999
43	1.0	0.998
44	1.0	0.998
45	1.0	0.984
46	1.0	1.000
51	1.0	0.988
52	1.0	0.965
53	1.0	0.986
54	1.0	1.000
55	1.0	0.977
56	1.0	1.000
57	1.0	1.000

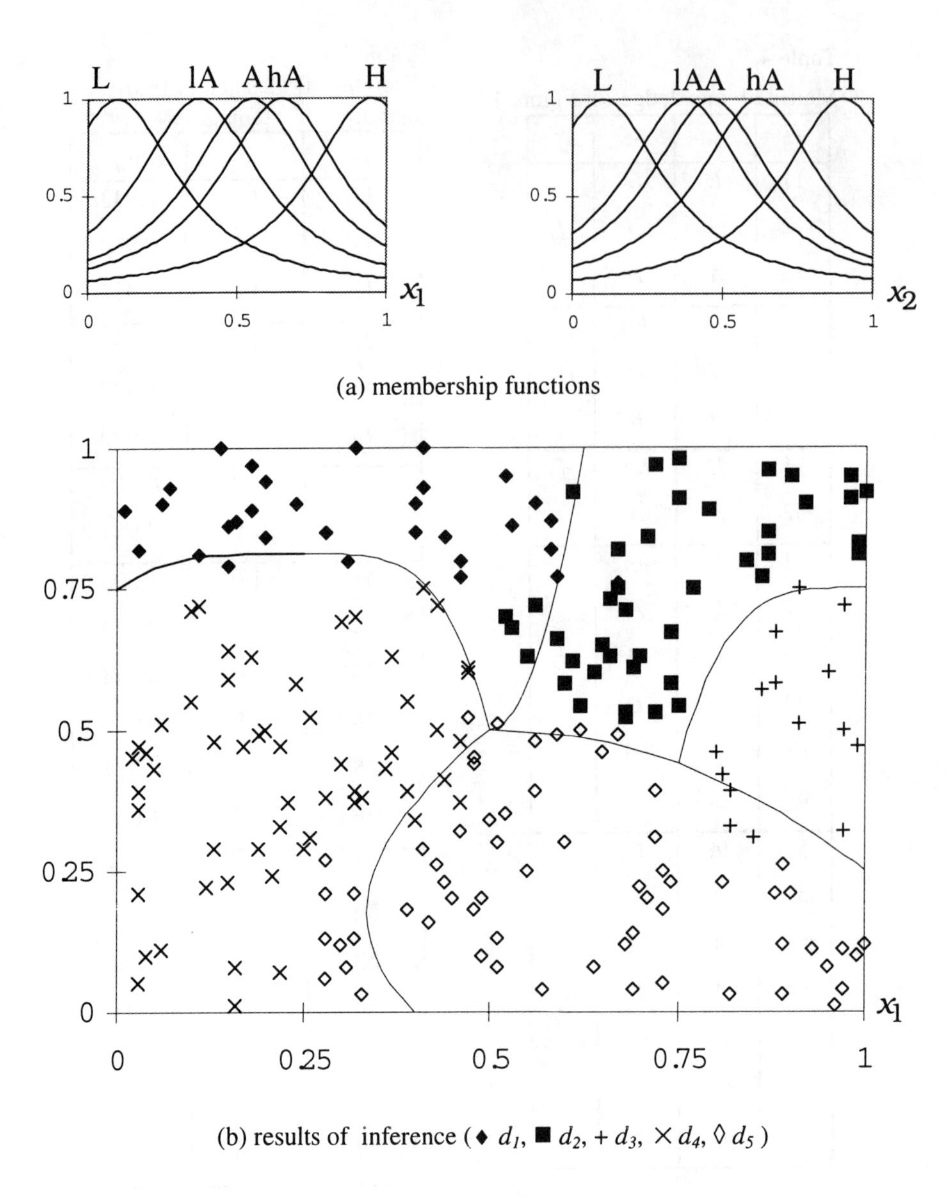

(a) membership functions

(b) results of inference (♦ d_1, ■ d_2, + d_3, × d_4, ◊ d_5)

Figure 15. The fuzzy model with discrete output, before tuning.

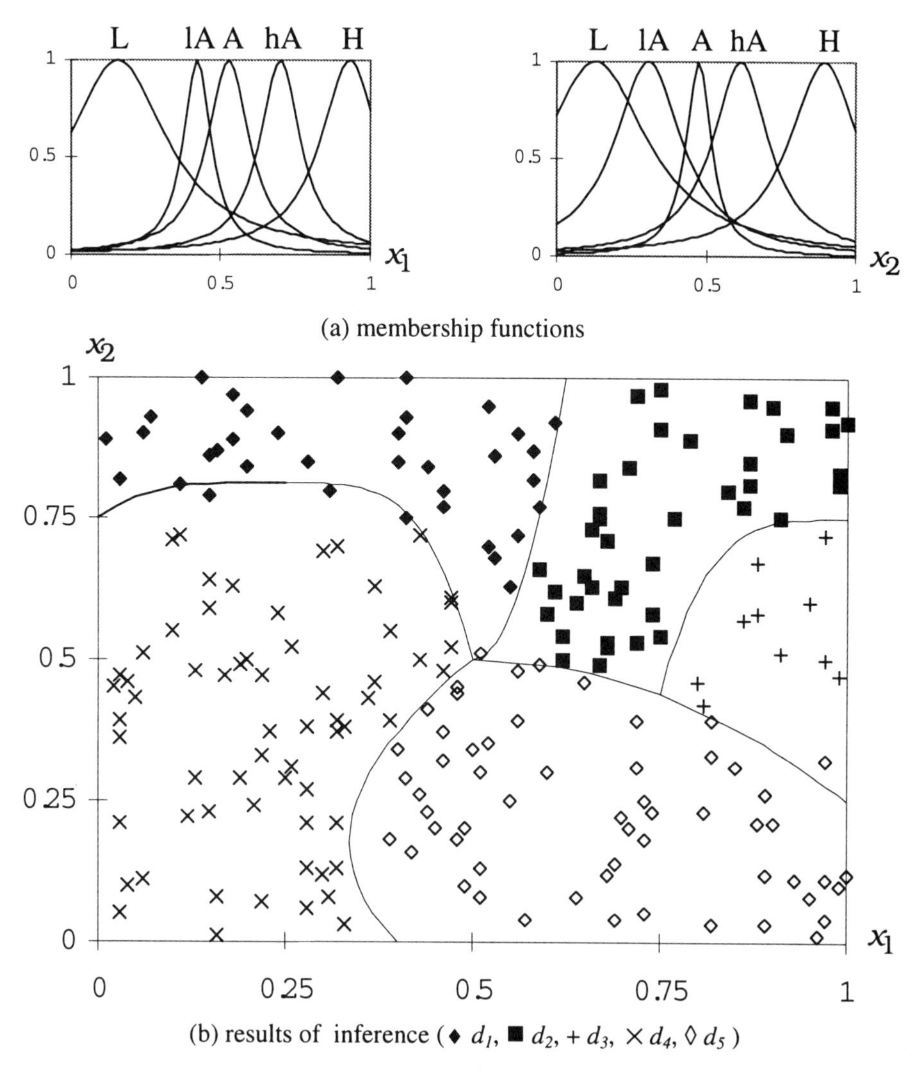

(a) membership functions

(b) results of inference (♦ d_1, ■ d_2, + d_3, × d_4, ◊ d_5)

Figure 16. Fuzzy model with discrete output after tuning.

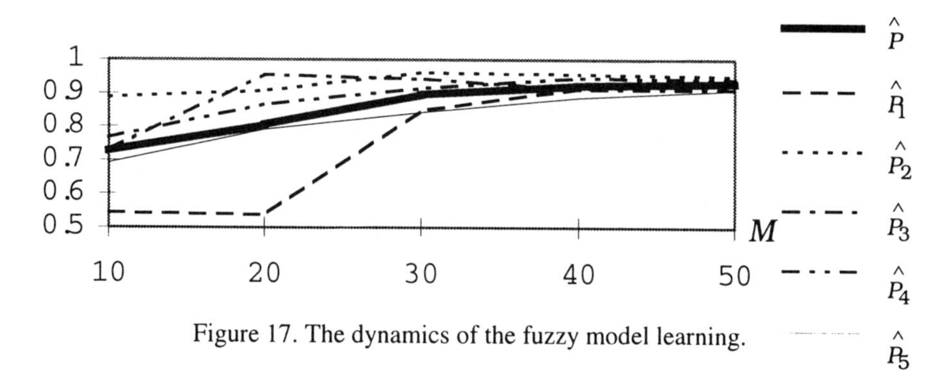

Figure 17. The dynamics of the fuzzy model learning.

6. APPLICATION TO DIFFERENTIAL DIAGNOSIS OF ISCHEMIA HEART DISEASE

Ischemia heart disease (IHD) is one of the most widespread sources of disability, and it has a high death rate among adults. The success of IHD treatment is defined by the achievement of some differential diagnosis, that is, a classification as one of the complication levels accepted in clinical practice: cardiac neurocirculatory dystonia or stenocardia. The quality of medical diagnosis strongly depends on the qualification of the diagnostician. Therefore, a computer support system for diagnostic decision making in such conditions is of particular significance. This section shows how to build the fuzzy rule-based system for the differential diagnosis of IHD by the proposed methodology.

6.1. DIAGNOSIS TYPES AND PARAMETERS OF PATIENT'S STATE

According to the current clinical practice [4,7], the complication of IHD will be defined at the levels as follows (from the lowest to the highest):

d_1 - neurocirculatory dystonia (NCD) of light complication,
d_2 - NCD of average complication,
d_3 - NCD of the heavy complication,
d_4 - stenocardia of the first functional disability degree,
d_5 - stenocardia of the second functional disability degree,
d_6 - stenocardia of the third functional disability degree.

The above mentioned levels $d_1 \div d_6$ are considered the types of diagnosis which should be identified. While making the diagnosis of IHD of a specific patient, we should take into consideration the next main parameters defined by laboratory tests (possible variation ranges are indicated in round brackets where c. u. is a conventional unit).

x_1 is the age of the patient (31÷58 years),
x_2 is the double product (DP) of pulse and blood pressure (128÷405 c.u.),
x_3 is the tolerance to physical loads (90÷1200 kgm/min),
x_4 is the increase of DP per kg of the patient body weight (0.6÷3.9 c.u.),
x_5 is the increase of DP per kg of load (0.09÷0.56 c.u.),
x_6 is the adenosine-triphosphoric acid - ATP (34.48÷69.49 mmol/l),
x_7 is the adenosine-diphosphoric acid - ADP (11.9÷29.4 mmol/l),
x_8 is the adenosine-monophosphoric acid - AMP (3.6÷27.1 mmol/l),
x_9 is the coefficient of phosphorylation (1.0÷5.7 c.u.),
x_{10} is the max. oxygen consumption per kg of patient weight (7.4÷40.9 mlitre/min × kg),
x_{11} is the increase of DP in response to submaximal load (46÷352 c.u.),

x_{12} is the ratio factor of milk and pyruvic acid (3.9÷30.2 c.u.).

The aim of the diagnosis is to translate a set of specific parameters $x_1 \dots x_{12}$ into a decision d_j ($j = 1 \dots 6$).

6.2. Fuzzy Rules

The structure of the model for differential diagnosis of IHD is shown in Figure 18, which corresponds to the following hierarchical tree of logic inferences:

$$d = f_d(x_1, y, z) , \tag{25}$$

$$y = f_y(x_2, x_3, x_4, x_5, x_{10}, x_{11}) , \tag{26}$$

$$z = f_z(x_6, x_7, x_8, x_9, x_{12}) \tag{27}$$

where d is the danger of IHD measured by levels $d_1 \div d_6$, y is the instrumental danger, and z is the biochemical danger.

The fuzzy if-then rules which correspond to the relations (25)-(27) are represented in Tables 6 to 8 using fuzzy terms as: L – low, lA – lower than average, A – average, hA – higher than average, H – high.

6.3. Fuzzy Logic Equation

Using Tables 6 to 8 and operations • (AND – min) and $\vee$ (OR – max), it is easy to write the system of fuzzy logic equations which connect the membership functions of diagnosis and parameters of the patient state as

$$\begin{aligned}
\mu^{d_1}(d) &= \mu^{L}(x_1)\cdot\mu^{L}(y)\cdot\mu^{L}(z) \\
&\vee \mu^{L}(x_1)\cdot\mu^{lA}(y)\cdot\mu^{lA}(z) \\
&\vee \mu^{lA}(x_1)\cdot\mu^{lA}(y)\cdot\mu^{L}(z) \\
&\dots \\
\mu^{d_6}(d) &= \mu^{H}(x_1)\cdot\mu^{H}(y)\cdot\mu^{H}(z) \\
&\vee \mu^{hA}(x_1)\cdot\mu^{H}(y)\cdot\mu^{hA}(z) \\
&\vee \mu^{A}(x_1)\cdot\mu^{H}(y)\cdot\mu^{hA}(z)
\end{aligned} \tag{28}$$

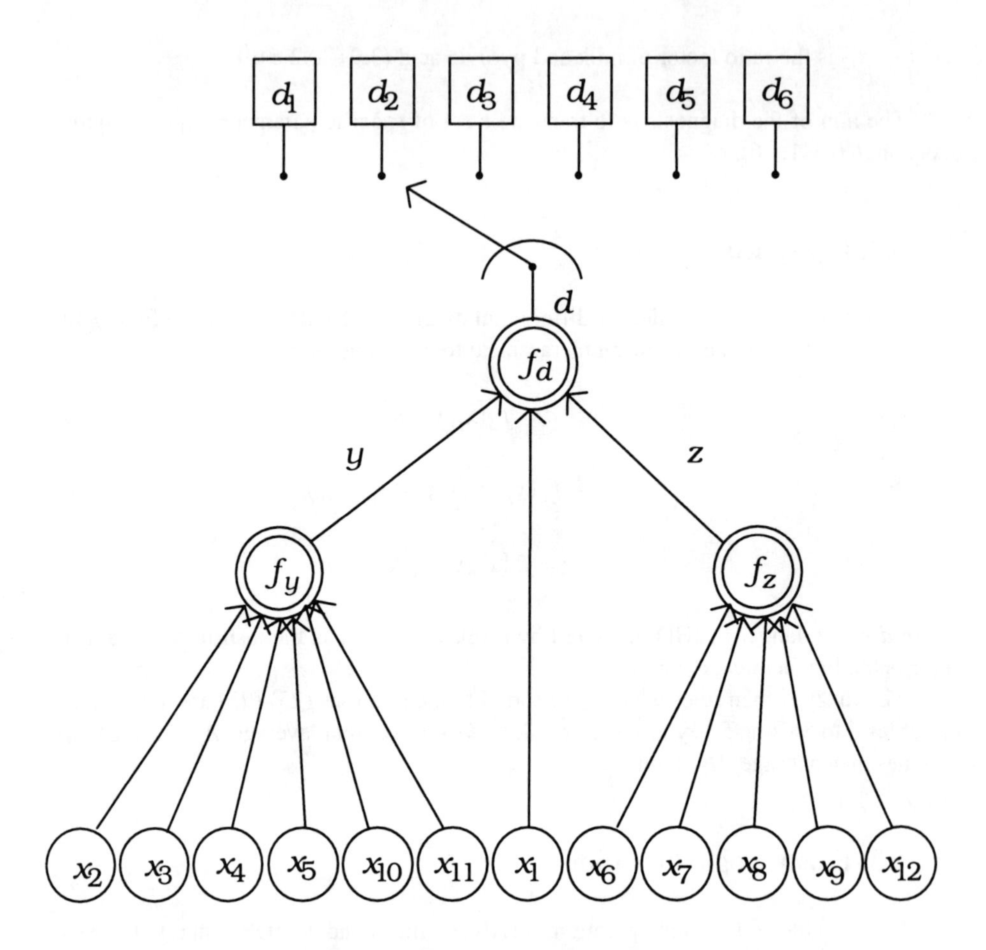

Figure 18. Diagnostic model structure.

$$
\begin{aligned}
\mu^{L}(y) = {} & \mu^{H}(x_2)\cdot\mu^{H}(x_3)\cdot\mu^{H}(x_4)\cdot\mu^{L}(x_5)\cdot\mu^{H}(x_{10})\cdot\mu^{H}(x_{11}) \\
& \vee\,\mu^{H}(x_2)\cdot\mu^{hA}(x_3)\cdot\mu^{H}(x_4)\cdot\mu^{lA}(x_5)\cdot\mu^{H}(x_{10})\cdot\mu^{H}(x_{11}) \\
& \vee\,\mu^{hA}(x_2)\cdot\mu^{H}(x_3)\cdot\mu^{hA}(x_4)\cdot\mu^{L}(x_5)\cdot\mu^{H}(x_{10})\cdot\mu^{H}(x_{11}) \\
& \ldots \\
\mu^{H}(y) = {} & \mu^{L}(x_2)\cdot\mu^{L}(x_3)\cdot\mu^{L}(x_4)\cdot\mu^{hA}(x_5)\cdot\mu^{L}(x_{10})\cdot\mu^{L}(x_{11}) \\
& \vee\,\mu^{lA}(x_2)\cdot\mu^{L}(x_3)\cdot\mu^{lA}(x_4)\cdot\mu^{H}(x_5)\cdot\mu^{L}(x_{10})\cdot\mu^{lA}(x_{11}) \\
& \vee\,\mu^{L}(x_2)\cdot\mu^{lA}(x_3)\cdot\mu^{lA}(x_4)\cdot\mu^{hA}(x_5)\cdot\mu^{L}(x_{10})\cdot\mu^{L}(x_{11})
\end{aligned}
\tag{29}
$$

$$
\begin{aligned}
\mu^{L}(z) = {} & \mu^{H}(x_6)\cdot\mu^{H}(x_7)\cdot\mu^{H}(x_8)\cdot\mu^{H}(x_9)\cdot\mu^{H}(x_{12}) \\
& \vee\mu^{hA}(x_6)\cdot\mu^{H}(x_7)\cdot\mu^{hA}(x_8)\cdot\mu^{hA}(x_9)\cdot\mu^{hA}(x_{12}) \\
& \vee\mu^{H}(x_6)\cdot\mu^{hA}(x_7)\cdot\mu^{H}(x_8)\cdot\mu^{A}(x_9)\cdot\mu^{hA}(x_{12}) \qquad (30) \\
& \dots \\
\mu^{H}(z) = {} & \mu^{L}(x_6)\cdot\mu^{L}(x_7)\cdot\mu^{L}(x_8)\cdot\mu^{L}(x_9)\cdot\mu^{lA}(x_{12}) \\
& \vee\mu^{lA}(x_6)\cdot\mu^{L}(x_7)\cdot\mu^{lA}(x_8)\cdot\mu^{L}(x_9)\cdot\mu^{L}(x_{12}) \\
& \vee\mu^{L}(x_6)\cdot\mu^{lA}(x_7)\cdot\mu^{lA}(x_8)\cdot\mu^{L}(x_9)\cdot\mu^{lA}(x_{12})
\end{aligned}
$$

The total number of fuzzy logic equations (28)–(30) is 16. Note that we do not use the weights of rules in (28)–(30) because all weights before tuning are equal to one.

Table 6. Knowledge about relation (25)

x_1	y	z	d
L	*L*	*L*	
L	*lA*	*lA*	d_1
lA	*lA*	*L*	
lA	*lA*	*lA*	
A	*lA*	*lA*	d_2
lA	*lA*	*A*	
A	*lA*	*A*	
hA	*hA*	*lA*	d_3
hA	*A*	*A*	
hA	*A*	*hA*	
A	*hA*	*hA*	d_4
lA	*hA*	*hA*	
A	*H*	*A*	
hA	*hA*	*H*	d_5
H	*hA*	*hA*	
H	*H*	*H*	
hA	*H*	*hA*	d_6
A	*H*	*hA*	

Table 7. Knowledge about relation (26)

x_2	x_3	x_4	x_5	x_{10}	x_{11}	y
H	*H*	*H*	*L*	*H*	*H*	
H	*hA*	*H*	*lA*	*H*	*H*	*L*
hA	*H*	*hA*	*L*	*H*	*H*	
hA	*hA*	*H*	*lA*	*H*	*hA*	
H	*H*	*hA*	*A*	*H*	*H*	*lA*
hA	*hA*	*H*	*lA*	*hA*	*hA*	
A	*A*	*A*	*A*	*A*	*A*	
hA	*hA*	*A*	*lA*	*hA*	*A*	*A*
A	*hA*	*hA*	*A*	*hA*	*hA*	
lA	*A*	*lA*	*hA*	*lA*	*lA*	
lA	*lA*	*A*	*A*	*L*	*lA*	*hA*
A	*lA*	*lA*	*hA*	*lA*	*A*	
L	*L*	*L*	*hA*	*L*	*L*	
lA	*L*	*lA*	*H*	*L*	*lA*	*H*
L	*lA*	*lA*	*hA*	*L*	*L*	

Table 8. Knowledge about relation (27)

x_6	x_7	x_8	x_9	x_{12}	z
H	*H*	*H*	*H*	*H*	
hA	*H*	*hA*	*hA*	*hA*	*L*
H	*hA*	*H*	*A*	*hA*	
hA	*hA*	*A*	*A*	*hA*	
A	*hA*	*A*	*hA*	*H*	*lA*
A	*H*	*hA*	*hA*	*hA*	
A	*A*	*A*	*hA*	*hA*	
hA	*hA*	*A*	*A*	*A*	*A*
hA	*A*	*hA*	*hA*	*A*	
lA	*A*	*lA*	*A*	*A*	
hA	*lA*	*A*	*lA*	*lA*	*hA*
L	*A*	*A*	*lA*	*A*	
L	*L*	*L*	*L*	*lA*	
lA	*L*	*lA*	*L*	*L*	*H*
L	*lA*	*lA*	*L*	*lA*	

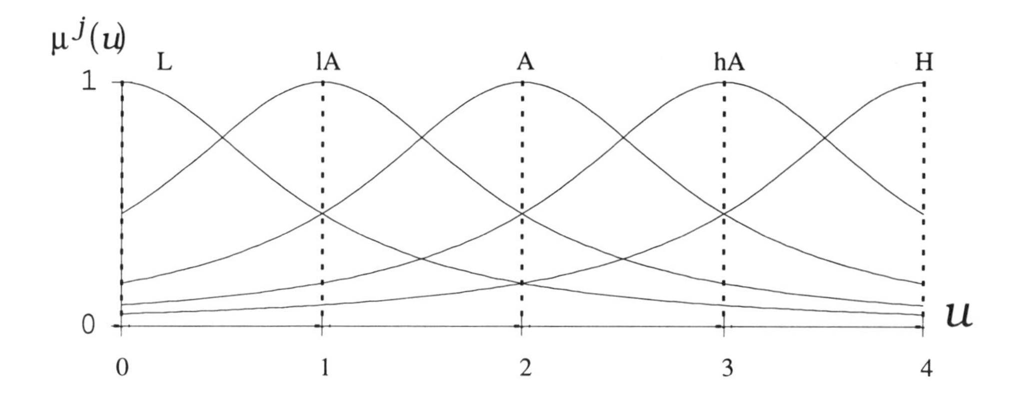

Figure 19. Rough membership functions.

6.4. Rough Membership Functions

Generally, all parameters x_1 to x_{12} have their own membership functions of the fuzzy terms (*L, lA, A, hA, H*) used in equations (28) – (30). To simplify the modeling, we can use only one shape of the membership functions for all parameters $x_1 \div x_{12}$, as shown in Figure 19

$$\mu^j(x_i) = \tilde{\mu}^j(u), \quad u = 4\frac{x_i - \underline{x_i}}{\overline{x_i} - \underline{x_i}}, \quad j = L, lA, A, hA, H$$

where $\left[\underline{x_i}, \overline{x_i}\right]$ is an interval of parameter x_i changing, $i = \overline{1,12}$.

Analytical expressions of the functions in Figure 19 are of the form:

$$\tilde{\mu}^j(u) = \frac{1}{1+\left(\frac{u-b}{c}\right)^2} \tag{31}$$

where the parameters b and c are given in Table 9. Selection of such curves is stipulated by the fact that they are approximations (5) of membership functions gained by the expert method of pairwise comparison, as described in Section 3.4.

Table 9. Parameters of rough membership functions

T	*L*	*lA*	*A*	*hA*	*H*
b	0	1	2	3	4
c	0.923	0.923	0.923	0.923	0.923

6.5. Algorithm of Decision Making

Fuzzy logic equations (28) – (30) with membership functions of fuzzy terms allow us to make the decision about the level of IHD according to this algorithm:

Step 1. Registration of parameters x_i^* ($i=\overline{1,12}$) value for a specific patient.

Step 2. Using the model (31), we define the values of the membership functions $\mu^j(x_i^*)$ when parameters values are fixed.

Step 3. Using logic equations, we calculate membership functions $\mu^{d_j}(d)$ for all diagnoses d_j, $j=\overline{1,6}$. In doing so, according to [12], *min* and *max* are substituted for the logic operations AND $(\cdot)$ and OR $(\vee)$, respectively.

Step 4. Let us define the decision d^*, for which: $\mu^{d^*}(d)=\max_j\left[\mu^{d_j}(d)\right]$, $j=\overline{1,6}$.

Example 2. Let us represent the next value of the parameters of a patient corresponding to her/his state:

$x_1^*=53$ years,	$x_2^*=175$ c.u. ,	$x_3^*=507$ kg/min,
$x_4^*=2.4$ c.u. ,	$x_5^*=0.25$ c.u.,	$x_6^*=60.7$ mmol/l,
$x_7^*=26.14$ mmol/l,	$x_8^*=10.4$ mmol/l,	$x_9^*=3.9$ c.u.,
$x_{10}^*=22.4$ mL/min·kg,	$x_{11}^*=172$ c.u.,	$x_{12}^*=26.1$ c.u.

Using model (31), we find the membership functions values at point x_i^*, $i=\overline{1,12}$ for all fuzzy terms and represent them in Table 10.

Table 10. Membership functions values in differential diagnosis of IHD

x_i^*	u^*	$\mu^L(x_i^*)$	$\mu^{lA}(x_i^*)$	$\mu^A(x_i^*)$	$\mu^{hA}(x_i^*)$	$\mu^H(x_i^*)$
53	3.259	0.074	0.143	0.349	0.927	0.608
175	0.679	0.649	0.892	0.328	0.136	0.072
507	1.503	0.274	0.771	0.775	0.275	0.120
2.4	2.182	0.152	0.379	0.963	0.560	0.205
0.25	1.362	0.315	0.867	0.676	0.241	0.109
60.7	2.996	0.087	0.176	0.462	1.000	0.458
26.14	3.255	0.074	0.142	0.343	0.911	0.635
10.4	1.157	0.389	0.972	0.545	0.201	0.095
3.9	2.468	0.123	0.283	0.795	0.751	0.266
22.4	1.791	0.210	0.576	0.951	0.368	0.149
172	1.647	0.239	0.670	0.872	0.318	0.133
26.1	3.376	0.070	0.131	0.310	0.857	0.687

Substituting the membership functions obtained from equation (29), we find

$$\mu^{L}(y)=0.072\cdot 0.120\cdot 0.205\cdot 0.315\cdot 0.149\cdot 0.133$$
$$\vee 0.072\cdot 0.275\cdot 0.205\cdot 0.867\cdot 0.149\cdot 0.133$$
$$\vee 0.137\cdot 0.120\cdot 0.560\cdot 0.315\cdot 0.149\cdot 0.133=0.120.$$

Similarly, we find: $\mu^{lA}(y)=0.137$, $\mu^{A}(y)=0.328$, $\mu^{hA}(y)=0.241$, $\mu^{H}(y)=0.210$.
According to equation (30), we find

$$\mu^{L}(z)=0.458\cdot 0.635\cdot 0.095\cdot 0.266\cdot 0.687$$
$$\vee 1.000\cdot 0.635\cdot 0.201\cdot 0.751\cdot 0.857$$
$$\vee 0.458\cdot 0.911\cdot 0.095\cdot 0.795\cdot 0.857=0.201.$$

Similarly: $\mu^{lA}(z)=0.545$, $\mu^{A}(z)=0.343$, $\mu^{hA}(z)=0.176$, $\mu^{H}(z)=0.087$.

According to equation (28) we ultimately find that

$$\mu^{d_1}(d)=0.074\cdot 0.120\cdot 0.201$$
$$\vee 0.074\cdot 0.137\cdot 0.545$$
$$\vee 0.143\cdot 0.137\cdot 0.201=0.137$$

Finally, we find: $\mu^{d_2}(d)=0.137$, $\mu^{d_3}(d)=0.328$, $\mu^{d_4}(d)=0.176$, $\mu^{d_5}(d)=0.210$, $\mu^{d_6}(d)=0.176$. Because the largest membership value corresponds to decision d_3, we select NCD with heavy complication as the patient's diagnosis.

6.6. Fine tuning of The Fuzzy Rules in Medical Applications

We used real data related to diseases with verified diagnoses as the training data for the fine tuning of fuzzy rules for differential diagnosis of IHD. The optimization problem (26) was solved by the combination of a genetic algorithm and gradient descent. The time required for the fine tuning of the fuzzy model is about two hours (Intel Pentium, 100 MHz).

Figure 20 shows an example of screen shoot with the knowledge table. The results of the fine tuning of the fuzzy model are shown in Figure 21 and Tables 11 to 13. The parameters b and c of the linguistic terms used in the fuzzy rules, after tuning, are shown in Table 14. The comparison of the inferred and correct diagnosis for 65 patients is shown in Appendix 1. There is only one case (**) in which the inferred decision (d_4) is too far from the real decision (d_2). In 8 cases (*) we have the real and inferred decisions on a boundary between classes of diagnoses. For the rest of the patients, there is full matching of real and inferred decisions. These results are quite satisfactory from a practical point of view, and thus the expert system can be used as a decision support system for the differential diagnosis of IHD.

Table 11. Weights of the rules before (w_b) and after (w_a) tuning in Table 6

d	w_b	w_a
d_1	1.000	0.934
	1.000	0.500
	1.000	0.419
d_2	1.000	0.500
	1.000	0.500
	1.000	0.764
d_5	1.000	0.428
	1.000	0.500
	1.000	0.724
d_4	1.000	0.663
	1.000	0.449
	1.000	0.449
d_5	1.000	0.499
	1.000	0.500
	1.000	0.770
d_6	1.000	0.500
	1.000	0.524
	1.000	0.915

Table 12. Weights of the rules before (w_b) and after (w_a) tuning in Table 7

y	w_b	w_a
L	1.000	0.500
	1.000	0.500
	1.000	0.734
lA	1.000	0.500
	1.000	0.632
	1.000	0.500
A	1.000	0.757
	1.000	0.470
	1.000	0.473
hA	1.000	0.527
	1.000	0.480
	1.000	0.664
H	1.000	0.499
	1.000	0.806
	1.000	0.499

Table 13. Weights of the rules before (w_b) and after (w_a) tuning in Table 8

z	w_b	w_a
L	1.000	0.500
	1.000	0.744
	1.000	0.500
lA	1.000	0.500
	1.000	0.500
	1.000	0.400
A	1.000	0.500
	1.000	0.500
	1.000	0.565
hA	1.000	0.771
	1.000	0.500
	1.000	0.500
H	1.000	0.500
	1.000	0.500
	1.000	0.500

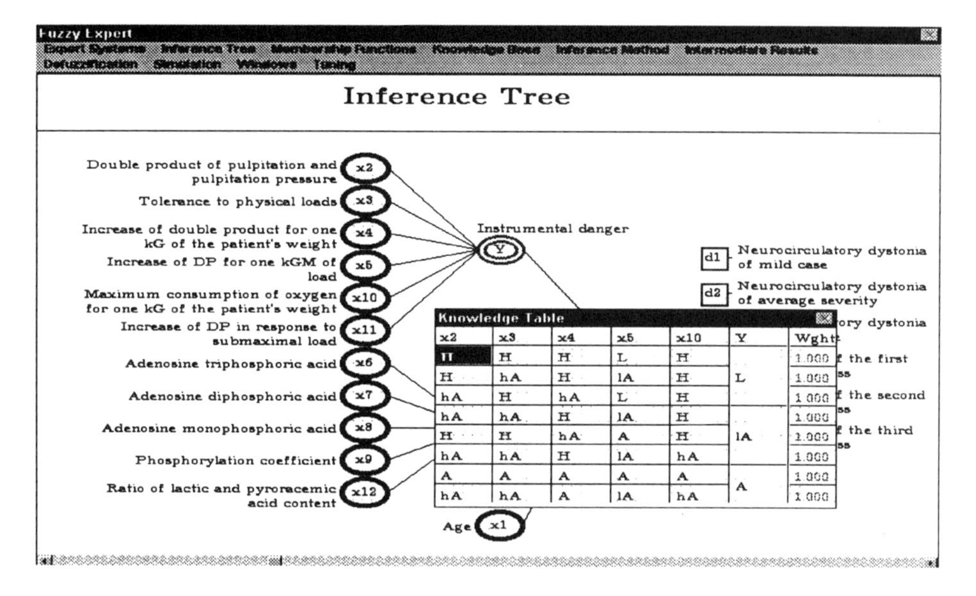

Figure 20. An example of a screen showing the knowledge table.

Table 14. Parameters b and c of the membership functions, after tuning

	L		*LA*		*A*		*hA*		*H*	
	b	*c*	*b*	*c*	*b*	*c*	*b*	*c*	*b*	*c*
x_1	32.58	23.32	38.21	9.79	43.38	11.92	51.06	16.00	56.74	22.61
x_2	128.00	57.31	186.39	87.84	235.23	80.38	332.76	109.61	389.32	162.74
x_3	182.46	807.90	509.01	242.90	648.07	575.60	922.17	261.00	1105.75	568.26
x_4	0.60	0.76	1.28	0.98	1.84	0.38	2.85	1.42	3.90	0.06
x_5	0.11	0.05	0.21	0.02	0.32	0.17	0.42	0.07	0.52	0.06
x_6	34.48	7.87	47.34	40.20	51.94	8.89	59.48	21.89	69.49	8.07
x_7	11.90	4.03	16.27	4.03	21.56	11.30	25.02	4.03	27.98	13.19
x_8	3.60	5.42	8.60	10.79	15.92	0.23	19.03	25.09	27.10	5.42
x_9	1.00	1.08	2.17	1.08	3.18	3.36	4.02	1.20	5.70	1.08
x_{10}	9.01	18.61	16.69	8.49	21.63	17.92	32.44	8.88	36.24	10.14
x_{11}	46.00	30.15	144.32	157.80	200.09	147.95	270.43	5.96	335.17	193.27
x_{12}	3.90	6.06	10.47	6.06	19.46	24.04	22.80	10.86	30.20	6.06

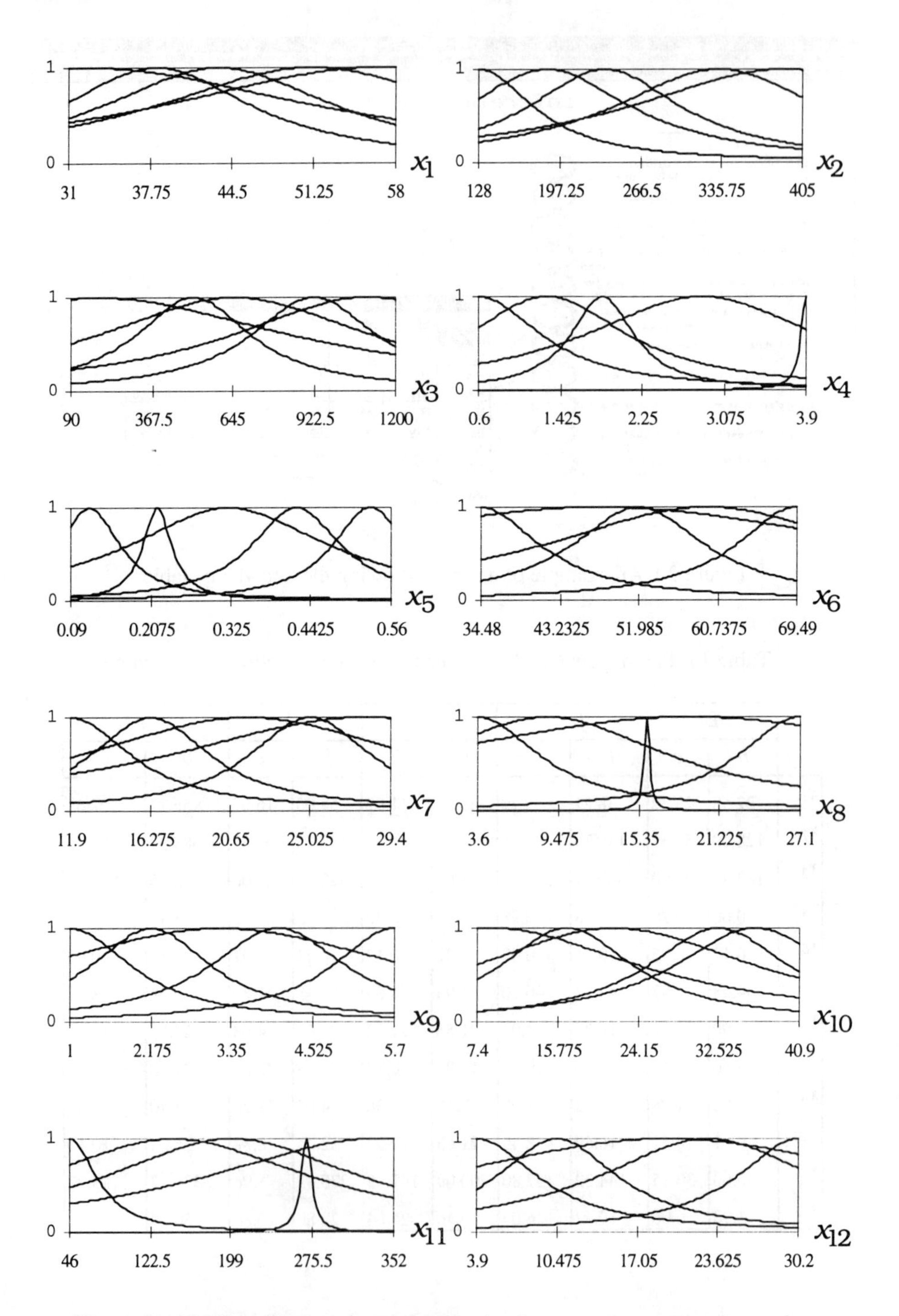

Figure 21. Membership functions of the patient's state parameters after tuning.

7. Conclusions

We have proposed an approach to the design and tuning of fuzzy rules for medical diagnosis. The proposed approach includes two fuzzy rules tuning stages, rough and fine. The rough stage consists of fuzzy rule generation by a medical expert and selection of the membership functions of the fuzzy terms by the method of pairwise comparison. The fine-tuning stage consists of finding the parameters of membership functions and the weights of rules, by using training data and solving the stated optimization problems. The stated optimization problems allow tuning of the fuzzy rules both with continuous and discrete consequent. The effectiveness of the tuning models was demonstrated by computer simulation. An important feature of the proposed method is that for design and tuning of the fuzzy rules, we use only one simple form of the expert knowledge matrix for both continuous and discrete consequent.

Using the proposed approach, we created a fuzzy expert system for differential diagnosis of ischemia heart disease. Enough correlation exists between the inferred and real diagnosis to recommend the created system for practical application as a decision support system in hospital usage. Other applications are presented in Appendix 2.

References

[1] Möller, D.P.F., Fuzzy Logic In Medicine, in *Proc. of Fourth European Congress on Intelligent Techniques and Soft Computing EUFIT-96*, 3, H.-J. Zimmermann (Ed.), Verlag Mainz, Aachen, 1996, 2036.

[2] Zahan, S., Michael, C., Nikolakeas, S., Fuzzy Expert System for Myocardial Ischemia Diagnosis, (in this volume).

[3] Pilz, U., Engelmann, L. Integration of Medical Knowledge in an Expert System for Use in the Intensive Care Medicine, (in this volume).

[4] Rotshtein, A. P., *Fuzzy Logic-Based Medical Diagnostics*, Continent-PRIM Publish. Comp., Vinnitsa, 1996. (In Russian.)

[5] Rotshtein, A., Zlepko, S., Zhupanova, M., Fuzzy Expert System for Medical Diagnostics, in *Proc. of 7th IMECO TC-13 Int. Conf. on Model Based Biomeasurements*, Stara Lesna, Slovakia, 1995, 231-233.

[6] Rotshtein, A., Katelnikov, D., Goldenberg, L., Design and Tuning of Fuzzy Expert Systems for Medical Diagnostics, in *Proc. of Int. Workshop on Biomedical Engineering & Medical Informatics (BEMI'97)*, Gliwice, Poland, 1997,106-110.

[7] Rotshtein, A., Katelnikov, D., Fuzzy Rule-Based Expert System for Differential Diagnosis of Ischemia Heart Disease, in *Proc. of Int. Conf. on Fuzzy Logic and Applications (FUZZY'97)*, Zichron-Yaakov, Israel, 1997, 367-372.

[8] Rotshtein, A., Modification of Saaty Method for Construction of Fuzzy Sets Membership Functions, in *Proc. of Int. Conf. on Fuzzy Logic and Applications (FUZZY'97)*, Zichron-Yaakov, Israel, 1997, 367-372.

[9] Rotshtein, A., Fuzzy Reliability Analysis of Labour (Man-Machine) Systems, in T. Onisawa, J. Kacprzyk, (Eds.) Studies in Fuzziness, Vol. 4, Reliability and Safety Analyses under Fuzziness, Physica - Verlag, Germany, 1995, 245-270.

[10] Saaty, Th., L., Exploring the interface between hierarchies, multiple objectives and fuzzy sets, *Fuzzy Sets and Systems*, 1978, 1, p. 57-68.

[11] Miller, C. A., The Magic Number Seven Plus or Minus Two: Some Limits on Our Capacity for Proceeding Information, *Psychological Review*, 63, 81, 1956.

[12] Zadeh, L. Outline of a New Approach to the Analysis of Complex Systems and Decision Processes, *IEEE Trans. Syst. Men. Cybern.*, 3, 28, 1973.

[13] Zimmermann, H.-J., *Fuzzy Set Theory and Its Application*, 3, Kluwer Academic Publishers, Dordrecht, 1996.

[14] Nechiporenko, V. I., *Structural Analysis of Systems (Effectiveness and Reliability)*, Sov. Radio, Moscow. 1977. (In Russian.)

[15] Casti, J., *Connectivity, Complexity and Catastrophe in Large-Scale Systems,* John Wiley & Sons, New York, 1979.

[16] Gen, M., Cheng, R., *Genetic Algorithms and Engineering Design*, John Wiley & Sons, New York, 1997.

[17] Reklaitis, G. V., Ravindran, A., Ragsdell, K. M., *Engineering Optimization. Methods and Applications*, John Wiley & Sons, New York, 1983.

Appendix 1
Comparison of real and inferred decisions for 65 patients

(1)	Parameters of state												Diagnosis	
	x_1	x_2	x_3	x_4	x_5	x_6	x_7	x_8	x_9	x_{10}	x_{11}	x_{12}	Real	Model
1	31	324	980	2.8	0.12	50.07	22.76	8.05	3.7	34.2	266	19.3	d1	d1
2	36	330	900	2.9	0.14	56.52	24.33	9.02	4.1	29.7	242	21.0	d1	d1
3	39	260	800	2.3	0.18	51.73	25.62	8.53	4.2	28.5	194	23.8	d2	d2
4	42	272	867	2.5	0.28	59.31	28.44	8.53	4.0	28.7	198	19.4	d2	d2
5	48	287	491	2.2	0.24	52.77	21.61	8.53	3.5	25.3	156	20.5	d3	d3
6	53	175	507	2.4	0.25	60.70	26.14	10.40	3.9	22.4	172	26.1	d3	d3
7	45	247	728	2.0	0.34	62.06	26.14	5.55	2.3	26.5	144	22.9	d4	d4
8	52	231	768	1.5	0.36	62.77	23.01	6.83	2.5	20.0	158	23.8	d4	d4
9	32	151	610	1.3	0.42	54.49	23.91	5.55	2.4	19.8	104	25.7	d5	d5
10	45	177	542	1.6	0.48	62.06	26.14	5.55	2.3	21.7	120	28.1	d5	d6 *
11	38	128	349	1.4	0.48	67.03	24.46	5.20	1.9	13.9	92	30.2	d6	d6
12	38	145	304	1.2	0.56	64.15	25.62	7.11	2.6	14.4	74	25.5	d6	d6
13	40	327	930	2.2	0.24	59.31	25.62	7.56	3.3	35.4	347	18.9	d1	d2 *
14	38	348	952	1.8	0.20	34.48	20.79	9.56	5.7	34.2	352	21.6	d1	d1
15	34	307	800	1.9	0.21	57.90	25.08	6.83	2.9	30.1	304	19.3	d2	d4 **
16	48	284	738	2.0	0.26	62.06	25.08	8.53	3.4	29.7	339	20.4	d2	d2
17	35	174	600	1.7	0.32	55.18	24.46	8.56	3.8	27.2	312	22.0	d3	d3
18	49	229	515	2.1	0.30	61.34	22.20	6.83	2.4	22.4	300	23.4	d3	d4 *
19	58	265	421	2.0	0.26	60.07	22.76	4.08	1.8	17.7	258	23.8	d4	d4
20	49	330	650	1.5	0.25	69.49	25.08	6.83	2.5	20.3	244	22.0	d4	d4
21	48	187	475	1.4	0.34	60.39	23.31	5.55	2.1	21.4	204	22.7	d5	d5

(1) Patient number

(continued)

(continued)

l	Parameters of state												Diagnosis		
	x_1	x_2	x_3	x_4	x_5	x_6	x_7	x_8	x_9	x_{10}	x_{11}	x_{12}	Real	Model	
22	42	224	400	1.5	0.39	55.18	21.05	7.11	2.7	20.4	215	22.5	d5	d5	
23	32	195	100	1.2	0.48	60.70	21.61	7.52	2.7	22.6	191	25.9	d6	d6	
24	51	192	292	1.3	0.45	62.77	23.70	5.55	1.6	19.2	188	24.4	d6	d6	
25	36	347	952	2.9	0.10	62.40	23.70	12.50	4.3	35.7	298	19.6	d1	d1	
26	48	314	902	3.2	0.14	59.40	24.20	10.50	4.2	33.5	287	18.8	d1	d1	
27	42	352	875	3.2	0.16	52.30	22.70	9.50	3.9	38.2	322	19.0	d1	d1	
28	40	323	1040	2.7	0.20	59.60	25.20	8.80	3.2	30.4	290	18.2	d1	d2	*
29	41	377	988	2.9	0.09	60.40	24.30	10.20	3.4	32.5	275	17.7	d1	d1	
30	34	309	932	3.2	0.15	60.80	25.40	9.40	4.4	31.5	312	18.5	d1	d1	
31	52	279	1056	2.7	0.09	59.90	21.30	8.80	3.7	33.4	334	18.7	d1	d1	
32	44	376	895	2.7	0.18	61.50	23.60	9.50	3.6	30.4	312	20.1	d2	d2	
33	46	304	929	2.6	0.22	58.20	25.10	10.70	3.8	32.5	346	19.2	d2	d2	
34	46	292	904	2.2	0.24	56.00	27.90	10.10	4.0	29.3	290	18.5	d2	d2	
35	42	276	885	2.4	0.25	61.40	29.40	11.20	3.6	27.8	226	20.8	d2	d2	
36	31	311	930	2.7	0.19	62.50	23.80	9.80	2.9	25.6	249	21.0	d2	d1	*
37	44	335	992	2.4	0.22	61.60	24.70	9.90	3.3	24.6	255	20.3	d2	d2	
38	47	346	873	2.3	0.18	57.70	22.50	10.60	3.7	28.7	267	18.8	d2	d2	
39	48	288	804	2.4	0.27	60.00	22.20	11.50	3.5	20.9	275	19.5	d3	d3	
40	50	316	875	2.1	0.31	61.40	24.00	9.30	2.8	22.5	302	21.2	d3	d4	*
41	51	292	774	2.0	0.28	62.50	25.90	8.80	3.0	26.7	277	22.5	d3	d4	*
42	54	315	766	2.2	0.22	53.70	26.20	8.70	2.7	21.4	265	20.5	d3	d4	*
43	40	300	865	2.1	0.25	59.40	25.80	9.30	3.5	21.9	303	21.4	d3	d3	
44	36	270	777	2.1	0.28	61.00	26.10	9.70	4.1	22.3	316	21.3	d3	d3	
45	34	275	859	2.3	0.30	62.50	27.00	9.60	4.2	24.0	295	22.5	d3	d3	
46	52	261	776	1.7	0.36	65.00	22.50	8.40	2.7	20.4	204	23.8	d4	d4	
47	41	258	785	1.5	0.36	62.70	23.80	7.60	2.5	19.8	225	24.0	d4	d4	
48	53	290	845	1.8	0.39	57.10	24.00	7.20	2.5	18.7	268	22.5	d4	d4	
49	39	203	723	2.0	0.40	58.50	23.70	6.20	2.8	17.1	209	24.7	d4	d4	
50	45	244	802	1.7	0.35	62.00	25.30	6.30	3.0	18.5	212	24.9	d4	d4	
51	46	233	795	1.9	0.39	57.90	24.90	5.20	2.4	17.4	251	23.5	d4	d4	
52	54	262	805	1.8	0.38	57.90	24.50	7.70	2.2	19.2	244	22.1	d4	d4	
53	51	245	595	1.3	0.44	64.20	26.40	5.60	2.1	16.5	204	24.7	d5	d5	
54	40	209	772	1.5	0.45	60.20	27.80	5.90	2.4	14.7	195	25.0	d5	d5	
55	42	198	621	1.4	0.42	58.80	25.20	6.10	2.6	12.2	225	24.5	d5	d5	
56	44	245	523	1.5	0.39	57.50	23.30	6.50	2.2	14.1	207	26.9	d5	d5	
57	50	237	652	1.6	0.45	63.70	24.70	6.40	2.1	11.9	262	24.2	d5	d5	
58	56	202	744	1.3	0.45	61.80	25.70	5.70	2.4	12.3	226	22.6	d5	d5	
59	51	247	723	1.2	0.38	62.50	26.90	5.60	2.3	10.4	230	25.8	d5	d5	
60	48	192	516	1.1	0.52	60.10	22.70	5.50	2.0	9.9	200	22.9	d6	d6	
61	39	188	446	1.2	0.48	59.00	23.50	5.20	2.4	9.5	212	26.7	d6	d6	
62	49	212	406	0.9	0.56	61.70	26.00	5.30	1.9	8.2	225	29.4	d6	d6	
63	45	247	527	0.7	0.51	62.60	27.40	5.10	2.0	7.4	197	28.5	d6	d6	
64	44	206	448	0.8	0.55	57.40	22.10	6.30	2.1	7.4	188	30.1	d6	d6	
65	42	228	512	1.0	0.52	53.90	25.60	5.40	2.3	7.8	204	29.5	d6	d6	

Appendix 2
FUZZY EXPERT shell and its application

The mathematical models and algorithms suggested in this chapter were implemented by A. Rotshtein and D. Katelnikov using the fuzzy expert system shell FUZZY EXPERT for medical diagnosis (C++, Windows 95). This system can be used in two modes: 1 – fuzzy knowledge base formation; 2 – fuzzy knowledge base utilization.

The first mode provides the following possibilities:

- collection of the information about medical diagnosis (output variable),
- collection of information about parameters of a patient's state (parameter's quantity, parameter's names, ranges of parameter's change, quantity and names of linguistical terms for each parameter, tree of hierarchical parameter classification),
- formation of fuzzy knowledge base about concerning parameters of the patient's state and diagnosis for each node of hierarchical tree,
- selection of membership functions for each linguistic term used in the hierarchical fuzzy rule-based system,
- tuning of membership functions and weights of fuzzy if-then rules.

The second mode provides the following functions for the user:

- presentation of the list of parameters of patient states in a given field of medicine,
- fixation of values for each quantitative and qualitative parameters corresponding to a given patient,
- execution of fuzzy logic evidence and presentation of information about patients diagnosis,
- observation of the fuzzy logic evidence results in the each node of the hierarchical knowledge tree,
- observation of diagnosis change during the variation of patient's state parameters,
- explanation of fuzzy logic evidence.

Two more screenshots of FUZZY EXPERT are presented in Figures A1 and A2.

The FUZZY EXPERT shell has been extensively used by medical doctors of Vinnitsa Medical University (Ukraine) for creating several fuzzy rule-based systems. The systems implemented are the following:

- differential diagnosis of ischemia heart disease (with V. Sheverda),
- forecasting of teenager's health disorders (with I. Sergeta),
- forecasting of bleeding danger in childbirth (with B. Mazorchuk),
- forecasting of womb myoma appearance risk in conditions of radioelectronic production (with N. Masibroda),
- forecasting of the quality of man-operator activity based on psychological and physiological parameters (with I. Sergeta),
- diagnosis of risk after appendicitis operations in the Chernobyl area conditions (with V. Loiko).

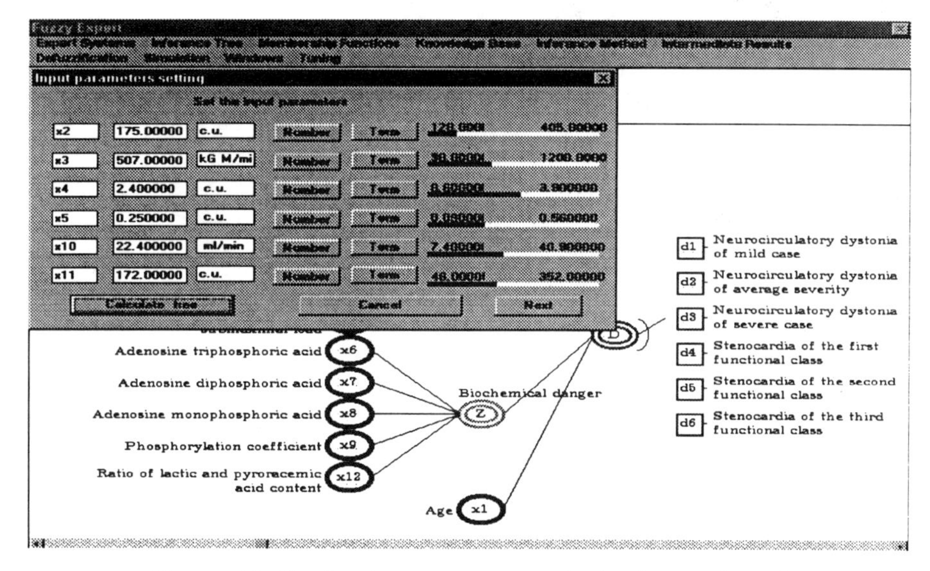

Figure A1. Setting input parameters in the form of number, linguistic term or min-max scale mark.

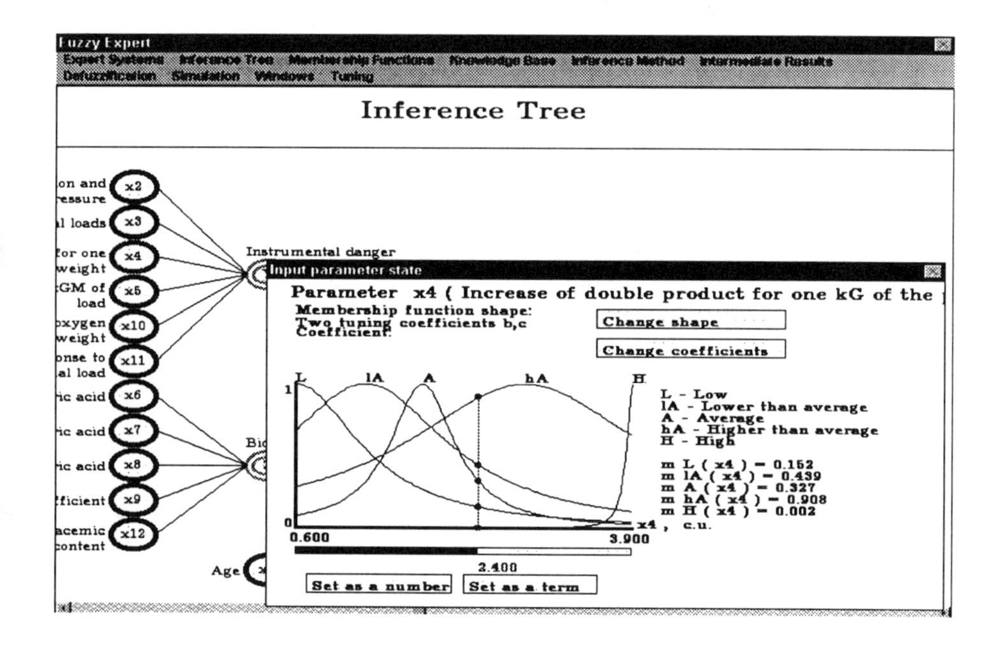

Figure A2. Membership functions after tuning.

Chapter 10

Integration of Medical Knowledge in an Expert System for Use in Intensive Care Medicine

Uwe Pilz and Lothar Engelmann

This chapter describes a fuzzy logic-based tool which is currently in use in clinical practice. Diagnostic decision in intensive care medicine requires medical knowledge adapted to the preferences of the physician and enriched with his experience. The knowledge-based system FLORIDA provides knowledge processing strategies, which enable a computer to use this knowledge at the bedside. With this system, it is possible to compute numerical values for physiological conditions like "heart failure" or "shock" from clinical signs. The physicians' knowledge is interpreted with the help of fuzzy logic, which enables subtle differentiation of a given physiological state. The system is robust against missing and faulty input.

1. INTRODUCTION

In intensive care units and other "high end medicine" fields, large amounts of data are collected to determine the patient's state. Such data can be monitored values, lab data, and clinical observations. Physicians need these data to describe and discuss the patient's physiological state. In intensive care, this state is usually described by physiological conditions like heart failure or shock.

Furthermore, it is important to evaluate the patient's progress. Therefore, any changes of physiological condition need to be followed over several hours or even several days. This task is quite difficult because of the large amount of data that needs to be taken into account. Usually, experienced physicians tend to consider almost all data for the current state but reduced data for the assessment of the patient's progress, while inexperienced physicians have problems determining the physiological state in an adequate form.

0-8493-9806-1/98/$0.00+$.50

Because a large amount of data exists in computer readable form, formalization of the data concentration process is required. It is necessary to determine the various classes of physiological conditions and to formalize them, so that a computer can correctly use this type of information. Assuming a numerical result, we get a trend for the state in the form of frequently calculated physiological conditions. The most problematic part of this task is the transformation of the medical knowledge from experts (i.e., the physicians) into a knowledge base that can be used in a computer program [1-4]. In this chapter, we describe our solution to this problem and the knowledge-processing program FLORIDA. A restricted version of FLORIDA is available and can be down loaded electronically [5].

2. SOFTWARE DESIGN PRINCIPLES

The task of FLORIDA is to process diagnostic knowledge. The input for such calculations can be found in patient monitoring systems or patient data management systems. Such systems usually offer an interface for dialogue with external programs and for presentation of their results. It is a preferable to integrate the knowledge processing ability in such systems [6], [7], [8].

To have a computer program that can be integrated into an arbitrary computational environment, we must determine the requirements for the necessary hardware and software.

- FLORIDA uses the computer language C++. It can be compiled with every C++ compiler that accepts the ANSI standard.
- **Interface:** FLORIDA itself contains a simple ASCII interface, which works with files. The files can easily be processed by computer programs but they can also be easily understood by a human user. Presentation of the results has to be done by a host system.
- **Patient database:** To follow the patient's progress, many FLORIDA results must be available to a single screen. In very fast computers, these can be calculated for presentation. Slower machines need a database with historical results. Creating and storing this database is the task of the host system.

FLORIDA's abilities are extremely well suited to reduce knowledge processing. Therefore, the following requirements and principles were established.

A senior physician has the knowledge needed to determine a patient's state. Often, she/he is unable to explain how she/he got their result. However, if the physicians could determine their thinking process they could probably explain how they came to their results. This knowledge is adequate for diagnostic purposes. We use this knowledge in the computer without transforming it into other forms, such as tables or networks. The main techniques used are fuzzy logic and the automatic rule expansion mechanism [9], [10].

In most cases, not all input is available for the diagnosis. With less input, the result is less reliable. FLORIDA gives values for each result, which describes its reliability.

FLORIDA is able to explain its results. The explanation is automatically generated from the knowledge base and the input.

3. MEDICAL KNOWLEDGE IN INTENSIVE CARE MEDICINE

3.1. Structure of the knowledge

FLORIDA was developed to describe the physiological state of patients in unstable situations. In such cases, it is important to detect changes in the situation rapidly and correctly. The base for the assessment consists of measured data, which have ranges of normality. If a measurement is outside of this range, a physiological dysfunction is indicated. The indication is more evident if the deviation from the normal range is large. Consequently, the intended working field for FLORIDA is the evaluation of unstable patients, on the basis of measured data. Such patients can be found in intensive care and anesthesiology.

Diagnostic knowledge suitable for an expert system can be obtained in different ways. One could observe the progress of a large number of patients, and use this empirical knowledge in a knowledge base. The advantage of this method is in improving the knowledge base with the help of patients whose data are used by the system. However, there are several disadvantages to empirical knowledge.

- Until it is validated by a huge amount of patient data, the knowledge base remains hypothetical.
- It is time consuming to obtain the knowledge to document sufficient data on patient's progress.
- Assessments obtained from a knowledge base that consists of empirical knowledge are expected to be of a lesser quality than the physician's opinion for the following reasons:
  - The knowledge is obtained from data only. No rules are available. Therefore, the assessment of the knowledge-based system is not comprehensible to the human user.
  - Rarely occurring but severe states are not reflected with adequate accuracy.
  - Physiological dysfunctions influence each other. Many of the combinations of dysfunctions are relatively rare occurrences. They cannot be examined with an acceptable amount of observable data.

Therefore, we have decided to use the "concentrated" knowledge used by physicians in the daily routine. To come closer to the physicians' knowledge, we must talk with them about their reasoning processes. The medical knowledge is obtained in colloquial language, mostly in the form of rules. These rules contain the specific terms and categories that are used by the experts. Such a rule could be

If a patient goes into a shock situation this can be recognized by a low blood pressure, a rise in heart rate, hyperventilation, and a high serum lactate level.

The categories in this colloquial rule are "shock situation," "blood pressure," "heart rate," "hyperventilation," and "serum lactate level." They can belong to linguistic classes (as determined by linguistic quantities) like "high" or "very low." Our "result" is *shock situation*. "Shock" can exist in different intensities, which may vary over time. Intensity can be described by "severe," "beginning," or similar categories. Variables with categories can be handled adequately with fuzzy logic, which is therefore used in FLORIDA. Examples will be used to demonstrate its operation.

3.2. Meaning of colloquial rules

The physicians' knowledge is "validated" in the sense that they use it successfully in their daily practice. Transferring this knowledge into knowledge-based systems usually requires a transformation of the knowledge, e.g., into tables or networks. The transformation result does not represent the original human knowledge, and cannot be presumed valid. To increase the quality and readability of the knowledge bases, we decided to keep the knowledge in nearly its original form. The interpreter needs to "know" the meaning of a rule to interpret it correctly.

To return to our simple "shock" example, in the linguistic rule there are four parameters (blood pressure, heart rate, hyperventilation, and serum lactate level) pointing to our result *shock*. If all of these indicate "shock," the situation is quite clear. However, we have to establish if it is necessary that all the four partial rules be fulfilled. Alternatively, is it sufficient for a single parameter to indicate shock?

To arrive at an answer we need to look at our result variable *shock*. There exist severe definitions, such as insufficient oxygen supply to the tissue. This leads to failure of the organ functions, and thereby to shock self-intensification. Oxygen supply can be insufficient to different degrees and, therefore, shock is also present to a specific degree. This oxygen supply cannot be seen or easily measured. Only its effect on the organism is visible, and can be detected by measured data and by clinical signs. We can conclude that the physiological conditions or dysfunctions themselves do not really exist in a mathematical sense. They cannot be measured directly or expressed in a special unit. Nevertheless, the physicians have an understanding of such conditions. These images are compiled from the measuring and observing, which can contradict each other to a certain extent.

What we wish to achieve with the expert system is a numerical value to detect small changes in the patient's progress. However, this value must be calculated with methods similar to the physician's "composition."

FLORIDA calculates the results by generally assuming the measured values to be independent components of the result. The human expert determines if the components form a "harmonic" image or not. FLORIDA tries to simulate this, and gives a value for the degree of contradiction between the partial rules.

Let us analyze a partial rule, such as

"Low blood pressure indicates a shock situation."

This colloquial rule contains a variety of meanings which will be "understood" by humans without further explanation. For use in a computer, we have to specify it. We arrive at the more accurate rule:

"Low blood pressure indicates a shock situation. The lower the blood pressure, the stronger the shock. Normal blood pressure does not indicate shock, whereas high blood pressure points against shock."

One of FLORIDA's paradigms is to accept rules in their linguistic form. The knowledge processing must "know" about the variety of meanings of symptoms, clinical signs, and data, and take them into account. The main instrument is therefore the automatic rule expansion, which will be explained in the next section.

3.3. Rule processing and result calculation

FLORIDA uses fuzzy logic for rule processing. More detailed information about fuzzy logic and its concepts can be found in the literature [11], [12], [13]. Fuzzy variables are used for input and as well for result. A variable in FLORIDA consists of five fuzzy sets which are able to describe a normal range, a high and a low range, and a very high and very low range using fuzzy concepts.

Intermediate sets (e.g., "*normal*" ... "*high*") are covered with fuzzy logic. The fuzzy sets for the middle three sets are triangular and each is described by three values: a lower value, a central value, and an upper value. Figure 1 shows a fuzzy set for low, normal, and high blood pressure, respectively.

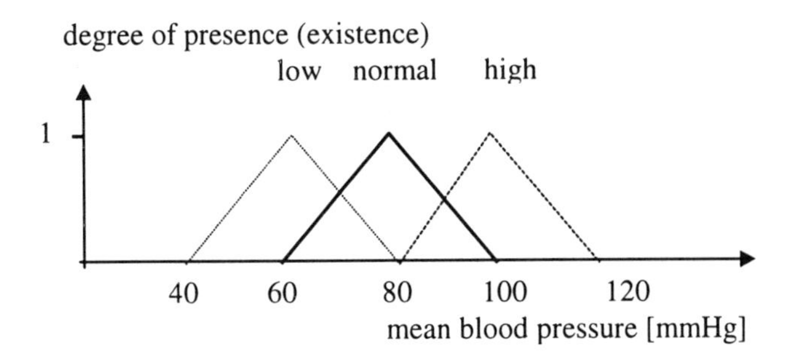

Figure 1. Fuzzy variable blood pressure, middle sets.

The sets should overlap in a way that, at every point of the "sharp" blood pressure, the sum of the presence (membership) degrees is 1, and at most two sets are valid for that point. The extreme sets need to be trapezoidal to ensure that the membership degree is 1 at all points in the definition range (Figure 2).

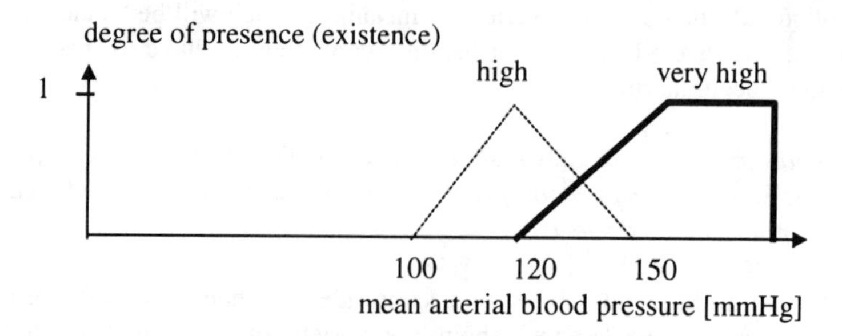

Figure 2. Fuzzy variable blood pressure, set very high.

Values outside of the definition range will be shifted to the border by FLORIDA's input processor, while triggering a warning. Thus, it is not necessary to extend the border values to possible but extremely rarely occurring values.

A result variable like *shock* will be implemented in a similar manner. Here we have no "real" values underlying the linguistic descriptions. However, the computation of a value is the goal of FLORIDA. We introduce an intensity ranging between – 100% and 100%. 100% is the strongest existence of a physiological state and –100% is the maximal rejection of this state. Our five sets are usually named *distinct, present, possible but rather unlikely, unlikely*, and *impossible.* These linguistic terms are used for linguistic output. Every input has an individual life span. This life span is connected with the age of the measured data. Older data count less than current ones. If the data reaches their life span, the data are assumed not to have been measured. Rules have a different degree of specificity. The "blood pressure" rule is more specific in comparison to the "heart rate" rule in our "shock" example. In FLORIDA, four degrees of specificity exist: *very specific*, *specific*, *unspecific*, and *very unspecific.*

During the precompilation process, FLORIDA normally expands rules. This means that a rule in the knowledge base will be expanded into several internal rules. For example, from the rule

Low blood pressure → shock

FLORIDA generates

1) *very low blood pressure* → *distinct shock*
2) *low blood pressure* → *shock is present*
3) *normal blood pressure* → *shock is possible but unlikely*
4) *high blood pressure* → *shock is unlikely*
5) *very high blood pressure* → *shock is impossible*

If FLORIDA receives an input, two of these rules will normally “fire.” Figure 3 shows that rules 1 and 2 fire to a different degree if the numerical value for mean blood pressure is 45 mmHg.

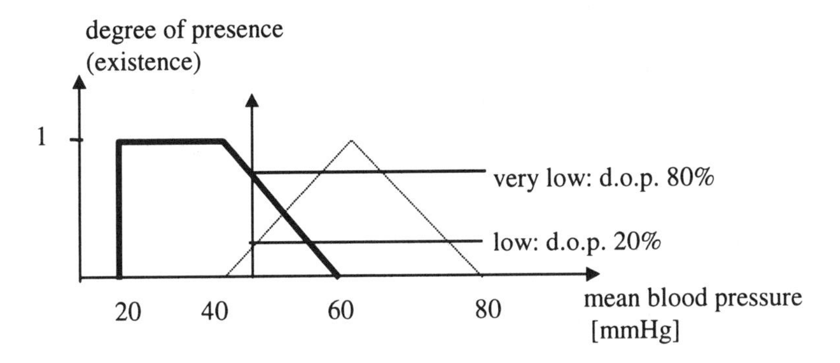

Figure 3. Fuzzification.

The change from numerical data to degree of presence (membership degree, which is often interpreted as a “degree of existence” of a symptom, or of a state) is called fuzzification.

The result for “If the blood pressure is very low” is “shock is distinct.” The result has the same validity of its presumption, in our case 80%. The result “shock is present” has a validity of only 20%. FLORIDA combines the results from these partial rules to a result for the original linguistic rule by summation of both partial results (Figure 4).

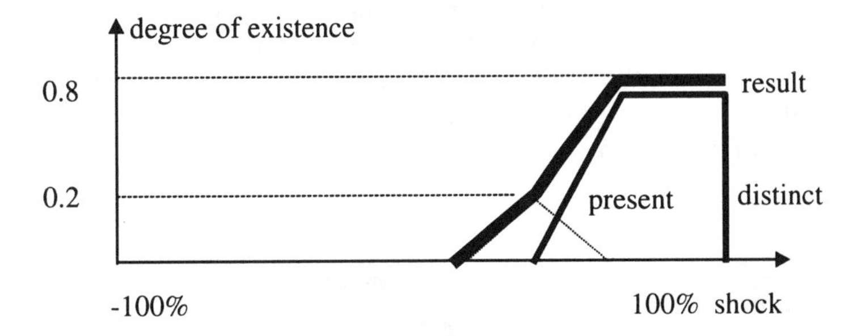

Figure 4. Calculating the result of a single rule: blood pressure.

3.4. Combining different rules

Different rules are assumed to be independent components of the total result. The total result consists of the intensity of the computed physiological state, its certainty and its consistency. Here, intensity is the main result and certainty and consistency reflect the trustworthiness of the result.

To calculate the intensity of a result, we need to process all rules with available input. The result area is the scaled sum of the partial areas. Scaling is done by

- the specificity
- the age and span of the input

Let us assume we have two input values for calculating *shock*: a mean blood pressure of 45 mmHg and heart rate of 70 min^{-1}. Serum lactate level and breathing frequency are assumed not to have been measured. The result is an area as shown in Figure 4 for the blood pressure rule, and an area as shown in Figure 5 for the heart rate rule.

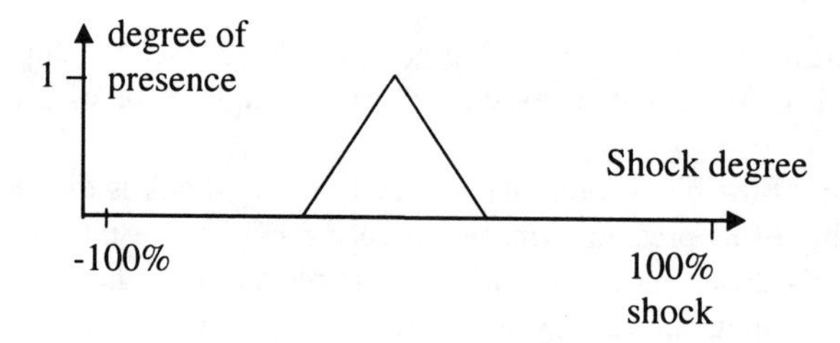

Figure 5. Result *shock* from the input heart rate.

The heart rate rule is less specific and, therefore, the area is reduced to 40% of its original value. If we have a span for heart rate of 1 hour and the value is 30 minutes old, the area reduces further to 40% * 50% = 20%. If we assume that the blood pressure is a current value, the partial area of this rule remains unreduced. The summation and scaling process is shown in Figure 6.

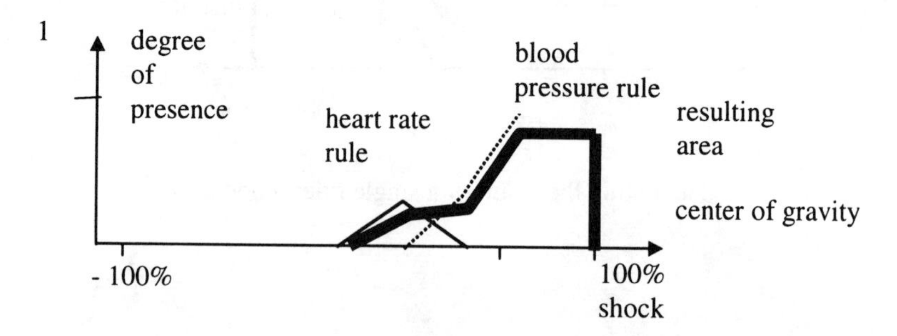

Figure 6. Summation and scaling process.

The result for "shock" is then computed as the x value of the center of gravity of the resulting area.

Certainty will be calculated numerically from the portion of input available in comparison to the possible input. Our example would give

Rule	Possible	Available
Blood pressure	100%	100%
Heart rate	40%	20%
Serum lactate level	100%	0
Breathing frequency	40%	0
Total	280%	120%

The certainty is then 120% /280% = 42.8%.

The meaning of consistency is difficult to understand. Center of gravity defuzzification cannot distinguish between the two cases in Figure 7. The same "intensity" determined by the center of gravity results from very differently formed areas (Figure 7).

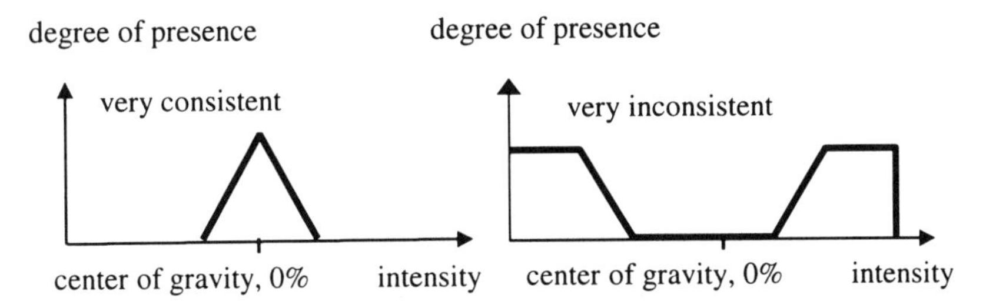

Figure 7. Different degrees of consistency.

The situation on the left is more reliable because all rules come to the same conclusion. In the situation on the right, the rules come to very different conclusions, showing that strong contradictions exist between them. To get a numerical value for this effect, we used the fact that area portions near the center of gravity indicate a high consistency and those area portions far away from the center of gravity point to low consistency. The integral:

$$C_i = \alpha \int_{-100\%}^{100\%} \mu_i(x) \,|\, x - c_i \,|\, dx \quad + \quad \beta$$

takes this fact into account. It calculates the consistency C_i for the result variable i. Here μ_i is the membership degree at x and c_i the center of gravity for this variable. The constants α and β are chosen for C_i to be 100% in the left case of Figure 7 and 0% in the right case, respectively. The result of our "shock" example is a consistency of 90%.

4. TRANSFORMATION OF KNOWLEDGE INTO FLORIDA COMMANDS

4.1. Introduction

Transformation from knowledge in a colloquial form to a knowledge base requires a computerized notation. In FLORIDA, this notation is a type of programming language. The knowledge needs to be written in this language with the help of an editor. The DOS program Edit and Windows' Notepad can be used for this task. The editor has to store the knowledge base in one or more files. We recommend the extension .flo for these files.

A knowledge base for FLORIDA requires the declaration of (fuzzy) variables for input and for results, and the notation of knowledge – mostly in the form of rules.

All commands and declarations must be noted in a single line.

4.2. Comments

Knowledge bases should have comments. Two types of comments exist in FLORIDA, single line comments and block comments.

A single line comment starts with a "#" and ends at the end of the line.

```
# This is a single line comment.<NL>
```

Block comments start with /* at the beginning of a line and end with */ at the beginning of a line. C programmers should note the difference to C block comments, which can start and end everywhere.

```
/* This is a <NL>
block comment. It reaches, until '*/' <NL>
at the  beginning of <NL>
line occurs.  <NL>
*/ <NL>
```

Block comments may contain single line comments.

4.3. Modules

FLORIDA supports modularization to some degree. The knowledge base can be divided into modules consisting of a single file each. However, there are no local variables. If necessary, they need to be realized with name conventions.

Separate modules exist for each physiological dysfunction. Such modules are commonly independent from each other, and can be changed without influencing other parts of the knowledge base.

The knowledge base has a module that contains input variables for laboratory values and monitoring parameters.

4.4. Linguistic variables

For every input and for every result, a variable has to be declared. The name of a variable can be formed arbitrarily. The maximum name length is 20 characters.

The main content of a variable is the names and limits of the five fuzzy sets. In some cases, the limits need to be adapted to the patient or the situation. The adaptation can be done with one of two methods: shifting or scaling. The method must be declared for each variable.

The syntax for a fuzzy variable is

```
feature name_of_the_feature
lower_limit_1    central_value_1 upper_value_1 setname_1
lower_limit_2    central_value_2 upper_value_2 setname_2
lower_limit_3    central_value_3 upper_value_3 setname_3
lower_limit_4    central_value_4 upper_value_4 setname_4
lower_limit_5    central_value_5 upper_value_5 setname_5
adaptation_method <NL>
```

For the adaptation method, either the sign + or * can be used for shifting or scaling, respectively. Scaling will be explained in more detail later.

The fuzzy variables for mean blood pressure and heart rate, as used in the previous section, can be defined as

```
# HR - heart rate [1/min] <NL>
feature HF
  very_low    0   25  50
  low        25   50  75
  normal     50   75 100
  high       75  100 125
  very_high 100  125 200
  * <NL>

# MAP - mean arterial pressure [mmHg] <NL>
feature MAP
  very_low   20   40  60
  low        40   60  80
  normal     60   80 100
  high       80  100 120
  very_high 100  120 150
  * <NL>
```

For every fuzzy variable connected to an input, a life span should be defined with the "duration" command. After this command, the variable name and span in minutes have to be given

```
duration MAP 120 <NL>
duration HR  120 <NL>
```

A result variable will be defined in a similar way, but no span is necessary

```
feature shock
  impossible             -100 -66 -33
  unlikely                -66 -33   0
  possible_but_unlikely -33   0  33
  present                  0  33  66
  distinct                33  66 100
  + <NL>
```

To force FLORIDA to add the result of *shock* to the result file, the result needs to be marked with the "result" command:

```
result shock <NL>
```

4.5. The FLORIDA calculator

Sometimes, it is necessary to calculate a variable from other ones. If we have input values for systolic and diastolic arterial pressure (SAP and DAP), then we can calculate the mean arterial pressure by

$$MAP = DAP + 1/3\ (SAP-DAP)$$

For such purposes, FLORIDA contains a calculator which can be used through the "calc" command. Its syntax is

calc *result_featurename RPN_expression* =

The formula must be typed in reverse polish notation (RPN) which notes the operands before the operators, a system, which is also used in some pocket calculators. To store the operators, a stack is available with four registers, usually called *x, y, z,* and *t.*

An example shows how RPN works. If we compute (12+8) * (7+9), we need to note:

input	explanation	*x*	*y*	*z*	*t*
12	first operand	12			
8	second operand	8	12		
+	sum of 12 and 8	20			
7	third operand	7	20		
9	fourth operand	9	7	20	
+	sum of 7 and 9	16	20		
*	result	320			

The notation for MAP in FLORIDA could be

```
calc MAP   DAP 1 3 / SAP DAP - * + = <NL>
```

The equivalent form

```
calc MAP   SAP DAP - 3 / DAP + <NL>
```

is shorter and needs less stack.

4.6. Rules: the knowledge itself

FLORIDA is able to explain its conclusions from rules and from input. However, it is a good idea to explain the knowledge behind the rules in linguistic form. Therefore, the "//" command is used: It contains free text to the result variable and is printed ahead of the automatic explanation. After the "//" symbol, FLORIDA expects the name of the variable and then free text.

```
// shock Shock: <NL>
// shock    - Low blood pressure <NL>
// shock    - tachycardia and hyperventilation <NL>
// shock    - high serum lactate level <NL>
```

The "//" command may be used instead of command lines.

A FLORIDA rule has the task of making human-like knowledge, expressed in a colloquial rule, available to the computer. Rules operate on variables with the help of fuzzy logic as described in the previous section. The syntax of a "rule" command is

rule
expansion_method
destination_variable setname
fuzzy_RPN_expression =
specificity <NL>

To get all partial rules as explained in the previous section, the expansion method "-" is necessary. The use of other expansion methods will be explained later.

The fuzzy RPN expression may contain fuzzy variables/fuzzy sets and fuzzy operators. Valid operators are & (fuzzy AND), | (fuzzy OR) and the unary negation operator ~. A fuzzy variable must be followed by its "value," which denotes a set name. So MAP can be followed by *low* or *very_high.*

Four degrees of specificity exist – *very_specific, specific, unspecific, very_unspecific.*

The rules for the example

MAP low	→ shock	(very specific)
serum lactate level high	→ shock	(very specific)
HR high	→ shock	(unspecific)
breathing frequency high	→ shock	(unspecific)

need to be noted as

```
rule - shock present MAP      low  = very_specific <NL>
rule - shock present lacatate high = very_specific <NL>
rule - shock present HR       high = unspecific <NL>
rule - shock present BR       high = unspecific <NL>
```

which is close to their linguistic form.

To increase the specificity of heart rate and breathing frequency, it is useful to combine both parameters

```
rule & shock present HR high BR high & = very_specific<NL>
```

The fuzzy AND operator is used in a RPN manner. If you combine two parameters with the fuzzy AND, the "&" expansion method must be used. The linguistic rule:

HR high AND BR high → *shock present*

generates the 25 following rules:

HR very high AND BR high → *shock distinct*
HR very high AND BR normal → *shock distinct*
HR very high AND BR low → *shock present*
HR very high AND BR very low → *shock possible but unlikely*
HR very high AND BR very high → *shock possible but unlikely*

HR high AND BR very high → *shock distinct*
HR high AND BR high → *shock present*
HR high AND BR normal → *shock possible but unlikely*
HR high AND BR low → *shock possible but unlikely*
HR high AND BR very low → *shock possible but unlikely*

HR normal AND BR very high → *shock present*
HR normal AND BR high → *shock possible but unlikely*
HR normal AND BR normal → *shock possible but unlikely*
HR normal AND BR low → *shock possible but unlikely*
HR normal AND BR very low → *shock unlikely*

HR low AND BR very high → *shock possible but unlikely*
HR low AND BR high → *shock possible but unlikely*

HR low AND BR normal	→ *shock possible but unlikely*
HR low AND BR low	→ *shock unlikely*
HR low AND BR very low	→ *shock impossible*
HR very low AND BR very high	→ *shock possible but unlikely*
HR very low AND BR high	→ *shock possible but unlikely*
HR very low AND BR normal	→ *shock unlikely*
HR very low AND BR low	→ *shock impossible*
HR very low AND BR very low	→ *shock impossible*

It is possible to form rules that will not be expanded. A rule with the expansion method "0" is used exactly as noted.

Rules that are more complex have to be expanded manually. However, it is necessary to tell FLORIDA that some rules are equivalent to a single linguistic rule. It is not valid to write down a list of rules with the expansion "0," because these rules count multiple for the calculation of certainty. Instead the expansion method "_" has to be used for all but the last rule. The last rule must contain the expansion method "0."

The following example is equivalent to the blood pressure rule – but less readable:

```
rule _ shock impossible
  MAP very_high   = very_specific <NL>
rule _ shock unlikely
  MAP high        = very_specific <NL>
rule _ shock possible_but_unlikely
  MAP normal      = very_specific <NL>
rule _ shock distinct
  MAP very_low    = very_specific <NL>
rule 0 shock present
  MAP low         = very_specific <NL>
```

4.7. Changing the normal value

In some cases the meaning of *normal* or *high* needs to be adapted to the patient's condition. As an example, the central venous pressure increases during mechanical ventilation. The command

set_normal *destination_variable source_variable*

modifies the fuzzy sets of the destination variable in a way that the numerical content of the source variable is used as the new central value of *normal.* The value can be read from the input or calculated with "calc."

It depends on the definition of the expansion method in the destination variable how the new normal value will be achieved. With the method "+" all set limits are shifted; with the method "*" all set limits are scaled.

5. INVOCATION OF FLORIDA

To determine the physiological state of a patient, FLORIDA must be integrated in an environment where measured data are available. To make FLORIDA easier to integrate in host systems we tried to assume as little as possible on the operating system, which will be host to FLORIDA. Therefore, we decided to design it as a command line application with an ASCII interface. If you want to use the DOS/ Windows® version of FLORIDA, you can get it electronically [5].

At the prompt, one can invoke FLORIDA by typing its name. This name must be followed by the file name of a knowledge base. To start FLORIDA with the example of the previous Section, you type in: `florida example.flo`.

FLORIDA then performs a syntax check of example.flo and returns to the prompt. If your knowledge base consists of several modules, you need a file which contains the names of all modules, one name per line. Assuming this file is named "all;" the syntax check can be invoked by: `florida @all`.

If the name of the knowledge base is followed by an input file name FLORIDA calculates all result variables and writes the results on the screen. The input file contains one input per line in the following order: *name value age_in_minutes*. If such a file is stored with the name values.inp, an evaluation can be done by

`florida example.flo values.inp` or `florida @all values.inp`.

6. EXPLAINING MORE OF FLORIDA'S FUNCTIONALITY – THE KNOWLEDGE BASE *INFLAMMATION*

6.1. Structuring the knowledge

The goal of FLORIDA is to make medical knowledge available to computers. If one talks with physicians about physiological dysfunction, one normally gets answers about typical parameter combinations and their linguistic values. These are valid for a strong dysfunction. With the help of FLORIDA, we try to describe the beginning, too, where the combination is only partially fulfilled.

If some of the parameters are changed and others are not, they compensate each other. This compensation is intended and is a tool to calculate dysfunctions at different degrees.

The parameters deviate from normal values based on different physiological effects. For example, in the case of inflammation, we measure low serum level of AT-III, low platelet count, and we detect the reaction products TAT and D-Dimer. This is an expression of the consumption of coagulation components during an inflammation.

Normally physiological effects are well covered by measured data but some are not. In partial or less dramatic dysfunction only some of these effects may be present and the parameters connected with them may be deviated. In such a case, the compensation process does not work very well. Effects well covered with parameters tend to overcompensate effects that are harder to measure. The situation gets even worse if some of the parameters are not measured. The asymmetry may increase even in this case.

Our recommendation is to create the knowledge base and first calculate the effects from the parameters. The combination of the effects gives the result needed for the output. In our inflammation example, we decided to consider the following parameters:

- body temperature
- white corpuscles
- heart rate
- breathing frequency
- fibrinogen
- C reactive protein
- AT-III
- platelets
- TAT
- D-Dimer

These parameters are routinely measured in our unit. Their combination causes the following physiological effects:

- fever
- leukocytosis or leukopenia
- tachycardia or tachypnoe
- synthesis of acute phase proteins in the lever
  - C reactive protein
  - fibrinogen
- consumption of coagulation components
  - AT-III
  - platelets
  - TAT, D-Dimer may detected

To make our knowledge base easier to understand we should add the description of these effects to the explanation with the help of the "//"-command (v. 6.6).

We recommend the reader follow the example by running the real program that can be obtained at [5]. Therefore, you can see how close to the linguistic form the rules in the knowledge base are. To see that the results are adequate, one can use the knowledge base with different sets of input parameters.

6.2. Rules for fever

Fever is connected only to body temperature so the introduction of a new variable is not necessary. However, problems occur with the definition of temperature that we have in our general-purpose parameter file.

```
# T - body temperature [°C] <NL>
```

```
feature T
  very_low  24 35 36
  low       35 36 37
  normal    36 37 38
  high      37 38 40
  very_high 38 40 41.5 + <NL>
duration T 360 <NL>
```

The central value of *high* is 38°C, and that is not yet a fever. The expansion mechanisms do not work very well with this type of definition. The problem can be solved with the help of a new variable and with the "copy" command, which is used to copy the temperature value from input to this variable:

```
# increased temperature is not fever yet <NL>
feature temperature <NL>
  low          35 36 37
  normal       36 37 38
  high         37 38 39
  very_high    38 39 40
  extreme_high 39 40 41 + <NL>
duration temperature 360 <NL>
copy T temperature <NL>
```

If fever exists, we have to take into account another effect. High temperature points to infection to some degree, but normal (or low) temperature is incompatible with infection to a larger degree. For such cases the expansion method "--" (or more generally, doubling the expansion sign) is the solution. The part for fever then consists of a single rule:

```
# Fever <NL>
rule -- inflammation present
        temperature very_high
        = very_specific <NL>
```

We probably don't want a result for inflammation if the temperature is not measured. The necessary_for command ensures that

```
necessary_for inflammation temperature
```

6.3. Rules for leukocytosis/leukopenia

During inflammation, we can find high or very low leukocyte counts (LKCS). The prepared expansion methods cannot be used in this case; hence, the rule expansion must be done manually. Similar to fever, high and low leukocyte counts have different meanings for inflammation – inflammation is not compatible with normal or low (but not very low) leukocyte count.

```
# Leukocytosis / leukopenia
rule __ inflammation unlikely
        LKCS normal = very_specific <NL>
rule __ inflammation possible_but_unlikely
        LKCS low = very_specific <NL>
rule __ inflammation distinct
        LKCS very_high = very_specific <NL>
rule __ inflammation present
        LKCS very_low = very_specific <NL>
rule 00 inflammation present
        LKCS high = very_specific <NL>
```

6.4. Rules for tachycardia/tachypnoe

For the calculation of tachycardia/tachypnoe, we need a new variable

```
feature tachycardia/tachypnoe
  impossible             -100 -66 -33
  unlikely                -66 -33   0
  possible_but_unlikely -33   0  33
  present                   0  33  66
  distinct                 33  66 100
  + <NL>
```

The rules are simple.

```
rule - tachycardia/tachypnoe present
   HR high = very_specific <NL>
rule - tachycardia/tachypnoe present
   BR high = very_specific <NL>
```

This result must be added to *inflammation:*

```
rule - inflammation present
tachycardia/tachypnoe present = specific <NL>
```

6.5. Rules for synthesis of acute phase proteins

For the synthesis of acute phase proteins, we need a new variable. The set names should be chosen so that the explanation is easy to understand.

```
feature acute_phase_proteins
        absent              -100 -66 -33
        in_traces            -66 -33   0
              low_concentration  -33   0  33
            present                0  33  66
        high concentration  33   66 100 <NL>
```

The consideration of fibrinogen is quite simple.

```
rule - acute_phase_proteins present
     fibrinogen high = very_specific <NL>
```

The C reactive protein points to an inflammation even at a serum level near the detection limit. The variable CrP takes this into account. The value *very_low* cannot be reached and *low* means "nothing detected" (below the detection limit).

```
# CrP - C reactive protein [mg/l] <NL>
feature CrP
very_low -5    -2.5    0
low       -2.5  0       2.5
normal     0     2.5    5
high       2.5   5     50
very_high 5    50    100 * <NL>
duration CrP 2880 <NL>
```

The rules must to be adapted to this definition. A manual expansion is necessary. CrP is also less specific in comparison to fibrinogen, so *unspecific* was used.

```
rule _ acute_phase_proteins absent
  CrP low = unspecific <NL>
rule _ acute_phase_proteins low_concentration
  CrP normal = unspecific <NL>
rule _ acute_phase_proteins present
  CrP high = unspecific <NL>
rule 0 acute_phase_proteins high_concentration
  CrP very_high = unspecific <NL>
```

The linguistic value *very_low* for *CrP* is not possible. Therefore, a rule is not needed for it.

The newly calculated variable *acute_phase_proteins* needs to be added to our result *inflammation.*

rule - inflammation present
acute_phase_proteins present = very_specific <NL>

6.6. Rules for consumption of coagulation components

A variable for the consumption is necessary. The set name should again be easily understood.

```
feature component_consumption
  not_any -100 -66 -33
  not_much -66 -33   0
  some      -33   0  33
```

```
  present     0  33  66
  severe     33  66 100 + <NL>
```

The rules are straightforward:

```
rule - component_consumption present
  AT-III low = very_specific <NL>
rule - component_consumption present
  platelets low = very_specific <NL>
rule - component_consumption present
  TAT high = very_specific <NL>
rule - component_consumption present
  D-Dimer high = very_specific <NL>
```

The variable is eventually added to our result. The specificity is less than the one of temperature or LKCS.

```
rule - inflammation present
  component_consumption present = specific <NL>
```

6.7. Improvement of explanation

The introduction of variables for physiological effects has the disadvantage that not all input appear in the explanation. Our knowledge base could give an explanation like this:

```
Inflammation:
- Fever <NL>
- Tachycardia and tachypnoe <NL>
- Acute phase proteins <NL>
   + Fibrinogen <NL>
   + CrP <NL>
- Consumption of coagulation components <NL>
   + AT-III <NL>
   + platelets <NL>
   + reaction products (TAT, D-Dimer) <NL>

Report: Inflammation is possible_but_unlikely ..
present.
        Intensity=13%, certainty=36%, consistency=45%

Explanation:
LKCS is very_high.                          [28.30, 2.8h old]
temperature is normal .. high.              [37.25, 0.5h old]
tachycardia/tachypnoe is present.                 [99.87%]
acute_phase_proteins is present                   [15.01%]
component_consumption is severe.                  [90.76%]

Missing input: BF D-Dimer
                  (FLORIDA 4.8#0,Inflammation__1.00)
```

To see the AT-III and platelets directly, we can add them with the help of the "explain with" command.

```
explain inflammation with AT-III
explain inflammation with platelets
```

Two lines with the numerical values of *AT-III* and *platelets* are added to the explanation.

7. DIFFERENTIATION OF DYSFUNCTIONS

Occasionally we want to divide a dysfunction in sub-dysfunctions. For example, *heart failure* can be distinguished between *left heart failure, right heart failure,* and *global heart failure.*

Normally, we need more input to calculate sub-dysfunctions. Therefore, the main dysfunction needs to be calculated to have at least this result in a situation with less input.

FLORIDA supports the calculation of sub-dysfunctions at high level. The main dysfunction and one of its sub-dysfunctions are connected the following way:

The sub-dysfunction is valid if

1. the main dysfunction is valid, or
2. the differentiation rules are valid.

The sub-dysfunction may be valid, at most, to the same degree as the main dysfunction. The same is true for certainty and consistency.

If we want to calculate the sub-dysfunction *left heart failure*, which is connected to *heart failure,* we have to take into account only the variables which point to the left heart. That means we should consider the parameters: central venous pressure (CVP), pulmonary artery pressure (PAP), left ventriculary ejection fraction (LVEF), and end diastolic volume index (EDVI).

PAP high at normal or low CVP → *left heart failure*
LVEF low → *left heart failure*
EDVI low → *left heart failure*

To express the fact that *left heart failure* is a sub-dysfunction to *heart failure*, we need the "depends_on" command.

```
depends_on heart_failure left_heart_failure
```

Notice that it is not necessary nor correct to add input like blood pressure or cardiac index to *left heart failure*. Florida automatically considers this when using the "depends_on" command.

8. VISUALIZATION OF THE RESULT

FLORIDA should be used in intensive care or similar units. Such an environment requires a fast overview of the patient's state for the physicians. The visualization must support this and show progress and state of the patient at a glance.

For routine use, FLORIDA results should be calculated as often as every 15 minutes. The results need to be shown over a time axis. A problem occurs because we must present three values for every result:

- intensity
- certainty
- consistency

We have tested several forms of visualization. Our last (and best) proposal is the following:

- The intensity, as the most important factor, is shown as the ordinate value of a marker.
- The consistency is shown in the form of the marker.
- The certainty is shown in the orientation of the marker.

Figure 8 shows the principle.

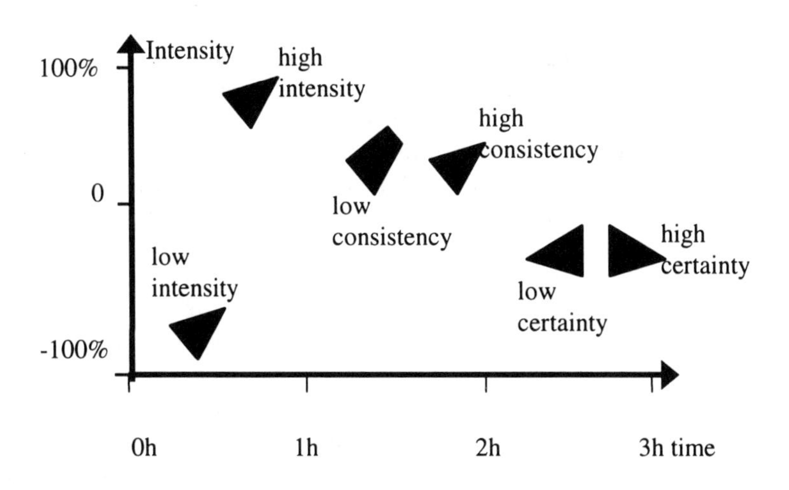

Figure 8. Visualization of the results.

In our unit, FLORIDA is integrated into a patient data management system. It provides the vital parameters, laboratory data, and medical information. Thereafter, FLORIDA is able to work autonomously, i.e., without human input.

For visualization, there is a screen with five sections. Each section can present up to eight physiological dysfunctions, which are presented by different colors. The time base can be changed between 1 hour and 8 days. Figure 9 shows a part of the FLORIDA screen.

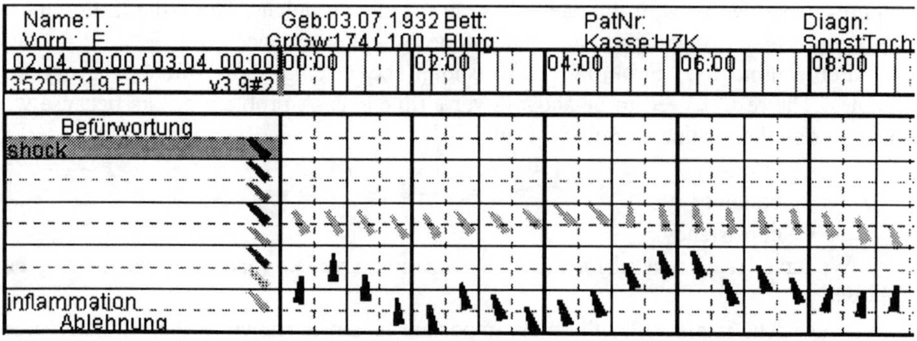

Figure 9. FLORIDA screen in the patient data management system.

9. DISCUSSION AND CONCLUSIONS

FLORIDA allows creation of knowledge bases for use in intensive care and anesthesiology. Because of its capabilities of processing knowledge, FLORIDA allows a straightforward transformation of human decision-making strategies into a computer environment. This works even with complex problems.

FLORIDA supports modularization. It is very easy to combine different knowledge data piece by piece with a single extended base. The program has well-defined interfaces to the operating system and can be adapted to each computer system with less expenditure.

REFERENCES

[1] Knebel F.G., Pilz U., Schirrmeister, W., Engelmann, L., Hommel, J., Krohn, K., Draws D.: Zielwertorientierte Therapieführung unter Anwendung des Wissensbasierten Systems (WBS) FLORIDA auf Basis des Patientendatenmanagementsystems COPRA zur Kostendämpfung in der Intensivtherapie entsprechend den Erfordernissen des Gesundheitsstrukturgesetzes (GSG). *Intensivmed* 32 Suppl. 1(1995) 192.

[2] Pilz, U., Engelmann, L., Werner, A.K.: Die Bewertung von Krankheitsverlauf und Therapieerfolg mit Hilfe des Expertensystems FLORIDA. *Intensivmed* 32 (1995) 511.

[3] Pilz, U.: Das Expertensystem FLORIDA zur Bewertung von Krankheitsverlauf und Therapieerfolg in der Intensivmedizin. In: Bocklisch S.F., Haass U.L., Bitterlich N., Protzel P.: *Fuzzy Technologien und Neuronale Netze in der Praxis*. Aachen: Shaker 1996.

[4] Knebel, F.G., Schirrmeister, W., Krohn, K., Pilz, U., Engelmann, L., Hommel, J., Draws, D.: A New System for Support in Intensive Care Units by Using Fuzzy Logic. European Society for Computing and Technology in Anaesthesia and Intensive Care (ESCTAIC) Fifth Annual Meeting Porto Carras Programme & Abstracts H07

[5] http://www.thuenet.de/Intensivehelp/f_dokumente.html or mailto:pilz@server3.medizin.uni-leipzig.de

[6] Pilz, U., Engelmann, L., Hommel, J., Schirrmeister, W., Knebel, F.: Erweitertes Monitoring auf der Intensivstation mit dem Expertensystem FLORIDA und der Wissensbasis EPSILON. *Medizin im Dialog*, April 1997, 16-9.

[7] Pilz, U., Schirrmeister, W., Knebel, F.G., Krohn, K., Hommel, J., Draws, D., Jahn O., Engelmann L.: FLORIDA - Intelligent Monitoring by a Knowledge Based System. European Society for Computing and Technology in *Anaesthesia and Intensive Care* (ESCTAIC) Fifth Annual Meeting Porto Carras Programme & Abstracts H08.

[8] Pilz, U., Werner, A., Engelmann, L.: Wissensbasiertes Monitoring: Design, Entwicklung und Validierung des Expertensystems FLORIDA.14. Arbeitstagung der Arbeitsgemeinschaft für Neurologische Intensivmedizin, 23-25.1.1997: Kurzfassungen der Vorträge und Poster. (Summary in the volume of the Symposium.)

[9] Pilz, U.: Struktur und grundlegende Funktionsweise des intensivmedizinischen Expertensystems FLORIDA. *Elektrie* 49 (1995) 399-406.

[10] Pilz, U.: Das Expertensystem FLORIDA - ein diagnostisches Hilfsmittel in der Intensivmedizin. 42. *Proc. Internationales Wissenschaftliches Kolloquium* Ilmenau 22.-25.9.1997, Vol. 2, 243-8.

[11] Zadeh, L.: Fuzzy Logic, Neuronal Networks and Soft Computing. *Communications of the ACM* 37 (1994) 377-83.

[12] Zimmermann, H.J.: *Fuzzy Set Theory*. Kluwer Academic Publisher Boston 1991

[13] Mayer, A., Mechler, B., Schlindwein, A., Wolke, R.: *Fuzzy Logic*. Addison Wesley Publishing Company, Bonn, 1993.

Part 3.

Neuro-Fuzzy Control and Hardware

Chapter 11

Hemodynamic Management with Multiple Drugs using Fuzzy Logic

Johnnie W. Huang, Claudio M. Held, and Rob J. Roy

Advanced clinical monitors have been proposed and tested in recent years for managing patient hemodynamics in operating rooms and intensive care units. These intelligent systems feature heuristic methods such as fuzzy logic for diagnosis and determining the optimal drug infusion rates for controlling the patient hemodynamic parameters. In this chapter, the concepts of fuzzy decision-making and fuzzy control are illustrated with an example of an automated multiple drug delivery system

1. Introduction

With the evolution of modern decision-making and control theory, biomedical engineers have been developing intelligent systems (Figure 1) for clinical monitoring and intervention and vital signs management in operating rooms (OR) and intensive care units (ICU). These control problems are often non-linear with time-delay, typifying the complex behaviors of the physiological systems. Such difficult tasks require high level reasoning and extensive past experience in the decision-making process for proper diagnosis and disciplined therapeutic approach.

The development of these intelligent systems started from the early closed-loop controllers [1], which were implemented in the single-input single-output architecture, such as the control of mean arterial pressure (MAP) through the infusion of sodium nitroprusside (SNP), a potent vasodilator. Considerable research efforts were later directed toward the development of systems with a multiple-input multiple-output architecture. The goal of these systems is the simultaneous management of hemodynamic variables [2], typically the MAP and the cardiac output (CO), with the infusions of SNP and dopamine (DPM), an inotropic drug (for terminology, see Appendix). The concept of combined drug therapy, for those clinical situations where an apparent reduction in the myocardial contractility is present (such as in the case of congestive heart failure (CHF) and post-heart surgery), was derived from the desire to

0-8493-9806-1/98/$0.00+$.50

minimize the myocardial oxygen consumption by relieving the end-diastolic pressure, while increasing or sustaining the CO. Conflicts in making drug choices and determining the corresponding dosages occur if the nature of the illness is not initially diagnosed for the formulation of a cohesive therapeutic approach. Capabilities in decision-making for selecting an appropriate control protocol are therefore essential when such an intelligent system is to cover a broader base of responsibilities in OR and ICU.

More recently, systems with controllers [3] equipped with an expanded arsenal of pharmaceutical agents were designed to first consider the secondary physiological parameters, such as the systemic vascular resistance, for diagnosing the patient's conditions before selecting a therapeutic approach in controlling those primary physiological parameters, such as the MAP.

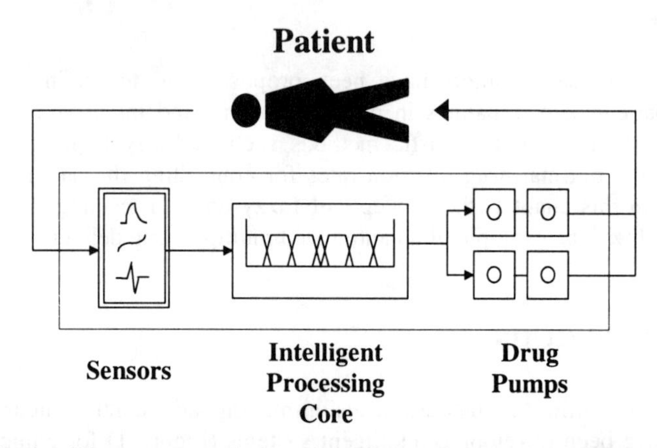

Figure 1. An intelligent system designed for fully automated drug delivery acquires the patient states through a collection of sensors. The information collected from these sensors may be the hemodynamic variables, blood gas level, anesthetic depth, ECG, and other health state signals. The information obtained is used to intelligently determine the optimal infusion rates of each pharmacological agent to be delivered to the patient through the infusion pumps. The pharmacological agents may be vasoactive drugs, anesthetics, inotropic agents, gases, and even fluids such as saline. (© 1988 IEEE, with permission.)

In this chapter, the application of fuzzy logic, in both decision-making and control, is demonstrated in the design of an intelligent system for handling a variety of critically ill patient cases in OR and ICU.

1.1. Progress in Decision Making

Prior to the introduction and the acceptance of fuzzy logic reasoning, models for medical diagnostic processes have been primarily based on classical decision theory [4]. Variations of Bayesian statistics, pattern classification, probabilistic approximation [5], decision trees [6], and other techniques have all been used to simulate the process of medical diagnosis. The development of knowledge-based expert systems for medical diagnosis started about a decade after Zadeh's introduction of fuzzy logic with the MYCIN system [7] in 1976. Although MYCIN was not based entirely on fuzzy logic, the concept of heuristic rules and the considerations of uncertainty factors were used. Since the introduction of MYCIN, fuzzy set theory has been applied to a variety of medical decision processes for diagnosis [8], [9]. Fuzzy decision-making is suitable for clinical diagnosis of the underlying ailment causing hemodynamic instabilities. This method emulates the thinking process of an anesthesiologist by providing a procedure for searching through the expert knowledge base. Various surveys of fuzzy expert diagnosis systems were conducted [10], [11], which have categorized most efforts into two schools of knowledge modeling: the equational models and the symptomatological models. These two categories were expressed in terms of their deep and shallow model dichotomy. Deep knowledge models are based on the sequential representation of cause-effect relationships, while shallow knowledge models provide relational dependence collaterally. Shallow knowledge representation is considered in this chapter, as the symptoms are inherently relational. Thus, the application of a method in which expert knowledge can be rapidly translated and prototyped as rules is justified.

1.2. Progress in Control

The task of adjusting for the appropriate titration of each drug is difficult. The environments, in which these controllers are expected to operate, are usually hostile. Disturbances such as the effects of anesthetic agents, bleeding, patient position changes, inter-subject drug sensitivity differences, and the highly non-linear behaviors of the circulatory system, have all made a simple proportional controller inadequate in handling this problem. Attempts to closing the loop with a feedback mechanism in the design of a drug delivery system were made by various researchers since the pioneering work of Sheppard [1]. Different controller algorithms were investigated and applied for hemodynamic management in the single-input/output architecture [12]. Stochastic adaptive [13], [14], model referenced [15], expert rule-based [16], [17], and artificial neural networks [18] were some of the methods implemented and achieved, with various degrees of success, with the single-input/output architecture. The introduction of the multiple-input/output concept, designed for simultaneous hemodynamic management with more than one drug [2], prompted additional system designs such as self-tuning moving average [19], fuzzy logic [20], and multiple-model referenced [21], [22]. Fuzzy control for managing hemodynamic variables is well suited clinically. It provides a method similar to that of an anesthesiologist, where the changes made in the titration of each drug, are based on knowledge and past experience.

2. System Development

The development of an automated drug delivery system is often complicated and requires the validation of a mathematical model such as the one by Yu *et al.* [23] (see Additional Resources) during the initial development phase. Subsequent animal and human experiments will require necessary hardware setups after obtaining protocol-approvals and subject consents. The eventual construction of a prototype machine will require proper equipment design for software integration and the user interface, with an emphasis on patient safety.

The task of an implemented controller in an intelligent system is to guide the targeted hemodynamic controlled variables to within some control state. As an example, we consider a system that handles three primary physiological parameters, the MAP, the CO, and the mean pulmonary arterial pressure (MPAP), to within the target box as illustrated in Figure 2.

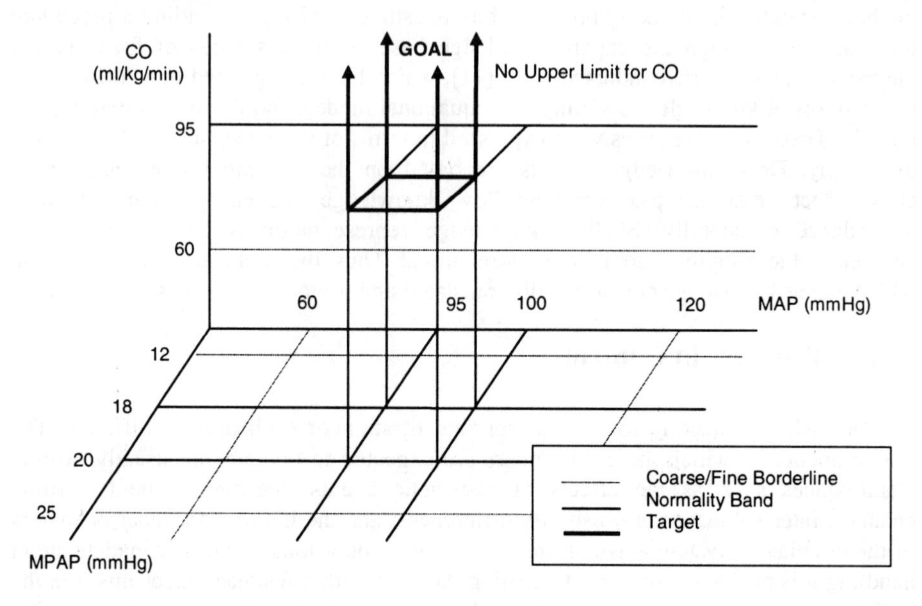

Figure 2. The MAP, the MPAP, and the CO control ranges shown in three-dimensional space. The intelligent system attempts to restore the three hemodynamic variables to within the target zone. When any of the three values falls outside the Coarse/Fine Borderline, coarse actions are taken by the supervisory commands to rapidly restore the hemodynamic variable back within the borderline. No upper limit is set for the CO, since satisfying only the minimal requirement for physiological functions is of concern. (© 1988 IEEE, with permission.)

The intelligent system uses four intravenous infusion drugs: dopamine (DPM) as an inotrope (infused at 5 to 10 µg/kg/min) to enhance cardiac performance, sodium nitroprusside (SNP) principally as an arterial vasodilator, nitroglycerin (NTG) as a venodilator for *preload* reduction, and phenylephrine (PNP) as an arterial vasoconstrictor for raising the MAP directly through its effects in tightening the systemic vascular tone. The objective of the system is to minimize the time required to achieve the task without causing major oscillations, and with the use of the least *number* and *amount* of pharmacological agents. The intention behind minimizing the drug usage is to reduce possible complications and drug accumulation in the patient. The design of such an intelligent system can generally be divided into three distinct modules, each responsible for decision-making, drug-titration control, and supervisory commands as depicted in Figure 3, further explained in the related textbox.

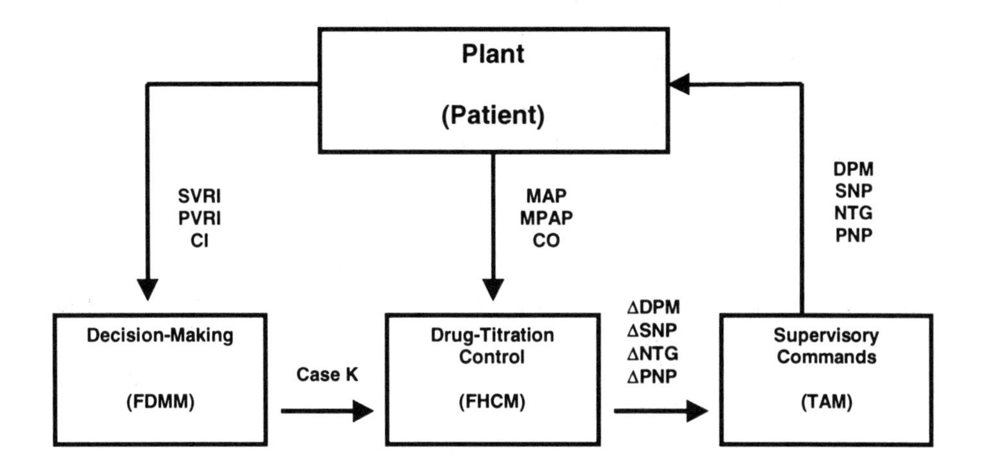

Figure 3. An intelligent system designed for automating drug delivery generally composed of three components: decision-making, drug-titration control, and supervisory commands. In the example of an intelligent system given in this chapter, the FDMM is a decision-making module that recommends a therapeutic strategy for the case based on the SVRI, the PVRI, and the CI from the patient. Changes of drug dosages are determined by FHCM, a drug-titration control module, from on the current states of the MAP, the MPAP, and the CO. The supervisory commands module TAM schedules the drug delivery based on the case K from the FDMM and the drug changes determined by the FHCM.

Decision-Making
Fuzzy Decision-Making Module (FDMM) is utilized to evaluate the status of the patient, and to designate an appropriate therapeutic strategy after the diagnosis.

Drug-Titration Control
Fuzzy Hemodynamic Control Module (FHCM) is implemented to determine the proper drug dosages based on the current state of the patient, by utilizing fuzzy-logic and other heuristic rules.

Supervisory Commands
Therapeutic Assessment Module (TAM) is devised to evaluate the effectiveness of the current remedial approach by the HMM, and to permit combinatorial infusions of additional drugs as necessary and allowed under the current therapeutic strategy as assigned by the FDMM.

2.1. Decision-Making: Fuzzy Decision-Making Module (FDMM)

As the development of intelligent systems advances, its embedded controllers are expected to be challenged by a variety of patient cases. Searching through discrete hierarchical rules as a method may not be sufficient to duplicate an anesthesiologist in determining the most suitable treatment. In addition, the complexity of the discrete rules and the non-availability or interruption of variable readings will further challenge the overall robustness of those systems.

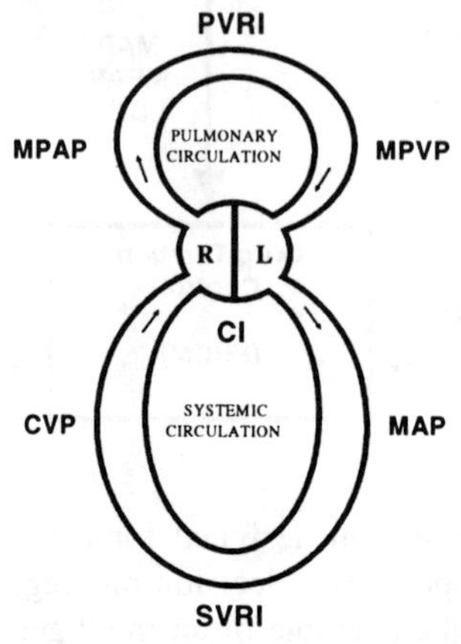

Figure 4. A model of the circulation system showing the relative location of the three controlled variables: the MAP, the MPAP, and the CO. The decision-making module of the intelligent system described uses the SVRI and the PVRI to assess the balance between the arterial and venous circulation systems, and CI to determine the health state of the heart. (© 1988 IEEE, with permission.)

2.1.1. Purpose

When assessing the patient, instead of sequentially checking the variables for exceeding some pre-defined thresholds, an anesthesiologist *concurrently* considers multiple variables, each with nondistinct borderlines. A fuzzy logic-based module is suitable for emulating the decision-making in determining the patient status, and in recommending an appropriate therapeutic strategy (e.g., Table 1). The controller determines the patient status as one of the several possible cases it has been designed to handle using the SVRI, the PVRI, and the CI as parameter inputs for fuzzy processing. The rationalization behind the usage of these three variables as the diagnostic determinants was derived from the desire to manage arterial and venous circulations independently.

As shown in Figure 4, which depicts the interrelationship of the physiological parameters in a circulatory system, the SVRI and the PVRI together provide a fair estimation of the balance between *preload* and *afterload*. The cardiac contractility is approximated by CI, which is critical in the evaluations of patients suffering from cardiac failure. The relationships of SVRI, PVRI, and CI with the effects of combined infusions of the four drugs are shown in the three-dimensional schematic diagram shown in Figure 5.

Table 1 - The intelligent system given in this chapter is capable of handling these six cases. Each case is represented by several possible pathological states reflecting the current patient status. The sequence of drugs to be delivered for each case typifies the standard procedures for hemodynamic regulations in emergency care with the available drugs.

Case	Patient Status	Possible Pathological States	Therapeutic Strategy
1	MAP Low CO Low MPAP High	Congestive Heart Failure, Myocardiac Infraction, Cardiomyopathy	Step 1: DPM Step 2: NTG Step 3: PNP (if MAP Low)
2	MAP High CO Low MPAP Normal-High	Post-Operative Hypertension, Post-Open Heart Surgery, Hypokinemia	Step 1: SNP Step 2: NTG (if MAP high) Step 3: DPM (if CO low)
3	MAP Low CO Normal/High MPAP Normal-High	Sepsis Shock, Spinal Shock	Step 1: PNP Step 2: DPM (if CO drops)
4	Map Low CO Low MPAP Low	Hemorrhagic Shock	Step 0: Fluid Loading Step 1: DPM Step 2: SNP Step 3: PNP (if MAP Low)
5	MAP High CO High/Normal MPAP Low	Post-Operative Hypertension	Step 1: SNP Step 2: NTG
6	MAP High CO High/Normal MPAP Normal/High	Hyperdynamic, Post-Operative Hypertension	Step 1: NTG Step 2: SNP

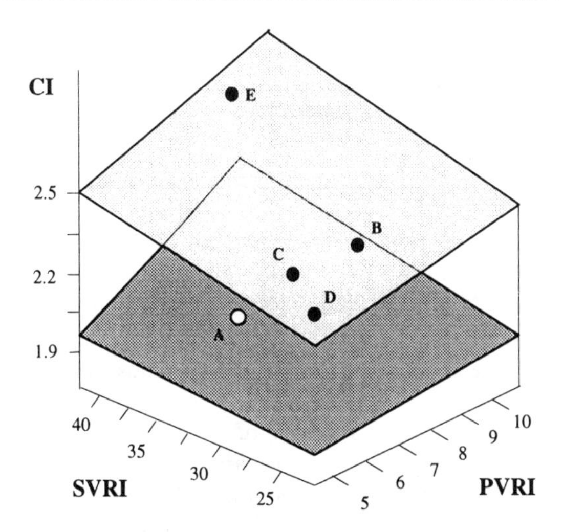

Figure 5. The relationships between the SVRI, the PVRI and the CI with the infusion of drugs are shown. A: Control. B: SNP + DPM. C: NTG + DPM. D: SNP + NTG + DPM. E: PNP + DPM. Two ventricular function planes are shown for representing the basal and the dopamine-augmented level of contractility as CI has been raised. Without any drugs injected, point A is in the lower plane. Points B-E have departed from the original state and are in the upper plane because of the combinatorial injections of different drugs. (© 1988 IEEE, with permission.)

Mathematical computation for the decision process is lessened by the reduction of input variables: MAP, MPAP, CO, the central venous pressure (CVP), and the pulmonary wedge pressure (PWP), to only SVRI, PVRI, and CI (see Terminology Appendix). The objective of the decision-making process handled by the fuzzy algorithm FDMM is to recognize the patient state and select the drug delivery sequence that would be most suitable for managing it. Future decision modules may include other vital signs, such as anesthetic level or blood gas measurements as additional attributes for distinguishing a larger variety of cases.

2.1.2. Operation

The Fuzzy Decision-Making Module (FDMM) is a fuzzy expert system that determines the propensity of the truth of a proposition based on possibilistic reasoning. The intelligent system has the ability to conduct the decision-making process by evaluating the accuracy of a certain descriptive statement quantified mathematically. The knowledge representation is flat rather than hierarchical, thus permitting solution searching in parallel for a multi-dimensional problem. A procedurally embedded monatomic inference engine is employed to derive a shallow explanation of the output via rule chaining. This rule chaining process systematically examines all rules by matching inputs with rule-antecedents, and deriving the outputs from rule-consequents by augmentation for those rules that apply. In the medical diagnostic module (FDMM), the symptoms determine the plausibility of the six cases for each patient with seven fuzzy membership functions [Figure 6-a] defined in the [–1,1] interval of real numbers: *Very True* (VT), *Rather True* (RT), *Slightly True* (ST), *Slightly False* (SF), *Rather False* (RF), *Very False* (VF), *Unknown* (UN) [24]. This unique term set allows the outputs to be polarized, which corresponds to the binary semantics that '*if something is probably true, then it is true and not false in nature*".[1] The symptoms are obtained from the current states of the CI, the SVRI, and the PVRI. These three secondary physiological parameters (SVRI, PVRI, and CI) are normalized using the corresponding sets of scaling factors, which establish their universes of discourse. These normalized values are the input vectors to be used for fuzzy matching in the state descriptions. The fuzzy term set for the inputs are composed of seven fuzzy membership functions [Figure 6-b]: Very High (VH), Rather High (RH), Slightly High (SH), Slight Low (SL), Rather Low (RL), Very Low (VL), Average (AV). These fuzzy states are defined by means of trapezoids and a triangles in the [–1,1] interval of real numbers.

The fuzzy rule set contains a series of *if-then* rules transcribed from expert knowledge (an anesthesiologist). The primary format for each rule is three "*if*" conditions as antecedents, which are the fuzzy linguistic secondary physiological parameters described earlier, and six "*then*" outcomes as the consequents, which are the fuzzy linguistic possibility of each case. Each output is represented by one membership, but, in order to reduce the number of rules, an input may have a range of fuzzy memberships. The Linguistic, the Surface, and the Deep structures of the rules (see Appendix) for inferences are as follows in the textbox:

---

[1] The polarization helps binary decision-making in the final stage of the decision process, for instance, "surgery" / "do not use surgery" (Editors' note).

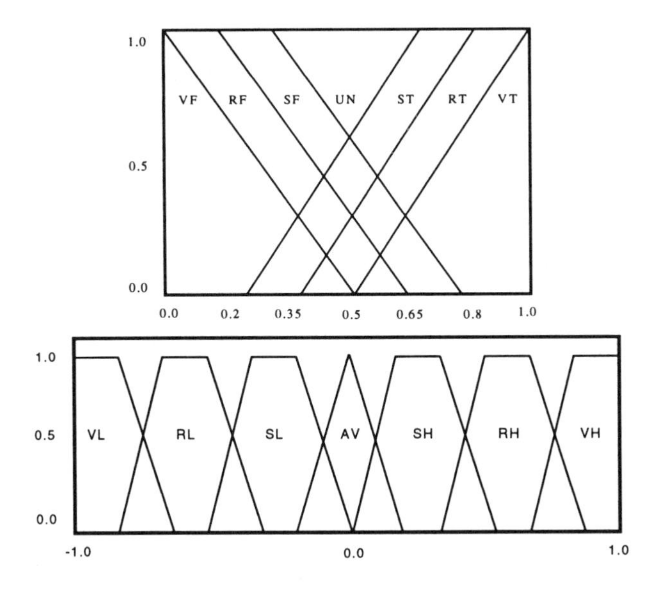

Figure 6. TOP: (a). Outputs from the FDAM are deduced from this true/false fuzzy term set. The fuzzy states used by the FDAM to describe plausibility for each case are: Very True (VT), Rather True (RT), Slightly True (ST), Slightly False (SF), Rather False (RF), Very False (VF), Unknown (UN). BOTTOM: (b). Inputs to the FDAM are fuzzified using this fuzzy term set. The inference process relies on Type II rules written and validated by experts (anesthesiologists). The fuzzy states used by the FDAM to describe the condition of the patient for each of the parameters SVRI, PVRI, and CI are: Very High (VH), Rather High (RH), Slightly High (SH), Slight Low (SL), Rather Low (RL), Very Low (VL), Average (AV). (© 1988 IEEE, with permission.)

***LINGUISTIC STRUCTURE*:**

IF the arterial side is very loose yet the heart is pumping out enough blood,
THEN heart performance is probably not the problem,
there appears to be enough fluid in the circulation,
right side of the circulation is normal,
some kind of shock is most likely, will need to tighten up the left side.

***SURFACE STRUCTURE*:**

IF the SVRI is critically low, AND
the PVRI is not very low, AND
the CI is sufficiently high,
THEN Case 1 is slightly impossible
Case 2 is rather impossible
Case 3 is most likely
Case 4 is slightly impossible
Case 5 is slightly impossible
Case 6 is indeterminate

***DEEP STRUCTURE*:**

IF (SVRI err = VL) AND (PVRI err = RL..VH) AND (CI Err = SH...VH)
THEN (K1 = SF), (K2 = RF), (K3 = VT), (K4 = SF), (K5 = SF), (K6 = UN).

The individual rule based inference process is conducted by computing the degree of match between the fuzzified input value *V* and the fuzzy sets describing the meaning of the rule-antecedent as prescribed in the ruleset R. The output μ is produced by clipping the fuzzy member describing the rule-consequent to the degree to which the rule-antecedent has been matched by V. The possibility distribution function is then found by finding the minimal of all μs:

$$\pi(V_{SVRI}, V_{PVRI}, V_{CI}) = \mu(V_{SVRI}) \wedge \mu(V_{PVRI}) \wedge \mu(V_{CI}) \tag{1}$$

As *u*s are aggregated on the true/false term set, the value of the overall output *U*, which is the minimized value of all *u*s determines the degree of applicability for each Case K. Each degree of applicability is a measure of the variable value at which the corresponding fuzzy membership set is at maximum. The rule-consequent is then inferred on the true/false fuzzy term set for each K. The defuzzification process utilizes the standard center of gravity method (COG):

$$DEFUZ_{COG}(X) = \frac{\sum_i u_i \mu x(u_i)}{\sum_i \mu x(u_i)} \tag{2}$$

It determines the output *X*, which is the abscissa of the center of gravity of the area describing the output of the inference engine in the true/false term set [25]. Then the case is chosen as the *K* with the highest operation value *X* obtained by COG. The entire decision process is illustrated in Figure 7.

2.2. Drug-Titration Control: Fuzzy Hemodynamic Control Module (FHCM)

2.2.1. Purpose

As the FDMM recognizes the underlying aliment causing hemodynamic instability in the patient, the Fuzzy Hemodynamic Control Module (FHCM) is employed to determine the dosage changes of each drug to be delivered, based on the diagnosis outcome by FDMM. Under the FHCM, the hemodynamic states are assessed according to a fuzzy rule-set composed of rules also written by an expert (anesthesiologist), and the infusion rates are changed when necessary.

2.2.2. Operation

Similar to FDMM, seven fuzzy membership functions defined by trapezoids in the [−1,1] interval of real numbers are used [Figure 8]: Negative Big (NB), Negative Medium (NM), Negative Small (NS), Zero (ZE), Positive Small (PS), Positive Medium (PM), Positive Big (PB). The fuzzified linguistic variables for the inputs to the fuzzy inference engine are the MAP error, the change in the MAP error (discrete derivative of the previous input variable), the CO error, the change in the CO error, the MPAP error, and the change in the MPAP error. The universes of discourse in the fuzzification process represent the MAP error as a band of 60 mmHg with an absolute range of [–40,20], the CO error as a band of 60 ml/kg/min with an absolute range of [–40,20], and the MPAP Error as a band of 8 mmHg with an absolute range of [–6,2]. As for the changes of error, a constant range of [–3,–3] for the MPAP, and of [–10,10] for the MAP and the CO are assigned.

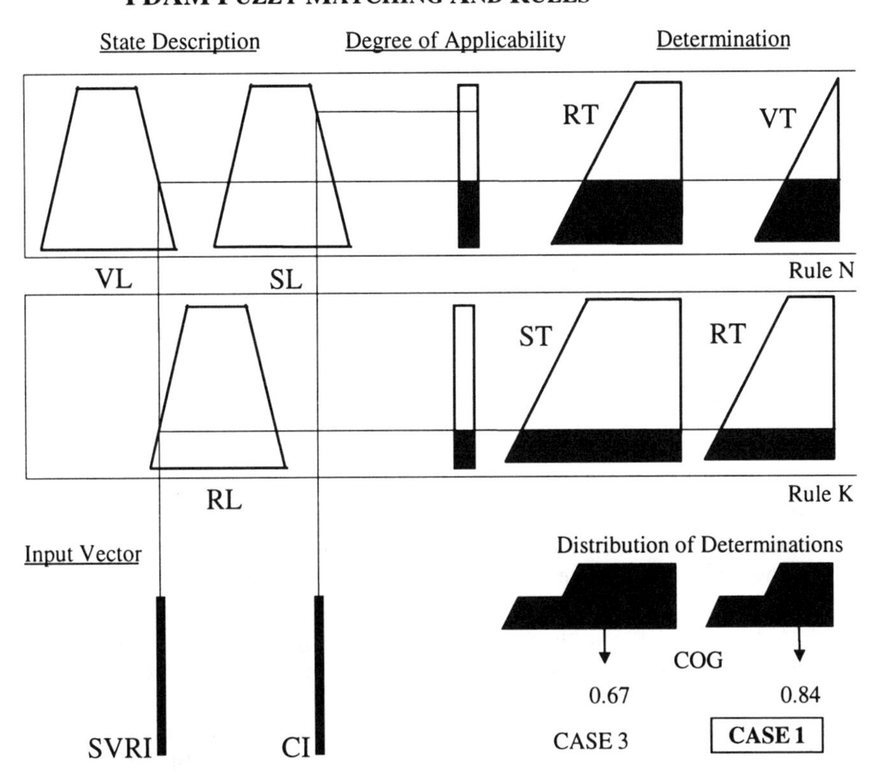

Figure 7. The process of fuzzy matching for decision-making is depicted. Only two input variables, SVRI and CI are shown in the inference; and only two rules, N and K, are used to demonstrate the determination of the degree of applicability. This shows FDMM has diagnosed the patient status as Case 1 since both SVRI and CI are rather low. A therapeutic strategy for Case 1 will be recommended for FHCM to proceed.

The fuzzified control outputs from the fuzzy inference engine are the changes in the infusion rates of DPM, SNP, PNP, and NTG. These outputs share the same membership term sets as the inputs shown in Figure 8. Similar to the FDMM defuzzification process, the COG method (Equation 2) is utilized to obtain the output values. The absolute ranges are [–7,8] μg/kg/min for SNP, [–0.8,0.5] μg/kg/min for DPM, [–6,5] μg/kg/min for PNP, and [–8,9] μg/kg/min for NTG. These universes of discourse were obtained through testing using a constant gain for the pharmacodynamics of each drug, and trying to achieve the fastest response without eliciting oscillations. Nevertheless, these ranges have been made adaptive, to account for the drug sensitivity differences found among different patients.

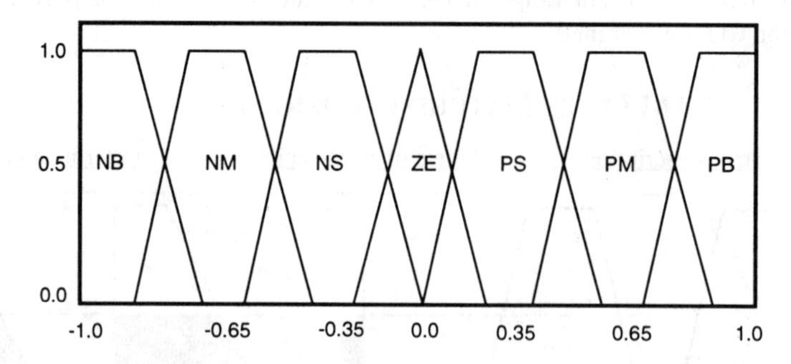

Figure 8. The FHCM utilizes this same fuzzy term set for fuzzifying the input variables (the MAP, the MPAP, and the CO) and defuzzifying to derive the output variables (the infusion rates of SNP, DPM, PNP, and NTG). The fuzzy linguistic symbols are Negative Big (NB), Negative Medium (NM), Negative Small (NS), Zero (ZE), Positive Small (PS), Positive Medium (PM), Positive Big (PB). (© 1988 IEEE, with permission.)

The fuzzy rule set contains a series of if-then rules. The primary format for each rule is similar to that of the FDMM, except that FHCM has six 'if' conditions as antecedents, which are the fuzzy linguistic hemodynamic variables described above, and four 'then' outcomes as the consequents, which are the fuzzy linguistic changes of drug infusion rates. Each output is represented by one membership, but in order to reduce the number of rules, an input may have a range of fuzzy memberships. The Linguistic, the Surface, and the Deep structures of the rules for inferences are shown in the textbox.

After the testing phase is completed, these rules may be tabulated and compiled which lessens the computation time required [26]. As with the FDMM rules, these FHCM rules are written and modified after extensive testing with expert advice.

2.3. Supervisory Commands: Therapeutic Assessment Module (TAM)

Both the FDMM and the FHCM implement fuzzy logic reasoning for emulating the parallel thinking process of human decision-making. The TAM represents the logical thinking process of humans for analytically deriving the conclusions. This module provides the supervisory commands for various linear tasks in an intelligent system. These tasks may range from drug scheduling to patient safety monitoring, which are essential in an intelligent system for upholding the rule exceptions and detecting unexpected events.

***LINGUISTIC STRUCTURE*:**

IF the blood pressure is rising rapidly from an already high value,
the cardiac output is stable,
pulmonary blood pressure is getting lower from a slightly low value,
THEN moderately dilate the left side of the circulation,
slowly turn off any inotrope injection,
do not alter the right side of the circulation,
slowly turn off any vasoconstrictor injection.

***SURFACE STRUCTURE*:**

IF the MAP is moderately high AND is getting higher very fast, AND
the CO is sufficient AND is not changing, AND
the MPAP is slightly low AND is getting lower slowly,
THEN increase SNP moderately,
decrease DPM slightly,
do not change NTG,
decrease PNP slightly.

***DEEP STRUCTURE*:**

IF (MAP Err = PM) AND (Chg MAP Err = PB) AND
(CO Err = ZE) AND (Chg CO Err = ZE) AND
(MPAP Err = NS) AND (Chg MPAP Err = PS)
THEN (Chg SNP = PM), (Chg DPM = NS), (Chg NTG = ZE), (Chg PNP = NS).

As for drug scheduling, the TAM follows a set of rules that execute the drug delivery sequence by evaluating several parameters, and considers each case individually as recommended by the FDMM. The TAM considers the time elapsed since last intervention, in order to mimic the rationale of an anesthesiologist in determining the therapeutic sufficiency of the current regimen. The system waits, adds a new infusion drug or eliminates an infusing drug, depending on the pharmacological delay and other properties of the infused drugs under the current therapeutic strategy.

As for patient safety monitoring, the TAM contains a collection of several algorithms designed to ensure reliable inputs to the system and to monitor for overall performance of this automated drug delivery system. For example, prolonged usage of SNP results in excessive thiocyanate formation, causing the patient to experience blurred vision, mental confusion, convulsions, or delirium. Any intelligent system must possess a simple drug counter as part of the supervisory command module to monitor for any possible acute dosages and the level of internal accumulation.

2.4. System Evaluations

The objective of an intelligent system for automating drug delivery is to monitor the patient for any hemodynamic instability, and manage that instability by the infusion of the equipped pharmacological agents. The system seeks to restore the hemodynamic variables to some defined ranges by infusing the minimal number and amount of the drugs.

2.4.1. Example One

The performance of a fuzzy logic based intelligent system developed by Held [27] for the simultaneous management of the MAP and the CO in dogs with DPM and SNP is shown in Figure 9. The system is designed to manage patients with CHF in ICU or in critical care units. It was validated with a mathematical model [20] and was subsequently tested with dogs as subjects. The decision-making module is not present in this system, since it assumes its subjects are all suffering from CHF. As part of the experimental protocol for evaluating the system response to a subject suffering from CHF, a higher level of the anesthetic agent halothane was infused into the airway loop to depress the hemodynamics, thereby simulating CHF symptoms. In the animal experiment, the end-tidal concentration of halothane (EtHal) was set to 1.0%. The initial arterial pressures before any system interventions were at 96/63 mmHg (MAP = 74 mmHg) and CO was 1.3 L/min (= 81.3 mL/kg/min). The controlled hemodynamic variable band for the MAP was set between 95 and 100 mmHg and for CO was a minimum of 125 mL/kg/min. The MPAP was not part of the controlled parameter set in this example. When the intelligent system was activated to restore the hemodynamic stability of the dog under serious cardiovascular depression, DPM was infused immediately to induce positive inotropic effects, thereby establishing a higher level of CO and preventing MAP from lowering any further. To account for the time delays associated with the pharmacological effects of DPM, the supervisory commands of the system prohibited any drug infusion changes in the first 3 minutes after DPM was first introduced. The infusion of SNP started after 3.5 minutes to prevent MAP from overshooting, as the elevated cardiac contractility by DPM also increased MAP. The DPM level was reduced after 10.5 minutes when the minimal CO level was achieved. This reduction in DPM would have also indirectly decreased the level of SNP required, consistent with the objective of minimizing the drug usage. After 25 minutes, hemodynamic stability was achieved in the dog and ample perfusion of the distal organs was ensured by maintaining a stable MAP with sufficient CO.

2.4.2. Example Two

The performances of an intelligent system [3] designed to regulate the three primary hemodynamic parameters, MAP, MPAP, and CO, are shown in Figure 10. The objective is similar to the earlier example, except now the MPAP will also be managed to within the normal range, shown in Figure 2. The results shown were obtained from a simulation with the mathematical model of a circulatory system by Yu *et al.* [23]. The system is now equipped with NTG and PNP in addition to DPM and SNP. In this intelligent system, the decision-making module FDMM first diagnoses the patient and selects a therapeutic strategy from Table 1 for the drug titration control module FHCM to follow. The FDMM continuously re-assesses the patient status during every intervention step (0.5 sec.), and, if necessary, re-adjusts the therapeutic strategy based on the new diagnosis after each examination. The ability and the efficiency of the FDMM to recognize any change of hemodynamic states in the patient are particularly important. Drugs being infused as recommended by the previous therapeutic strategy may be obsolete and counter-productive in the current state. In the simulation, a subject simulated by a modeled dog having 100% heart contractility was suffering from hypertension when the intelligent system first initiated the interventional steps.

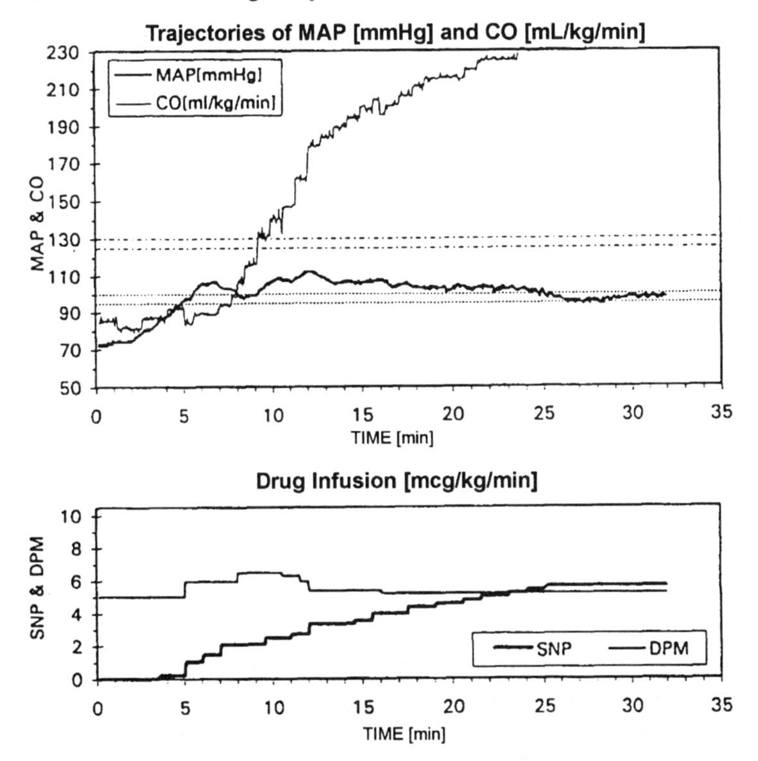

Figure 9. The experiment results of an intelligent system [27] designed for regulating MAP and CO with SNP and DPM are shown. The experiment was conducted on a dog with CHF simulated by an elevated level of EtHal. The system was able to restore the hemodynamic stability by a sophisticated sequence of drug infusions and dosage level adjustments. (© 1988 IEEE, with permission.)

The FDMM assigned Case 6 as the therapeutic strategy and NTG was subsequently infused to depress the MAP. However, the patient became hypotensive after 10 minutes because the controlled band for the MAP arbitrarily changed from 95-100 mmHg to 125-130 mmHg, and from a minimal of 95 ml/kg/min to 125 ml/kg/min for the CO, to represent a change of state in the patient. The FDMM immediately recognized this change during the next controller interventional step at 10.5 minutes where the therapeutic strategy was adjusted from Case 6 to Case 1. Consequently, DPM was infused immediately and maintained its infusion rate at 5 μg/kg/min for 2.5 minutes while NTG was progressively shut off. PNP infusion started at 15.5 minutes as determined by the supervisory commands module TAM for boosting the MAP after DPM successfully increased the CO. The infusion of PNP was followed immediately after NTG infusion was stopped. An earlier infusion of PNP would be counteracted by the effects of NTG, and might cause a MAP overshoot if PNP infusion was determined as not necessary after NTG was shut off.

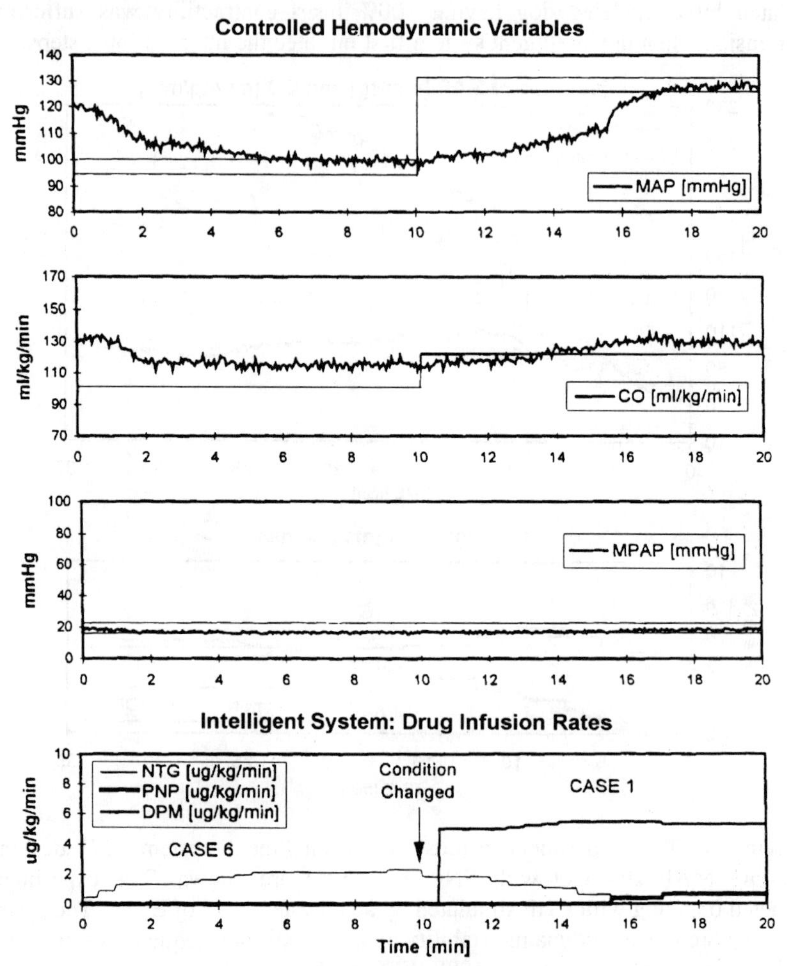

Figure 10. The simulation results of the intelligent system [3] on a mathematical circulatory model are shown. The simulation started with the patient, having hypertension initially, changed to hypotension after 10 minutes. The system is capable of detecting the changes in the patient status where the FDMM will provide the FHCM a different sequence of drugs to be delivered when necessary. NTG was not immediately shut off to minimize large perturbation in the relatively controlled states. PNP was used to provide a more secure control of the MAP after the CO had been satisfied. (© 1988 IEEE, with permission.)

3. Future Prospects

The feasibility of an intelligent system for automated drug delivery lies within its abilities to determine and react to a wide range of possible cases. Moreover, the robustness in regulating those hemodynamic variables according to the prognosis by the decision-making module is essential. Future patient monitoring techniques will be combined with patient interventional procedures by such intelligent systems to assist anesthesiologists maintain patient hemodynamic stability. Several important issues including limitations of an intelligent system should be discussed and evaluated.

3.1. Design Possibilities

Although the example given in this chapter only describes a system, which uses a limited number of hemodynamic variables for diagnosis and control, there are several other potential sources of information that an intelligent system may use for processing. Additional patient information obtained will enable the system to have a better estimation of the patient status, thus the diagnosis outcome will be more accurate and the control strategy will be more effective.

It is widely known that anesthesia depresses hemodynamics in the patient through its suppression of the central nervous system. An intelligent system can use the level of anesthetic currently applied as an additional parameter when processing. Other parameters such as heart rate (HR) may be used. This parameter is potentially useful since inotropic agents such as DPM are likely to produce tachycardia. Monitoring the trend of HR may prevent tachycardia thus avoiding a significant decrease in CO due to insufficient filling time during diastole. Since stroke volume (SV) is determined by dividing CO with HR, SV may also be used along with SVRI and PVRI during the diagnosis when evaluating the balance between preload and afterload.

The system may be equipped with more drugs thus enabling it to manage patient cases not previously discussed. Drugs such as beta-blockers or calcium channel blockers may be added to the regimen. Combining these drugs with information processed by an ECG analyzer module, in situations where a stable patient is suffering from atrial fibrillation with a rapid ventricular response, may be useful for rate control. Other drugs such as anesthetics and analgesics or even fluids such as saline may be considered. Nevertheless, more sophisticated processing algorithms may be required in addition to obtaining necessary information from the patient.

3.2. "Curse of Dimensions"

Although in flat knowledge representation, the expansion of the number of input or output variables will not increase the level of complexity in transcribing expert knowledge into rules. However, this expansion of the number of variables will greatly increase the number of rules involved. Any rules missing in the knowledge base may potentially cause devastating consequences in the outcome, although fuzzy systems are generally considered robust even if the knowledge may not be complete. This inherent hindrance may be overcome by having multiple FHCM units, each dedicated to a particular case as determined by the FDMM. In addition, multiple fuzzy engines may be stacked on top of each other when a "fuzzy hierarchical" system has evolved. Essentially, the combination of FDMM and FHCM as proposed in this chapter is a fuzzy hierarchical system designed to minimize the collateral rule confusion if only one fuzzy engine is to be implemented.

3.3. Machine Intelligence

As the intelligent systems are equipped with a wide selection of drugs currently available to most anesthesiologists for countering a variety of possible cases, the prognosis by some decision-making module becomes critical in prescribing the sequence of drugs that will be most effective. A knowledge-based decision-making algorithm for determining the patient status was presented in this chapter. This assessing capability using artificial intelligence will be essential in providing a more in-depth diagnosis compared to the threshold decision-tree algorithms. For an example, in a patient with extremely low MAP and a CO that is slightly low but very close to normal, a threshold decision algorithm may initially advise the use of DPM since both variables are below the target level. However, a fuzzy decision-making algorithm like FDMM will recognize the inadequacy in peripheral tonicity in the patient, and advise the use of PNP instead, which will be more effective in such scenario. This computational operation emulates that of an anesthesiologist in formulating a strategy, which may not be obvious if processed linearly. In reference to the Turing Test proposed by mathematician Alan Turing in 1950 for judging machine intelligence, the course which the system selects and its ability to preemptively alter its titration of drugs in expectation of potential instability, demonstrates consciousness when looked upon by a third person. Nevertheless, when evaluating the internal processing in which the intelligent system operates, fuzzy logic is a method based on the *interpolation* of the knowledge`base. It still does not provide the *interpretation* that a human is capable of in analyzing situations, which is necessary when unexpected events occur. The processing core of the intelligent system is merely a *Chinese Room*, a parable formulated by philosopher John Searle in 1980 for disparaging the notion of artificial intelligence. Nonetheless, a system, capable of monitoring and managing multiple parameters continuously in a clinical setting with proper titration of drugs mimicking actions of an anesthesiologist, is very valuable in a clinical setting.

ADDITIONAL RESOURCES

The nonlinear canine circulatory model developed by Yu *et al.* [23] emulates the dynamics of the circulation system by using eleven compartments in an electric circuit analog. This model is sufficient for approximating the hemodynamic responses of drug injections for testing the systems during the initial development phase. In the model, the heart is subdivided into two compartments representing the left and right ventricles. Each compartment is characterized by the maximum magnitudes of a varying elastance, which provides the energy necessary for blood flow. The left and right ventricles pump the blood through an ideal valve into the systemic and the pulmonary circulation correspondingly. The systemic circulation is serially linked by aorta, arterial, and venous compartments, while the pulmonary circulation is composed of pulmonary arterial and pulmonary venous compartments. These vasculature compartments are represented by capacitors, each with its own compliant element, and connected through resistors.

This canine circulatory model and the program codes of various intelligent systems designed for automating drug delivery are obtainable electronically with a hypertext browser from the Software Archive of the Biological Signal Processing and Control Laboratory, Department of Biomedical Engineering, Rensselaer Polytechnic Institute at http://www.rpi.edu/~royr/roy_sftwr.html. Additional information including resources for the development of other intelligent systems for OR and ICU settings are available.

APPENDIX: TERMINOLOGY

Afterload: the pressure in the artery during systole, which is correlated to the resistance in the circulation.

Cardiac Index (CI): defined as the cardiac output normalized by the body surface area (BSA), where BSA is estimated by $BSA = 0.007184 \times W^{0.425} \times H^{0.725}$, having W for body weight in kilogram and H for height of the subject in centimeter.

Congestive Heart Failure (CHF): a disease state having impaired heart function thus causing poor circulation of the blood.

Deep Structure: knowledge as rules symbolized for machine processing.

Degree of Applicability: measures the variable value at which the corresponding fuzzy membership set is at maximum.

Fuzzy Logic Controller: a system designed to process control of plants for which the state variables could not be precisely measured or where it is difficult to describe the course of actions.

Inotrope: any pharmacological agent that influences the force of cardiac muscular contractility.

Linguistic Structure: knowledge as rules perceived or given by an expert (anesthesiologist).

Preload: refers to the volume of blood or the pressure in the ventricle at the end of diastole when the ventricle is filled.

Pulmonary Vascular Resistance Index (PVRI): defined as $\frac{MPAP - PWP}{CI} \times 79.9$

Rule Chaining: systematically examining all rules by matching the inputs with rule-antecedent and deriving the outputs from rule-consequent by augmentation for those rules that applied.

Rule-Antecedent: the IF clause in an IF-THEN statement that represents the rule which necessitates a certain condition for the rule to be applicable.

Surface Structure: knowledge as rules fuzzified for rule prototyping.

Systemic Vascular Resistance Index (SVRI): defined as $\frac{MAP - CVP}{CI} \times 79.9$

Truth of a Proposition: accuracy of a certain descriptive statement quantified mathematically.

TERMINOLOGY

CHF	Congestive Heart Failure
CI	Cardiac Index
CO	Cardiac Output
COG	Center of Gravity Method
CVP	Central Venous Pressure
DPM	Dopamine
FDMM	Fuzzy Decision-Making Module
FHCM	Fuzzy Hemodynamic Control Module
MAP	Mean Arterial Pressure
MIMO	Multiple-Input Multiple-Output
MPAP	Mean Pulmonary Arterial Pressure
NTG	Nitroglycerin
PNP	Phenylephrine
PVRI	Pulmonary Vascular Resistance Index
PWP	Pulmonary Wedge Pressure
SISO	Single-Input Single-Output
SNP	Sodium Nitroprusside
SVRI	Systemic Vascular Resistance Index
TAM	Therapeutic Assessment Module

REFERENCES

[1] Sheppard, L.C., Kouchoukos, N.T., Kurtis, M.A., and Kirklin, J.W., Automated treatment of critically ill patients following operation. *Annals of Surgery*, 1968; 168: 596-604.

[2] Roy, R.J., Adaptive Cardiovascular Control Using Multiple Drug Infusions, *Proc. of the IEEE-EMBS Conference on Frontiers of Engineering in Health Care*, 459-464, 1982.

[3] Huang, J.W. and Roy, R.J., Multiple-drug hemodynamic control using fuzzy decision theory. *IEEE Trans. on Biomed. Eng.*, 1998; 45 (2): 213-228.

[4] Croft, D.J., Mathematical methods in medical diagnosis. *Ann. Biomedical Engineering*, 1974; 2.

[5] Gouvernet, J., Ayme, S., and Sanchez, E., Approximate Reasoning in Medical Genetics, *Proc. of Int. Congress on Applied Systems Research and Cybernetics*, Acapulco, Mexico, 1980.

[6] Gorry, G.A. and Barnett, G.O., Sequential diagnosis by computer. *Journal of American Medical Association*, 1968; 205: 849-54.

[7] Shortliffe, E.H., *Computer-Based Medical Consultations, MYCIN*. Elsevier/North Holland, New York, 1976.

[8] Bezdek, J.C., Feature selection for binary data: medical diagnosis with fuzzy sets, *Proc. of National Computer Conf.*, AFIPS Press, Montvale, New Jersey, 1976.

[9] Smets, P., Medical diagnosis: Fuzzy sets and degrees of belief, *Fuzzy Sets and Systems*, 1981; 5 (3): 259-66.

[10] Raoult, O., Survey of diagnosis expert systems, in *Knowledge Based Systems for Test and Diagnosis*, Elsevier Science Publishers B.V. North-Holland, 1989.

[11] Hudson, D.L. and Cohen, M.E., Fuzzy logic in medical expert system, *IEEE Eng. in Med. and Bio. Magazine*, Nov./Dec. 1994: 693-698.

[12] Isaka, S., Sebald, A.V., Control Strategies for Arterial Blood Pressure Regulation, *IEEE Trans. on Biomed. Eng.*, April 1993; 40: 353-63.

[13] Walker, B.K., Chia, T-L., Stern, and K.S., Katona, P.G., Parameter Identification and Adaptive Control for Blood Pressure, *Proc. IFAC Identification System Parameter Estimation*, Washington, D.C.; 1982: 1413-8.

[14] Arnsparger, J.M., McInnis, B.C., Glover, J.R., and Normann, N.A., Adaptive Control of Blood Pressure, *IEEE Trans. on Biomed. Eng.*, 1983; BME-30: 168-76.

[15] Kaufman, H, Roy, R.J., and Xu, X., Model Reference Adaptive Control of Drug Infusion Rate, *Automatica*, 1984; 20: 205-9.

[16] Hammond, J.J., Kirkendall, W.M., and Calfee, R.V., Hypertensive Crisis Managed by Computer Controlled Infusion of Sodium Nitroprusside: A Model for the Closed Loop Administration of Short Acting Vasoactive Agents, *Computers and Biomedical Engineering*, 1979; 12: 97-108.

[17] Martin, J.F., Schneider, A.M., Quinn, M.L., and Smith, N.T., Improved Safety and Efficacy in Adaptive Control of Arterial Blood Pressure Through the Use of a Supervisor, *IEEE Trans. on Biomed. Eng.*, April 1992; 39: 381-8.

[18] Sebald, A.V., Sebald, C.A., and Schlenzig, J., Use of Neural Net Control Strategies in Difficult Adaptive Control Problems: Closed Loop Control of Drug Infusion, *Proc. Asilomar Conf. on Circuits and Systems*, Nov. 1989: 342-4.

[19] Voss, G.I., Katona, P.G., and Chizeck, H.J., Adaptive Multivariable Drug Delivery Control of Arterial Pressure and Cardiac Output in Anesthetized Dogs, *IEEE Trans. on Biomed. Eng.*, 1987 Aug; BME-34 (8): 617-23.

[20] Held, C.M. and Roy, R.J., Multiple Drug Hemodynamic Control by Means of a Supervisory-Fuzzy Rule Based Adaptive Control System: Validation on a Model, *IEEE Trans. on Biomed. Eng.*, 1995 Apr; 371-85.

[21] Westvold, S.S., Kaufman, H., and Roy, R.J., Model Reference Adaptive Control of a Circulatory Model for Combined Nitroprusside-Dopamine Therapy, *Proc. Of ACASP 1992*, Grenoble, France 1992 Jul 1-3: 223-8.

[22] Yu, C., Roy, R.J., Kaufman, H., and Bequette, B.W., Multiple-Model Adaptive Predictive Control of Mean Arterial Pressure and Cardiac Output, *IEEE Trans. on Biomed. Eng.*, 1992 Aug; 39 (8): 765-78.

[23] Yu, C., Roy, R.J., and Kaufman, H., A Circulatory Model for Combined Nitroprusside-Dopamine Therapy in Acute Heart Failure, *Medical Progress through Technology*, 1990; 16: 77-88.

[24] Tazaki, E., Medical Diagnosis Using Simplified Multidimensional Fuzzy Reasoning, *Proc. IEEE Int. Conf. on SMC*, 1988.

[25] Lee, C.C., Fuzzy Logic in Control Systems: Fuzzy Logic Controller - Parts I and II, *IEEE Trans. on Sys. Man. Cyber.*, 1990 Mar./Apr.; 20: 404-35.

[26] Bonissone, P.P., A Compiler for Fuzzy Logic Controllers, *Proc. Int. Conf. on Fuzzy Eng. Sys.* IFES 91, Yokohama, Japan, Nov. 1991.

[27] Held, C. M., Closed-loop hemodynamic management by means of a rule-based control system using fuzzy logic, Ph.D. Thesis, Rensselaer Polytechnic Institute, Tory, NY, July 1995.

Chapter 12

Neuro-Fuzzy Hardware in Medical Applications

12 A. System Requirements for Fuzzy and Neuro-Fuzzy Hardware in Medical Equipment

Horia-Nicolai L. Teodorescu, Abraham Kandel, and Daniel Mlynek

In this chapter we deal with system and technological requirements for applications of fuzzy and neuro-fuzzy systems in medical equipment. The requirements of circuits and equipment in various classes of medical applications are reviewed. The main system parameters, such as errors, sensitivity, power of approximation (classes of approximated functions and related approximating precision), speed, reliability, etc. are discussed. Several issues related to hardware minimization are detailed and specific system architectures are presented.

1. Introduction

Since 1985, the field of dedicated hardware for fuzzy and neuro-fuzzy applications has seen constant development and significant progress. A good historical perspective on the development of fuzzy hardware can be found in [1], and a comprehensive reference is [2].

The use of either fuzzy or classic circuits in medical equipment is governed by several domain-specific requirements, rules, and operating conditions. In general, there is no restriction or limit for any circuit or part to operate in medical equipment, based solely on the operating principle. For example, there is no incompatibility in using ceramic or plastic film capacitors in such equipment. On the other hand, if the capacitor

0-8493-9806-1/98/$0.00+$.50

materials are noxious, or if the reliability of the capacitor is too low, then restrictions apply. Similarly, there is no specific preference for plastic or ceramic encapsulations for integrated circuits (ICs), for digital or analog circuits, or for fuzzy or crisp circuits – provided that the operating capabilities and properties do not pose a threat to the life of the patient and consequently ban some sub-class of specific circuits. The properties related to biocompatibility, reliability, precision, electromagnetic interference behavior, and cost govern the acceptance of a specific class of circuits in medical applications.

There are several classes of applications related to medical use, each with its specific requirements and constraints. The main classes of medical equipment are

- diagnosis equipment (measuring and imaging)
- "non-contact" equipment
- external (to the human body)
- internal (implanted)
- laboratory
- surgery equipment (critical) – anesthesia control, other life-support systems
- education – including Virtual-Reality related
- rehabilitation – several applications
- prosthetics (general: sense prostheses, limb prostheses; life-support prostheses and ortheses).

Each of the above classes has specific constraints at the system level, regarding the properties mentioned above.

Until now, the use of specific fuzzy hardware was reported in several of these classes, as described in the next sections.

2. SPECIFIC REQUIREMENTS OF MEDICAL APPLICATIONS

The general conditions for fuzzy and neuro-fuzzy integrated circuits (ICs) and software included in medical equipment do not differ from the conditions imposed on other electronic or software components of the equipment. These conditions can be found in the corresponding international ISO and IEC, or national standards. ISO standards for specific classes of medical equipment are extensively used. As an example, the standards ISO/FDIS 8835-1, ISO 8835-2:1993, ISO/DIS 8835-2 and ISO 8835-3:1997, ISO 11196: 1995 refer to medical equipment for anesthesia [3], [4], [5]. Several IEC/ISO standards refer to the general conditions for medical electrical equipment; standards in this class, dealing with specific aspects, are for instance [6], [7], [8]. Different countries have their own standards; most of the national standards are similar to IEC/ISO standards.

There are significant differences between the requirements imposed on medical equipment, depending on the class of application. Various classes have different requirements regarding safety, biocompatibility, reliability, cost, and precision.

It is beyond the purpose of this chapter to enter details and the designer of neuro-fuzzy systems for medical applications is referred to applicable standards.

2.1. General system and technological requirements

Cost

The cost of the equipment – and subsequently of the circuits – should be low for prosthetic devices, low to average for portable instrumentation, and average for general clinical equipment and rehabilitation equipment. The cost of small parts, including integrated circuits (ICs), is generally not essential for higher-end equipment such as imaging equipment, clinical life-support equipment, and high-end laboratory equipment. The cost range for various classes of equipment is shown in Figure 1.

Operating and storage temperature ranges

The requirements regarding temperature ranges are different for clinical, laboratory, and implanted equipment and for portable/ambulatory instrumentation. The last class should withstand adverse conditions, including temperatures possibly as low as –30 Celsius degrees (–22° Fahrenheit) up to 70° Celsius (158°F), when operated in some adverse ambulatory conditions. Circuits for such equipment generally require ceramic encapsulation and specific manufacturing technologies. Requirements that are more specific are given in national and international standards. However, the standards determine only minimal conditions; in some applications, the operating conditions may be more demanding.

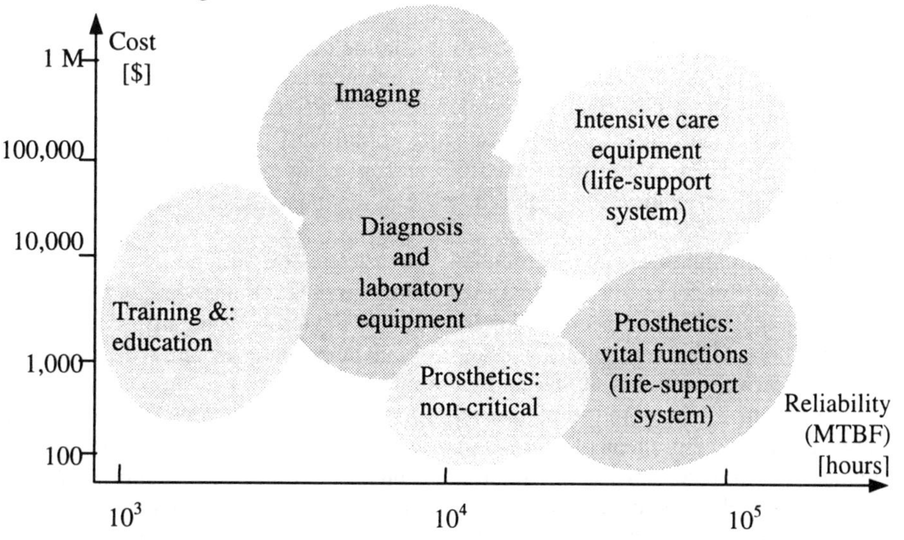

Figure 1. Reliability requirements and cost limits for medical equipment.

Noise immunity and electromagnetic compatibility

Depending on the application, the requirements for noise immunity and electromagnetic compatibility wildly vary for medical equipment. For instance, the laboratory equipment is supposed to operate under "mild" conditions, while equipment used in the surgery room should withstand adverse conditions such as high voltage applied to its inputs (hundreds of volts in the MHz frequency range, from electric surgery devices, or a few kilo-Volts from defibrillators.)

Power supply requirements

The power supply requirements vary in medical equipment from extremely low, in implanted devices, to not essential (high, non-critical) in clinical equipment. Techniques are readily available to manufacture fuzzy and neuro-fuzzy processors with low supply voltages for portable instrumentation. Power consumption is minimized using appropriate technology (MOS, CMOS) and designing a minimal hardware circuit for the application in hand (see subsequent chapters). Many of these problems are discussed in detail in Section B of this chapter.

Another problem related to power supply is the insensitivity of the circuits to the changes of the power supply voltage, such as voltage decrease due to battery aging. Several of the currently existing fuzzy and neuro-fuzzy circuits that operate in analog mode have poor stability to power supply changes and rely on the power supply conditioning circuits to achieve precision.

2.2. Reliability requirements

The reliability requirements for medical equipment varies from a mean time between defect of about 1000 hours, for training and education material, to 5 to 10 years (43,800 to 87,600 hours) required for no-defect operation, for life-support prostheses. Figure 1 illustrates the reliability requirements for various classes of medical equipment. In this figure, several subclasses are not included. For instance, endoscopy equipment (in the class of imaging equipment) has a significantly lower price, but higher reliability than classic imaging equipment (similar to surgery equipment). Surgery equipment is placed between diagnosis equipment and intensive care equipment, with a large price range.

Medical equipment poses higher reliability constraints on the circuits. The system designer has to remember to check that the existing circuits are intended and approved for medical applications, especially for life-support devices. (Generally, commercial circuits are not.) Moreover, the designer has to remember that there are still significant differences between various countries regarding the conditions to be satisfied for a device approved by the authorities for such critical applications. There are also large differences in the requirements for life-support systems and general-use applications in medicine (for instance, devices to measure the blood pressure). The authorization for use in critical application usually has to be secured from the circuit manufacturer.

The reliability issue determines the technology that is used to implement the fuzzy system in some applications. For instance, a general-purpose computer would be incompatible in the operation of a critical application. Its reliability in operation is tolerably low in an industrial or pre-clinical application (non-critical diagnosis). However, such a machine is intolerably unreliable in the operating theatre or in life-supporting applications (one may expect a general use computer program or a personal computer to frequently crash, while it should operate without any problem during tens of thousands of hours in a life-support application). Risk analysis and software validation should be accomplished for any equipment. The IEC/ISO standards related to these problems are IEC 601-1-4 si IEC/ISO-DIS 14971-1. A question to be answered is if (and how) the use of fuzzy and neuro-fuzzy systems improves the life cycle of software and reduces the risks of the equipment use.

2.3. Precision and sensitivity to parameters

From the engineering point of view, most pieces of medical equipment are either measurement or control systems, therefore the precision and sensitivity to parameters are essential issues in the design. Fuzzy logic and neuro-fuzzy circuits used in medical equipment must be submitted to sensitivity analysis.

In digital implementation, the sensitivity is not a critical issue. On the other hand, the sensitivity of analog fuzzy circuits to various parameters may be expected to be high. Several papers deal with the sensitivity analysis in analog fuzzy circuits, for example [10] – [17]. The reported results for the analogue fuzzy and neuro-fuzzy circuits show that the sensitivity of the output to temperature is in the range 0.1%/ Celsius degree. Sensitivity to the power supply voltage of the classic fuzzy logic implementations is high, especially for the voltage-mode circuits.

Table 1 summarizes some of the requirements, while Figure 1 shows the cost vs. reliability relationship for various classes of medical equipment.

Table 1. General requirements

Medical Application	Precision Requirement	Reliability	Sensitivity to External Conditions	Computation (Inference) Power	Flexibility (Programability)
Prosthetics	average or high	high	very low	low	low or average
Artificial organ	average or high	high to critical depending on the organ	very low	low, average	low or average
Portable medical instrumentation	very high or high	average to high (in emergency)	very low or low	low, average	average
Clinical device: diagnosis	very high	average (easily supplemented by other reliability enhancement solutions)	average or low (sensitivity to power supply can be compensated by increasing stabilizer; good screening available, etc.)	high	average or high
Laboratory equipment	very high	average	average	average	average or high
Rehabilitation	average	average	average or low	low, average	high
Imaging	very high	high	average	very high	high
Education and training	low to average	low	low	low to high	high

3. Analysis of Several Applications

3.1. Life-support applications

There are several applications of fuzzy and neuro-fuzzy systems in medical equipment reported in the literature. The applications described below are either in a development stage, or they are commercial products.

3.1.1. Artificial heart control

The use of neural and fuzzy controllers was reported [18] in a typical application, namely, a total artificial heart based on a moving actuator. The neural controller aims to regulate the aortic pressure by means of the actuator velocity, taking into account a predetermined peak level of the motor current. A second controller is a fuzzy logic system that monitors and regulates the pulmonary artery pressure to keep it at permissible level (not to exceed a given value). Moreover, the suction in the right heart is controlled, to prevent atrial suction. The controlled device is the actuator. The *in vivo* tests performed so far were successful.

Application analysis. The application is a device for life support, with high reliability (significantly higher than the electromechanical parts, namely the actuator). Power consumption should be low, while speed and computation (inference) powers are low. The system should withstand accelerations and the whole design should insure biocompatibility. To avoid the use of sensors – that pose problems of biocompatibility and decay in precision under the difficult conditions represented by the bio-chemical attack of blood – the designer decided to measure the blood flow indirectly, by means of the current needed by the actuator. The use of both a neural and a fuzzy system shows the designer split the problem into two sub-problems, both with significantly lower complexity, thus keeping the development and validation time and costs at a minimum, with a compromise over power consumption and chip area. In a large-scale production, this compromise is undesirable and one could expect the fusion of the two sub-systems (neuro and fuzzy, respectively) into a single neuro-fuzzy system.

3.1.2. Assisted ventilation

The use of fuzzy logic has been reported in an assisted control system of inspired oxygen in ventilated newborn infants [19]. The controller – that should operate in real time – regulates the mechanical ventilation, or more exactly, the inspired oxygen concentration in newborns. The fuzzy system uses the rules derived by neonatologists (no training of the fuzzy system). Clinical trials are reportedly under progress.

Application analysis. The application is a device for life-support, especially for infant life support (an extremum sensitive application from the point of view of the medical professional community, as well as of the public). This is a typical example of critical application, where high reliability of the equipment is essential. Indeed, such controllers should operate unsupervised, over long periods of time, with almost zero risk of failure, as any failure may result in the death of a child.

In a less demanding application in the same field of neonatal care, Bosque [20] reports on the "symbiosis of nurse and machine through fuzzy logic" through improved specificity of a neonatal pulse oximeter alarm. The aim is to eliminate false alarms – a problem that often plagues the monitoring systems and determines their rejection by the medical personnel.

3.2. Anesthesia related equipment

1. One of the applications in this field is reported in this volume (Chapter 11). Similar equipment was demonstrated as early as 1991 by Tsutsui and Arita [21].

Application analysis. The application is a device for life-support equipment. High reliability is, however, less a concern in this application, because almost in all countries, the law asks the anesthesiologist to assist and make the final decision, irrespective of the equipment used as a decision support. Power consumption is also irrelevant, as the equipment is not portable. The adverse conditions are related to the use of anesthesia gases (some of them are explosive, so the equipment should be designed with specific rules in mind; this has no specific implication on the system, apart from the fact that the system cannot be based on general purpose computers). Speed is irrelevant, as the controlled parameter (anesthetic flow) changes very slowly (one per minute, or less). The use of a fuzzy system is needed because of high variability of the human body response. In a large-scale production, the system could use a dedicated chip – to minimize costs – or a classic digital implementation.

2. In another approach, the problem of designing general purpose anaesthetic agent administration systems was addressed by Nebot [22], using fuzzy reasoning to find a general and reliable model for the patient response to anesthetics.

The authors separate the system-generic from patient-specific behavior in the frame of the model, by combining knowledge obtained from different patients. The authors claim to have been able to detect similar patterns for the patients undergoing similar operations, preserving the common characteristics, while filtering out the specific behavioral patterns of the individual patients.

3.3. Fuzzy and neuro-fuzzy-based equipment for prosthetics

Neuro-fuzzy systems are reputed for their capabilities in fusing data from a large number of different sensors and to extract the essential information. This advantage recommends the use of neuro-fuzzy systems in prosthetics and interfacing to the biologic neurons (nerves) in the "ascendant" way, i.e., "from sensors toward the central nervous system." On the other hand, the "descendent" way, i.e., the interfacing toward effector systems could also be considered; moreover, the replacement of the nervous control to effector muscles can be envisaged.

Several such applications were proposed. The use of neuro-fuzzy and fuzzy systems to control the muscles of disabled persons was introduced in [23], [24], [25]. Both groups proposed the generation of a specific stimulation pattern for electrical stimulation to control movements by functional electrical stimulation supervised by a fuzzy system.

Chen et al. used a robust closed-loop control scheme to improve cycling system efficacy for subjects with paraplegia. In their approach, the stimulation patterns are fixed and only the gain of the control loop patterns was adjusted. The system is a model-free, fuzzy logic controller.

A fuzzy controlled electrical stimulator, based on fixed or adjustable stimulation patterns, was developed to reduce the tremor movements [24]. The system determines the tremor movements by means of sensors and applies electrical stimuli to muscles to minimize the periodical low amplitude movements. In another version, devoted to muscular rehabilitation after an injury or disease, the system presents the patient a description of the movement, to help her or him to improve the movement characteristics.

In the field of prosthetics and neurology, Siemens developed neuro-electronic circuits to interface natural neurons to artificial neural systems in prosthesis and related applications. The direct interface between natural and artificial neurons – in a biological-electronic coupling – is an obvious application of neural networks and neuro-fuzzy circuits, with specific problems related to biocompatibility and biomaterials, microsystems (including micromachining), and sensorics [26].

Applications analysis. Prosthetics for limbs and movement-assisting devices and rehabilitation equipment are not critical applications. Hence, reliability is not an essential parameter. Cost, portability, low power, adaptability to various subjects, and operation under adverse environmental conditions may be the main requirements of the application. Biocompatibility is an issue of concern in cases of direct connection to the muscles and nerves. Because the equipment is in direct contact with the patient and electrically stimulates the body, patient safety is also a major concern.

3.4. General purpose devices

An automatic blood pressure monitor, that uses fuzzy logic technology, was recently introduced on the market for general use [27]. In this monitor, fuzzy logic is applied to adaptively determine the user's best inflation level and the best deflation level. The aim is to achieve higher accuracy and simplified operation and user increased comfort.

Application analysis. The application is a device for general purpose, ambulatory, non-critical use. The main requirements are high adaptability, fair robustness, fair reliability under non-critical conditions, and low cost. The main reason for using fuzzy logic is the adaptability and low cost (compared to a classic control that would be more expensive). As this is a commercial product aimed at a large market, the cost of this application is quite critical.

3.5. Other applications

Several real-life applications of fuzzy and neural systems were reported in image processing, including medical applications. For example, the SGS-Thomson Microelectronics fuzzy processor family ST-WARP [28], [29] is reported to have been used in medical applications, too [30].

Auxiliary equipment and health-related systems

In the field of biomedical applications, there are numerous applications much less demanding than the medical equipment. These include biologic process controllers and auxiliary equipment used in hospitals: laboratory systems, heating, air filtration, sterilization control, etc. All this equipment can work in a medical environment, but the requirements are closer to those for industrial equipment. An example of such equipment is the dust collecting electrical filter based on a microprocessor control complemented by an in-house manufactured fuzzy-logic controller (80C166), manufactured by Siemens [30]. The equipment is primarily intended to serve in industrial plants, still its use is health-related. Moreover, air filtration systems are common auxiliary equipment in hospitals, too. The fuzzy logic control is intended to insure adaptation to cope with large process fluctuations, such as a disruptive breakdown (for example), yet retaining a near-optimal operation.

4. GENERAL SYSTEM DESIGN ISSUES

4.1. Nonlinearity implementation – simulation power

In control and other applications, the main advantage of the fuzzy and neuro-fuzzy systems is the ease and power to simulate a nonlinear behavior. One of the questions the designer has to answer from the beginning is how complex should a fuzzy system be (at the hardware level) to implement a given nonlinearity.

The implementation of nonlinearities with fuzzy system is reknown to be easy. Examples in Figure 2 illustrate nonlinearities easily implemented with a four-chip configuration (discussed in the next section).

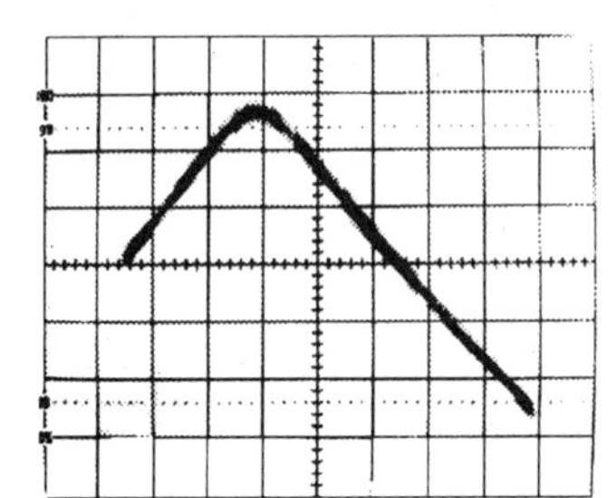

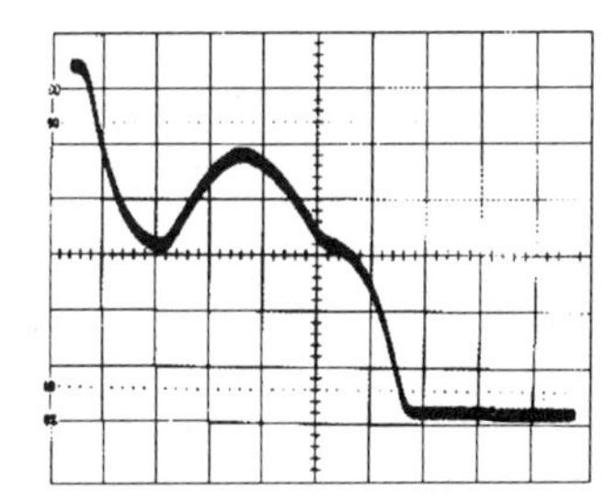

Figure 2. Examples of nonlinearities easily obtained by using fuzzy logic systems, but difficult to obtain by polynomials of lower order.

An important question – generally obscured in the literature – is the ability of fuzzy and neuro-fuzzy circuits to implement nonlinearities. Usually, this question is contoured by saying that "in general" these systems are able to perform any nonlinear mapping, or saying that they are universal nonlinear approximators. This is true when the number of input/output membership functions and the number of rules, or the number of neurons increase to infinity, but does not address the practical questions:

- How many input/output membership functions are needed to implement an n-th order interpolator (i.e., an interpolator for n fixed points in the input-output space)?
- How many input/output membership functions and rules are needed to implement an approximator for a given function with a predefined error?

- How does the sensitivity of the circuit impede its ability to perform as an interpolator or approximator?

Such questions should be asked at the initial stages of designing neuro-fuzzy systems for medical application as well. Classes of medical applications that benefit from the nonlinear characteristic of the neuro-fuzzy systems are:

i) The equipment uses a fuzzy system to correct the nonlinearity of a sensor; the fuzzy system will be used to perform the correction, in a simple manner, more or less as a human will fit some point assuming that the rest will follow close enough to the desired curve.

ii) The equipment uses a fuzzy system to perform a continuous nonlinear mapping or a continuous to discrete mapping representing either a control or a recognition/decision function.

iii) The equipment uses a fuzzy system to perform a combination of signal nonlinear processing, data fusion, and recognition/decision making processes into a single fused manner. This type of task completely justifies the cost of the fuzzy system.

Remember that an interpolating system is a system ideally performing a mapping that (exactly) goes through a pre-defined set of points (the interpolation points). We emphasize that a fuzzy system with triangular functions overlapping, at most, two by two is an interpolator. In this case, the tuning – as an approximator – asks for a change of the fixed points. By allowing more than two input membership functions to overlap in the input space, the system capabilities as an approximator increase. Consequently, fuzzy systems may be used either as interpolators or approximators, depending on the application.

Different types of fuzzy systems have different degrees of flexibility in changing the nonlinear characteristic function. The Mamdani-type fuzzy systems ask for the change of the parameters of both input and output membership functions. They have higher flexibility in the tuning process, with the drawback of a higher computational load in the defuzzification stage. Sugeno-Takagi type fuzzy systems are easier to develop, as the number of the adjusted parameters is lower (at least for lower orders of the polynomial determining the output of the system).

4.2. Dynamic errors

Analog implementations for fuzzy systems are mainly used to increase the speed and to reduce the chip area, at the cost of lower flexibility.

In such high-speed applications, the dynamic characteristics – including dynamic errors – are essential, as they add to the other types of errors. At present, there is no analysis for such errors, and no manufacturer provides characteristics in this respect.

5. HARDWARE IMPLEMENTATION ISSUES

5.1. Implementation choice: analog vs. digital fuzzy processors

In applications related to medicine, the use of fuzzy system analog implementations is preferred to digital implementations because analog circuits have lower power consumption and dimensions. Speed is hardly critical in all but image processing applications in medicine. Still, in such applications, the high precision could jeopardize the low silicon area/low consumption if the circuit complexity is to be increased to cope with high precision. One can conclude that the target applications for existing hardware solutions for fuzzy system are those of low or moderate precision, low power supply, low dimension, and low cost, or, at the other end of the range, very high speed, high cost (dedicated circuits), and still low or moderate precision.

Many other applications of fuzzy systems in medicine fall in the category of software-based applications, using either classic technology or a combination of classic digital technology and fuzzy accelerators.

5.2. Hardware minimization

For reasons that are often specific to the medical field, such as conjoint need for miniaturization, power reduction, and cost reduction in portable instrumentation and prosthetic devices, the hardware minimization in medical equipment is a more stringent problem than in industrial equipment.

The reduction of specific hardware in fuzzy systems uses various strategies, starting from the general problem analysis and ending to integrated circuit solutions, as discussed in the previous chapter.

The system minimization can be performed in two main ways:

- By minimizing the problem complexity, for instance, applying a principal component analysis to the input space, to reduce its dimension. Usually, the reduction is of about one order of magnitude in each variable. (Examples of such reduction, also using the symmetry of the problem or other properties of the input space to reduce the input space dimension, are shown in Chapters 3 and 7, in this volume.)
- By reducing the dimensionality of the problem through "problem splitting" techniques.

As far as fuzzy systems are concerned, the reduction of the number of fuzzy inputs is a major way to reduce problem complexity, and hence hardware complexity. The inference speed also increases when the rules number is decreased.

Moreover, the reduction of complexity is essential in keeping low the costs of the design and development phases. For systems aimed to small-series manufacturing, these costs can significantly contribute to the overall production costs.

5.3. Parallelism vs. number of rule blocks

Fuzzy control systems, as well as general fuzzy inference engines are built around some basic blocks, such as elementary rules inference blocks, truncation blocks, summing (MAX) blocks, and defuzzifiers (see Figure 3). Defuzzifier blocks are needed only one per complete system, for single-output systems. The elementary inference rules blocks are needed in a large number per system, namely, one block per implemented rule. Therefore, the number of elementary-rule blocks increases according to the product of number of inputs, more exactly, this number is given by the product of numbers of membership functions assigned to these inputs. For example, for a 3 input system, each input having assigned 9 membership functions, and one output with 9 membership functions, the number of rules is 9 x 9 x 9 (729 rules for a completely defined system). Such a system requires a large silicon area. When increasing the number of input variables, as needed in bigger problems, let us say to 10, the number of rules becomes huge (9^{10}).

To cope with this problem, simplification of the hardware is needed at an early stage of the design. The result could be a significant reduction of circuit complexity, with a tolerable loss of parallelism in computation.

A simple method to reduce the number of rules is to apply the hardware only to rules that yield a non-zero output. This method was first suggested by Ikeda et al. [32], in relation to speed increase in fuzzy circuits, and independently by the present author and Yamakawa [33], [34], [35], in relation to the minimization of the number of rule chips in large fuzzy systems. The method is briefly explained in this section.

The number of rule blocks may be reduced taking into account that only a few rules are active ("fired") at once. By determining which rules have to be applied at every moment and by using flexible rule blocks able to implement any rule, the number of rule blocks in a fuzzy processor can be drastically reduced. This principle requires only a few additional circuits, to determine what rule should be activated. The processing speed reduction is very low.

Consider, for example, a usual two-input, one-output fuzzy system described by rules of the type

$$R(h,i): \textit{ If } x \textit{ is } A_h \textit{ and } y \textit{ is } B_i, \textit{ Then } z \textit{ is } C_k$$

For the system performing the inference, the appropriate rule *R(h,i)* must be fired, and thus a mechanism inside the system has to map the two-dimensional real intervals corresponding to the membership functions to which the instant inputs actually belong, to the sub-set of corresponding functions. This is the basic mechanism of rules firing and it is illustrated in Figure 4. In this figure, one considers that all membership functions, for both inputs, are non-vanishing on finite intervals, and that these intervals overlap only two-by-two. These properties of the input membership functions give rise to a firing mechanism that fires only four rules at a time. (In Figure 4, the membership functions are triangular. However, any shape of the membership functions that satisfies the above properties leads to the same results.)

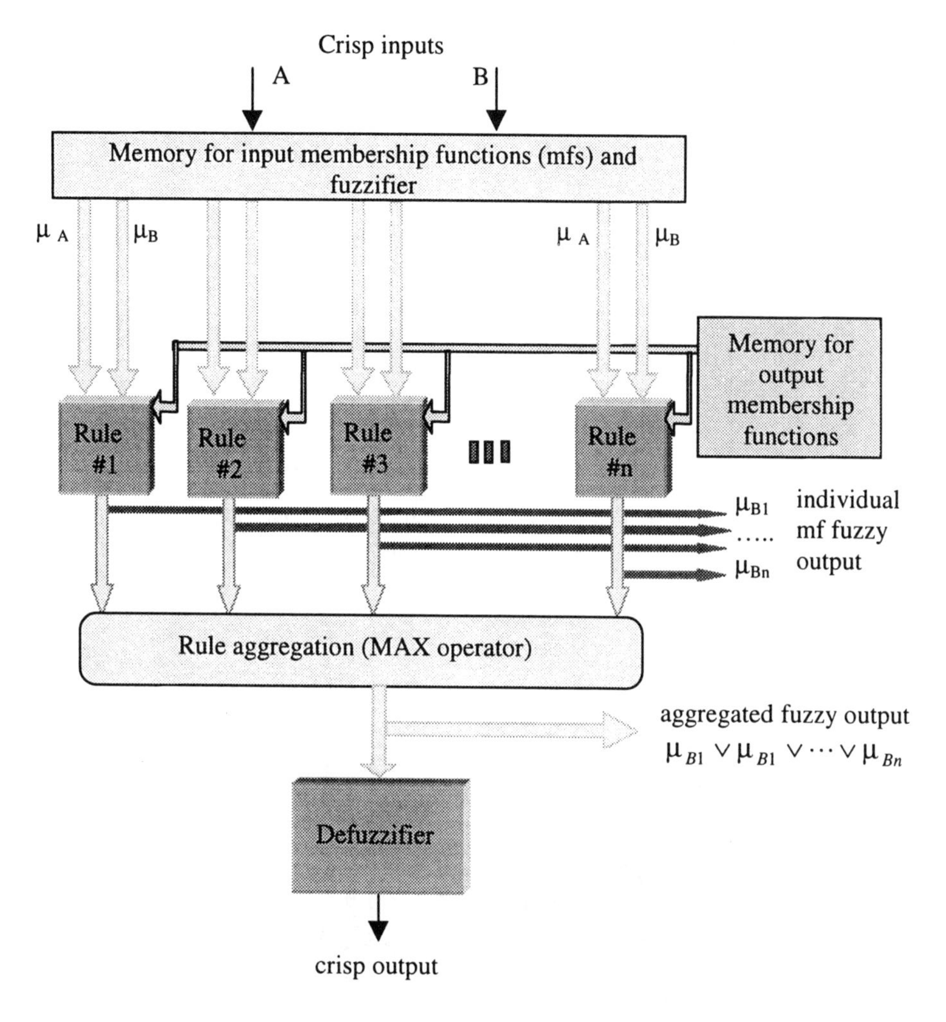

Figure 3. Block diagram of a classic fuzzy logic system.

For a system with n inputs, each having m membership functions, the complexity reduction is of the order n^m vs. 2^n (when the input membership functions overlap at most two by two). For example, for a system with two inputs, the number of rule blocks is 4, irrespective of the number of membership functions of each input. Such a system is shown in Figure 5.

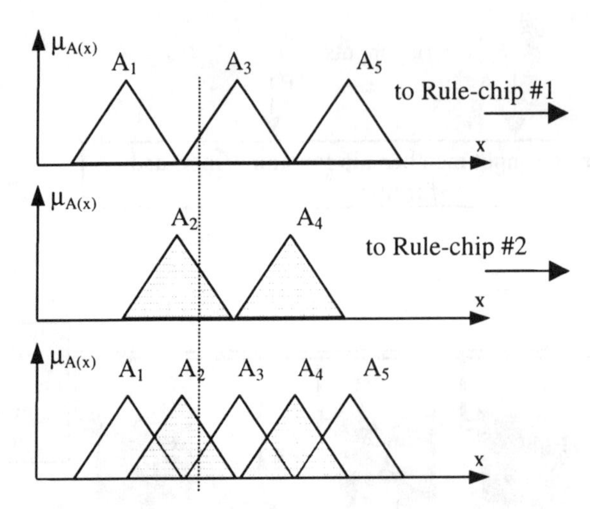

Figure 4. Explanation for the principle of selective rule firing (one input variable shown).

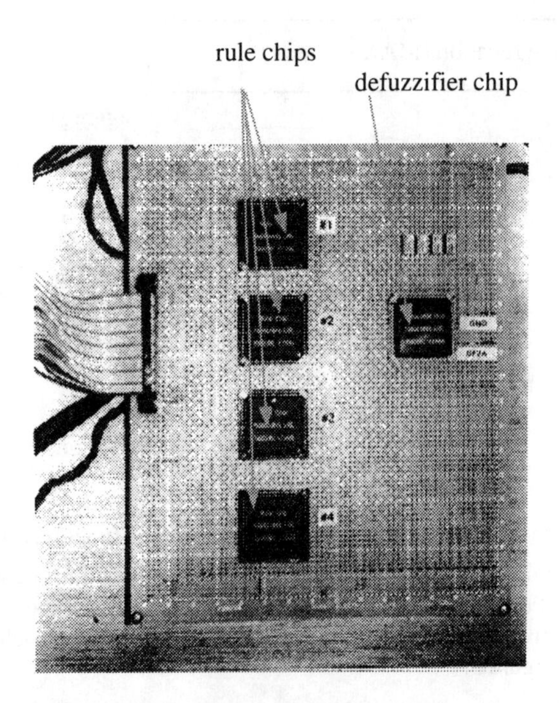

Figure 5. Example of circuitry used to implement a two-input, single-output fuzzy universal system, based on the *TG004MC* chips. The decoding (rules firing) circuitry is not shown.

5.4. A minimal system design

The above structures avoid any memory for recording the output membership functions generated by individual rules. The use of memory devices is generally avoided due to speed constraints. However, when the speed is less a concern, using memory cells may contribute to further reduction of the number of rule blocks. The maximal hardware reduction is obtained when a single rule block is used to perform the implementation of all the rules. To achieve this goal, the system has to store the partial results. If the fuzzy system is a Mamdani type system, the hardware reduction may be low, as the increase in the number of memory cells to store the output membership functions may balance the reduction in rule blocks. Consequently, the solution suggested here is recommended for Sugeno-type systems. For such systems, the output is a single value (singleton), hence the required memory is of low capacity. A simple analog memory that avoids the analog digital conversion is a capacitor and a set of switches. The architecture of this type of system (due to the first author) is presented in Figure 6.

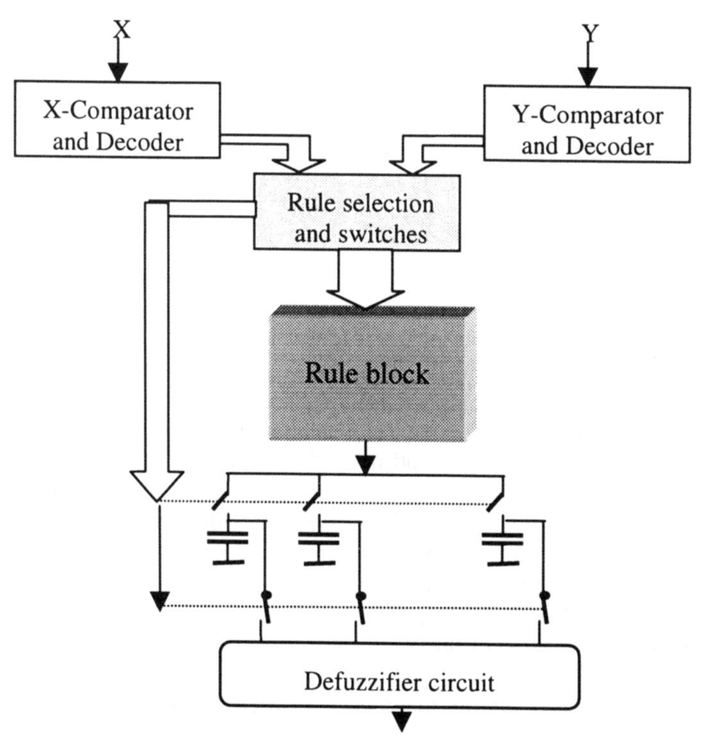

Figure 6. Scheme of the switched capacitor system (connections from inputs X and Y to the rule block not shown).

6. CHOOSING THE RIGHT DESIGN

The designer must first choose between various solutions of implementing a nonlinear control or classification system. The choice has to be done between neural networks, fuzzy systems, neuro-fuzzy systems, or technologies that are more conventional. The second major choice regards the technology to be used (digital vs. analogue, voltage or current mode, etc.). It is dependent on the application in hand (power supply requirements, operating temperature range, etc.) When a fuzzy logic or neuro-fuzzy approach is used, the type of inference, the I/O power (number of inputs, outputs, and type of input/outputs) and the inference power should be assessed.

When choosing the circuit, note that, in general, current implementations do not allow a fuzzy output, most of them concentrating on a crisp (defuzzified) output, as required in control applications. Therefore, such implementations are not usable in some decision-support applications, where the non-aggregated fuzzy output membership functions are needed.

Several existing commercial fuzzy and neuro-fuzzy chips are general-purpose, flexible enough systems and may be adapted to medical applications. There are several major companies manufacturing fuzzy and neuro-fuzzy chips. Smaller companies and university laboratories have a significant share of the market for dedicated circuits used in small series products; moreover, they have a large share in the research and development of such circuits. Detailed description of some of the current implementations may be found in [2], [37]-[45].

7. CONCLUSIONS

In this chapter we reviewed the main requirements for the fuzzy and neuro-fuzzy systems in medical applications and analyzed several of their system-related properties.

The state of the art in fuzzy and neuro-fuzzy hardware technology is at the border of a large-scale development era, with several already sound achievements. As for neural networks, the use of the fuzzy and neuro-fuzzy systems is to cover some of the slots where more classic circuits do not fit, for reasons of cost and lower flexibility. Another important reason for the use of fuzzy systems is the low cost of development: for experienced engineers in the field, the time to develop a "robust" yet easily adaptable application can be significantly lower than for a classic nonlinear system.

In the second part of this chapter, due to Schmid and Mlynek, more details are provided on the implementation of fuzzy and neuro-fuzzy systems at the circuit level.

Acknowledgments. The authors thank Steli Loznen for pointing out some errors and incomplete quotations.

This work was performed with the support of The Institute of Information Science of the Romanian Academy and a Grant from Swiss National Foundation. The first author is also thankful to the Fuzzy Logic Systems Institute and to Yamakawa Laboratory, Kyushu Institute of Technology for their hospitality during the years 1992, 1993, and 1994. The photographs in this chapter were obtained during his time there.

REFERENCES

[1] Russo, M.: Fuzzy Hardware Research from Historical Point of View. In: Kandel, A. and G. Langholz (Eds.): *Fuzzy Hardware. Architectures and Applications.* Kluwer Academic Publishers, 1998, Chapter 1.

[2] Kandel, A. and G. Langholz (Eds.): *Fuzzy Hardware. Architectures and Applications.* Kluwer Academic Publishers, 1998

[3] ISO/FDIS 8835-1, ISO 8835-2:1993, ISO/DIS 8835-2 and ISO 8835-3:1997 refer to "Inhalation anesthesia systems," while ISO 9703-1:1992, ISO 9703-2:1994 and ISO/DIS 9703-3 refer to "Anaesthetic and respiratory care alarm signals."

[4] ISO 11196: 1995: Anaesthetic gas monitors.

[5] IEC/ISO-DIS 14971-1: Medical devices - Risk management. Part 1: Application of risk analysis. Also IEC 601-1-4.

[6] ISO 60601-2-32:1994: Medical electrical equipment. Part 2: Particular requirements.

[7] ISO 60601-2-33:1995: Medical electrical equipment. Part 2: Particular requirements.

[8] IEC 60601-2-2nd Ed: 1991 Medical electrical equipment. Part 2: Particular requirements.

[9] IEC 60601-1-1 Ed 2b: 1988 Medical electrical equipment. Part 1: General requirements for safety.

[10] Sofron, E. and Teodorescu, H.N.: On the error in fuzzy gates, *Proc. Int. Conf. AMSE,* Cetinje, Yugoslavia, pp. 79-83, 1990.

[11] Sofron E., Teodorescu, H., and Gheorghe, I.: The influence of the technological parameters on features of the current sources in the fuzzy logic circuits, in *Fuzzy Systems and Artificial Intelligence*, Teodorescu, H.N., Yamakawa T., and Rascanu, A. (Eds.), Technical University of Iasi Press, pp. 221-228 Iasi, Romania, October 1991.

[12] Sofron, E. and Gheorghe, I.: Sensitivity analysis for the membership function circuit, *Proc. of The 5'th European Congress on Intelligent Techniques and Soft Computing, EUFIT'97,* Aachen, Germany, September 8-12, 1997.

[13] Balteanu, F., Gheorghe, I., Sofron, E., and Serban, G.: Sensitivity analysis for some CMOS neuro-fuzzy circuits. *Fuzzy Systems & A.I., Reports & Letters,* vol. IV, no. 3/4, pp. 75-80, Romanian Academy Publishing House, 1995.

[14] Gheorghe, I., Sofron, E., Balteanu, F., Serban, G., and Ciobanu, D.: The propagation of the error in fuzzy logic circuits, in *Proceedings First European Congress on Fuzzy and Intelligent Technologies, EUFIT'93,* pp. 1007-1015, Aachen, Germany, Sept., 1993.

[15] Hashem, S.: Sensitivity analysis for feedforward artificial neural networks with differentiable activation functions, In *Proc. of Int. Joint Conference on Neural Networks*, vol. I, pp. 419-424, New Jersey, U.S.A., 1992.

[16] Yamakawa, T.: A fuzzy inference engine in nonlinear analog mode and its application to a fuzzy logic control, *IEEE Transactions on Neural Networks*, vol. 4, no. 3, pp. 496-522, 1993.

[17] Gheorghe, I.: Contributions to the analysis of dynamic fuzzy systems. Ph.D. Thesis, Technical University of Iasi, Romania, March 1998.

[18] Lee, M., Ahn, J.M., Min, B.G., Lee, S.Y., Park, C.H.: Total artificial heart using neural and fuzzy controller. *Artificial Organs*, 20, (11), 1220-6, 1996.

[19] Sun Y., Kohane I., Stark A.R.: Fuzzy logic assisted control of inspired oxygen in ventilated newborn infants. *Proc. Annu. Symp. Comput. Appl. Med. Care*, 756-61, 1994.

[20] Bosque, E.M.: Symbiosis of nurse and machine through fuzzy logic: improved specificity of a neonatal pulse oximeter alarm. *ANS Adv. Nurs. Sci.*, 18(2), 67-75 1995.

[21] Tsutsui T. and Kuramoto T.: The composition of cerebrospinal fluid during enflurane anesthesia. (In Japanese, with English abstract) Masui (*Jap. J. Anesth.*); 1991; 40: 1477-80.

[22] Nebot, A., Cellier, F.E., and Linkens, D.A.: Synthesis of an anaesthetic agent administration system using fuzzy inductive reasoning. *Artif Intell Med,* 8(2), 147-66, May 1996.

[23] Chen, J.J., Yu, N.Y., Huang, D.G., Ann, B.T., and Chang G.C.: Applying fuzzy logic to control cycling movement induced by functional electrical stimulation. *IEEE. Trans. Rehabil. Eng.*, 5(2): 158-69 1997.

[24] Teodorescu, H.N. and Mlynek, D.: Research Proposal, Grant, Fonds National Suisse (National Swiss Foundation), July 1996; Research Report.

[25] Teodorescu H.N., Mlynek, D., Kandel, A., Ropota, I., Teodorescu, C., Posa, C., Brezulianu, A., and Ciorap, R.: Analysis of Chaotic Movements And Fuzzy Assessment of Hands Tremor in Rehabilitation. *Proc. KES'98 Int. Conf.,* Vol. 2, Adelaide, Australia, May 1998.

[26] http://w2.siemens.de/infoshop/umwelt/ear03_e.htm.

[27] http://www.omronhealthcare.com/

[28] Caponetto, R., Lavorgna, M, Occipinti, L, and Rizzotto G.G.: Fuzzy Cellular System: Characteristics and Architecture. In: Kandel, A. and G. Langholz (Eds.): *Fuzzy Hardware. Architectures and Applications.* Kluwer Academic Publishers, 1988, Chapter 14.

[29] SGS-Thomson Microelectronics: STFLWARP11 Data Sheet, STonLine, http://www.st.com.

[30] Personal communication by Luigi Occhipinti, during Int. Symposium on Fuzzy Systems, Zichron Yaakov, Israel, 21 May 1997.

[31] ANL-97-07, 1996 Siemens Aktiengesellschaft.

[32] http://w2.siemens.de/infoshop/umwelt/ear03_e.html.

[33] Ikeda, H., Kisu, N., Hiramoto, Y., and Nakamura, S.: A fuzzy inference coprocessor with an 'Active-rule-driven' architecture. In *Proc. IEEE Int. Conf. on Fuzzy Systems*, San Diego, CA., pp. 537-544, March 1992.

[34] Teodorescu, H.N. and Yamakawa, T.: Architecture of a fuzzy inference engine with minimal rule blocks and adaptive membership functions. Vol. *Fuzzy Systems. Proc. ISKIT'92 Conference*, Iizuka, June 1992. pp. 79-84.

[35] Teodorescu, H.N., Yamakawa, T.: Architectures for rule-chips number minimizing in fuzzy inference systems. Vol. *Proceedings of the 2nd Int. Conf. on Fuzzy Logic and Neural Networks.* Vol. 1., pp. 547-550 (Jono Printing Co., July 1992, Iizuka, Japan).

[36] Teodorescu, H.N.: Minimization of rule-chips in fuzzy control and fuzzy inference systems. Vol. *Proceedings of the 2nd Int. Conf. on Fuzzy Logic and Neural Networks.* Vol. 1., pp. 441-444 (Jono Printing Co., July 1992, Iizuka, Japan).

[37] Conti, M., Orcioni, S., and Turchetti, C.: Analog Neuro-Fuzzy Network for System Modeling and Control. In: F.C. Morabito: *Advances in Intelligent Systems.* IOS Press, 1997, pp. 496-501.

[38] Sibigrtroth, J.M.: New MCU is ideal for fuzzy applications. *Proc. European Congress on Intelligent Technologies and Soft Computing, EUFIT'96,* (Vol. 3), Aachen, Sept. 2-5, 1996, pp. 2235-2239.

[39] Viot G.: The fuzzy micro-architecture of the CPU12. *Proc. European Congress on Intelligent Technologies and Soft Computing, EUFIT'96,* (Vol. 3), Aachen, Sept. 2-5, 1996, pp. 2240-2244.

[40] Eichfeld, H. and Kunemund, Th.: Fuzzy Coprocessors. *Proc. European Congress on Intelligent Technologies and Soft Computing, EUFIT'96,* (Vol. 3), Aachen, Sept. 2-5, 1996, pp. 2245-2249.

[41] Yamakawa, T.: A fuzzy inference engine in nonlinear analog mode and its application to a fuzzy logic control, *IEEE Transactions on Neural Networks,* vol. 4, no. 3, pp. 496-522, 1993.

[42] Yamakawa, T.: Silicon implementation of a fuzzy neuron, *IEEE Trans. on Fuzzy Systems,* vol. 4, no. 4, pp. 488- 501, Nov. 1996.

[43] http://sunvlsi4.bo.infn.it/ papers_95.

[44] http://sunvlsi4.bo.infn.it/ papers_96.

[45] http://sunvlsi4.bo.infn.it/ papers_97.

Chapter 12 B.

Neural Networks and Fuzzy-Based Integrated Circuit and System Solutions Applied to the Biomedical Field

Alexandre Schmid and Daniel Mlynek

This chapter focuses on the hardware integration of neuro-fuzzy algorithms. We first highlight the properties that embedded biomedical systems are expected to exhibit. The development of autonomous, reliable, accurate integrated circuits, while keeping the objectives of low-power consumption, low-area, and a straightforward design process, is descibed. The main implementation families for neural networks, fuzzy and hybrid paradigms, are presented. Finally, a successful integration example is reported as a mixed analog/digital artificial neural network dedicated to intercardiac morphology classification.

1. INTRODUCTION

The extensive development of embedded microsystems has been the technological answer to the large market demand for portable, easy to handle, efficient, and cheap products. Microcontrollers and application specific integrated circuits (ASICs) have become the algorithmic core of microsystems, including such peripheral electronic circuits and devices such as sensors, actuators, filters, converters, multiplexors, and power devices [1]. Hence, the integration of neuro-fuzzy algorithms into silicon hardware ASICs has become a standard method, allowing the full exploitation of the speed offered by making parallel computation available.

Silicon is the hardware medium most frequently used in the electronics industry. The implementation of neuro-fuzzy algorithms is, however, not restricted to silicon integration. Applying material intrinsic properties (chemical, optical, and opto-electronic, etc.) is also a prospective field of research, which attempts to increase the number of parallel processing elements, and thus increase the processing speed and

0-8493-9806-1/98/$0.00+$.50

power provided, at the lowest energy cost. This chapter, however, focuses on silicon hardware developments.

This chapter is an overview of current technological trends, issues, and possibilities that have to be considered for successful integration of neuro-fuzzy algorithms. Section 2 covers the hardware specification requirements for the implementation of autonomous, safety-critical systems. The main hardware families implementing neural networks and fuzzy paradigms on-chip are described in Section 3.

From software to hardware

An efficient design methodology has to be applied to the development of any electronic system, after a careful study of the technical requirements and specifications [2]. Decisions taken at the early stages of system development define either software running on conventional or dedicated processors, or ASIC hardware, as the optimal implementation choice.

Several technical reasons influence defining hardware the most convenient choice for the implementation of a module or algorithm. Required performances in terms of processing speed, reduction of software-hardware communication, limitation in the available area, and maximum power consumption may be some important criteria. Nevertheless, economic parameters also have a relevant influence; the costs of development and production, as well as the expected time-to-market, have to be taken into account. A dedicated hardware solution is advantageous when applied to a large series of low-cost, highly integrated components, which integrate several functions on-chip. Moreover, some algorithms are inherently best suited for hardware implementation because of their intrinsic nature. ANN and fuzzy algorithms follow such a model, which exhibits a large degree of parallelism, ideally exploited by highly parallel machines or dedicated neuro-fuzzy hardware. On the other hand, any hardware solution is less flexible than software implementation, which can be upgraded to adapt to a new environment configuration, or according to up-to-date algorithms.

The top-down methodology, widely applied to the development of integrated circuits (ICs) is depicted in Table 1. Five levels of abstraction can be distinguished. The system level includes the definition of the project specifications, to which all further decisions are related. The development method and design-flow, CAD software tools, as well as a development technology which these tools support, have to be selected. The technical project participants involved include IC foundry and CAD crew who provides technical support. The macro-functions which are to be implemented can have several algorithmic descriptions, depending on the nature and range of data and availability of specific operators. The architecture defines the required hardware modules, as well as their interconnection and scheduling, to support the execution of the chosen algorithms. The electronic circuitry which actually implements the functions has to be selected at the circuit-level of the development. Finally, the integration technology that is provided by the IC foundries defines the actual on-silicon implementation of the designed structures.

Table 1: Hardware design top-down methodology.

High-level	SYSTEM LEVEL	Project specifications, conception methodology, design-flow to be applied, project participants.
↓	ALGORITHMIC LEVEL	Choice of the algorithms to be implemented to successfully achieve a specified function.
	ARCHITECTURAL LEVEL	Choice of the macroblocs to implement the algorithms, specification of their interconnection and scheduling.
	CIRCUIT LEVEL	Design techniques.
Low-level	TECHNOLOGY LEVEL	Implementation technology.

The successful development of any embedded system should include a software-hardware partitioning step. The hardware-software codesign [3] methodology aims to lowering inadequacies between software and hardware paths that lead to inefficient system behavior. Emphasis is given to a cooperative development of both modules. Every decision on the software module significantly influences the development of the hardware part, and vice versa. Applying a codesign partitioning approach helps define the nature of the modules (HW/SW) and information flows at board-level and integrated circuit level.

2. REQUIRED PROPERTIES FOR EMBEDDED MEDICAL SYSTEMS

This chapter focuses on technical aspects of embedding systems at the IC level. Emphasis will be given to the targeted biomedical field of application, which can be compared in many respects to other high-tech research areas such as telecommunication, space engineering, and robotics. Nevertheless, any successful development does not depend only on technical expertise; some parameters such as the patients' and doctors' acceptance of the new device, research and production costs, and potential market are of prime importance, as in any industrial development [4].

2.1. Embedding medical systems

Electronic systems follow a general downward trend in their overall dimensions and weight, while increasing their efficiency, reliability, and range of applications. Biomedical electronic systems follow this trend. Integrating biomedical dedicated algorithms into single ICs aims to build up portable embedded systems in order to improve support for emergency medical interventions, increase patient autonomy and comfort, and lower costs.

Ambulatory intervention is often of first importance to a patient's recovery quality and time. A fast and safe diagnostic, followed by efficient emergency treatment until arrival at a heavily equipped hospital, are the keys for a successful intervention.

Medical care is given with the help of portable instrumentation. Hence, efficient equipment has to be integrated into ambulatory intervention units.

Increasing the autonomy of patients is another issue that may be solved by embedding data acquisition systems, drug delivery systems and prosthetic systems into portable units. The impact of such systems, is not only in terms of improving the patients' comfort; it also helps to reduce the duration of hospital stays and lower costs.

Lowering the cost of medical systems is a crucial concern in order to increase the diffusion of up-to-date diagnosis and healing assistance systems. The development of highly integrated technology proves to be a clever approach to cost reduction, as it affects the production stage (less raw material, packaging), transportation (less unity weight, overall dimensions), and the amount of stocks (single systems integrate several functions).

2.2. Autonomy

The increasing demand for autonomous and portable systems requires the development of compact and low-power consumption systems. Reducing both of these, while increasing processing speed and power, is a challenging task. It requires finding the optimal trade-off between three strongly correlated parameters: area, speed, and power.

Low area issues

Several factors influence the silicon area required to achieve a given function. These factors can be significantly influenced at different levels of the development process.

The choice of the *design-flow* that will support the development of any IC dramatically affects the occupied silicon area. Full-custom design and HDL-based automated design are the two main coexisting design-flows. Full-custom design-flow requires an experienced designer, who manually defines all geometric information on the layers positions within a dedicated computer aided design (CAD) environment. This method leads to very high density designs, but at the cost of a long development time. On the other hand, automated design requires the designer to produce a hardware description language (HDL)-based description of the system, which may be derived from a graphical-based one. Logic synthesis is applied to this textual input to produce a gate-level description. The next step consists of applying a placement and routing tool, which selects and instantiates pre-designed, pre-characterized standard cells and interconnects them. The resulting layout implements the HDL-described functions in a less dense way than full-custom design would allow. The advantages of this approach include fast and flexible design techniques, increased reusability of the modules, and a technology independent process.

Architectural decisions have to be made very early in the development of an IC, as dictated by the circuit specifications. Architectural trade-offs include serialization-pipelining-parallelizing, test procedures, definition of the IC's input/outputs, and communication protocols. Decisions on the architectural level of an IC development affect not only the size of the functional structures (those which implement the actual functions), but also the size of peripheral circuitry such as chip control circuitry, memories, multiplexing circuitry, input-output devices, and test devices, either integrated or off-chip.

The choice of the integration technology is of utmost importance; the design rules associated with it determine the possible structural density. A technology is usually characterized by both its type (CMOS, BiCMOS) and the minimal feature size, given in microns. The Semiconductor Industry Association (SIA) roadmap (see Table 2) [5] gives technology targets in terms of minimal feature size and supply voltage, among many others.

Table 2: Targets of the SIA roadmap concerning minimal feature size and supply voltage

YEAR	1997	1999	2001	2003	2006	2009	2012
TECHNOLOGY [nm]	250	180	150	130	100	70	50
VDD [V]	1.8-2.5	1.5-1.8	1.2-1.5	1.2-1.5	0.9-1.2	0.6-0.9	0.5-0.6

Low-power issues

The design of low-power consumption systems has become a constant concern in attempts to address the issue of powering miniaturized autonomous devices. The overall weight and dimensions as well as the autonomy requirements are to be balanced with the energy/pound ratio of portable sources (batteries), which have a relatively slow evolution. The reduction of heat dissipation may be another motive for applying low-power design.

The successful development of low-power systems must be considered at all levels in the design flow. Considering this requirement at the highest integration levels produces higher efficiency. System-level specifications for low power apply to the nature and performance of the whole system, or its constituting macroblocks. Current electronic systems often feature automatic power-off; this is a consequence of a high-level directive that obviously has a dramatic effect on power consumption. The supply voltage also has to be determined carefully. Fine-tuning the algorithmic level requirements also has a relevant influence on power dissipation. Data representation, required precision, complexity and regularity of the algorithms, to mention only a few, are parameters to be carefully studied, and optimally adapted to the targeted application. The architectural level directives determine the necessary resources to process the algorithms. The degree of parallelism, the processors' instruction set, and the possible implementation of voltage scaling devices are such decisions. The efficient use of the instruction set is also necessary for power saving.

The compiler which generates the microcode has to be optimized for the efficient use of the available hardware architecture. The circuit logic level low-power criteria applies to the choice of an adequate logic family to achieve a given function, the optimal sizing of low-level structures (transistors, logic functions), and low-level activity reduction techniques such as clock gating. At the lowest technological level, the reduction of the MOS transistor threshold can have an influence on the overall circuit consumption; thus, some techniques involving multi-threshold designs have emerged.

2.3. Reliability – safety

Reliability of hardware is a key issue in the successful development of safety-critical biomedical systems. Hardware devices have to operate properly as soon as they are powered up, and perform faultlessly. Obviously, the failure of such a system may have severe consequences to a patients' health; moreover, the upgrade, maintenance, or mending of a hardware module may simply be impossible, because of time or space constraints. Improving the reliability of hardware is a systematic task that involves all levels in the development and production phases.

System level issues: The IC design-flow is widely supported by CAD software tools, which tend to ease and speedup the development process by providing automatization and checking utilities. Again, full-custom or high-level description language (HDL)-based methodologies can be selected, both of which involve automation as a means to achieve fast and failureless design [6]. Several frameworks support high-level synthesis of neuro-fuzzy systems: AMICAL [7], and AFAN [8] are two examples. Coherence among all developed models can be ensured by the strict application of design directives [9]. On the other hand, low-end silicon compilers tend to avoid the use of silicon-consuming standard cells by implementing a dedicated adaptable layout. Among several, let us mention the following: SCOFIC [10], IDEA [11], and [12].

Algorithmic level issues: The design of reliable neuro-fuzzy systems requires full confidence in the implemented algorithms, which have their foundation in mathematically proven theorems. Nevertheless, their implementation at the hardware level involves applying restrictions or modifications to cope with practical limitations (limited precision depth, function approximation through lookup tables, etc.). Thus, the hardware-friendly algorithms have appeared [13] as a means of shifting processing complexity from the hardware to the algorithmic (software computed) parts: weight perturbation algorithms [14], and derivativeless and multiplication free algorithms [15], [16]. Probably the most severe limitation stems from the quantified representation of the weight and input values; several authors have been experimenting in this field with various architecture implementations [17]. Feed-forward ANNs generally require a full 16-bit precision [18], [19]. The critical precision-depth operation is the back-propagation algorithm [20], which modifies the ANNs current weights according to a reward/penalty supply mechanism; the recall phase usually requires less – even very low – precision [21].

On-line topological modification is a possible way to increase the reliability of IC integration of neuro-fuzzy systems. Redundant on-chip implementation of critical circuits is a technique that may be applied using two approaches; the main drawback is an increase of both the silicon area and the power consumption. The first approach to exploiting redundant structures consists of conveniently selecting one of the modules as the allowed speaker, all the others (failure, not yet required) being ignored. Majority decision is the other approach which is based on the analysis of all the speaking modules, and convergence to a common, identical system answer. ANNs inherently use such paradigms. ANNs reconfigurability may be considered topological alteration, which also applies to hardware realizations. Several improved solutions are offered for the relatively low cost of extra switching circuitry: increase of the network's resolution, ease of testing, and the possible isolation of defects [22].

The optimal ANN topology (the number of layers, neurons in each layer, and the interconnections between neurons) for solving a given problem cannot be determined analytically. Undertraining and overtraining have to be avoided, as both cause inaccurate solutions (see Figure 1). Early training stopping methods attempt to detect an increase in a computed error term, which is considered the start of overfitting. Overtraining may also be avoided by limiting the degree of freedom in the network to its optimal value. Pruning algorithms [23] intend to converge toward an optimal network topology by successively removing neurons that are not believed to be vital for the solution. Discrimination of the elements to be removed involves the processing of sensitivity terms (sensitivity of an error term to the removal of one neuron), forcing unessential neuron weights to zero, and interactive pruning (constant observation of the network's behavior, to select the neurons without significant impact the solution).

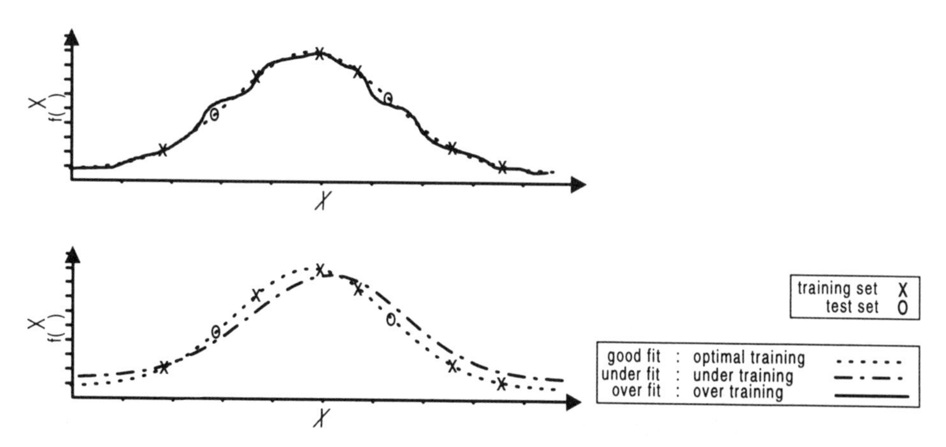

Figure 1: Possible results of function approximation through ANN

Artificial neural networks are inherently robust systems. The robustness is due to the distribution of knowledge throughout the whole network. Any loss of a processing element induces only a small degradation of the overall performance. Hence, it is possible to increase the robustness of a network prior to any degradation by applying training with robustness. This method consists of introducing, during the training phase, a failure the network may encounter during the recall phase [24]. Enhancing the fault tolerance of ANNs is promoted by applying synaptic weight noise during the training phase [25]; the generalization ability and the overall performance of the network are expected to be extended. It has been demonstrated that noise injection into the training set improves the ability of the network to generalize.

Architectural level issues: Architectural features that enhance reliability include resources that allow reconfigurability of the device. Reprogrammability allows the upgrade of a software module or of a software programmed hardware configuration to adapt to a new environmental situation, to correct a possible software error, or to apply software-driven modified algorithms to replace a failing hardware module, which is a relevant additional security element for autonomous systems.

Applying a design reuse methodology is another architectural consideration. The use of previously designed, characterized units is a relevant factor of speeding up the design process, while maintaining a high degree of confidence. Existing hardware material that may be instantiated multiple times by semi-automatic tools may consist of existing architectures, HDL modules, or low-level layouts. Frameworks supporting the hardware reuse methodology are available [26]. Within this context, let us mention the use of normalized devices and protocols as a way to produce safe and reliable electronic systems.

Embedding test structures into a design is a key issue in achieving reliability. It is the only systematic method for detecting failure, which may occur during the design, manufacturing, or packaging phases of an IC, and also during its lifetime. The cost is generally in terms of reduction in the area efficiency, as circuit testing implies the hardware integration of specialized circuitry, as well as extra input/output pins. The IEEE 1149.1 boundary scan standard is extensively used in the development of electronic systems, from board to IC level [27].

The possibility of system failure during the lifetime of an IC gave rise to the development of run-time mechanisms [28], to be applied to safety critical applications as dedicated extra circuitry for detecting or preventing a failure. System failures can actually result at each level in the flow of development. Indeed, the specifications may be incomplete or inconsistent, the algorithms may not be appropriate in the case of unexpected external conditions (they may diverge or overshoot some values), the IC peripheral devices may be defective (either hardware or software), or development mistakes may remain undiscovered. Finally, some devices may become inoperative or behave unexpectedly due to external physical conditions (temperature, circuit aging). The implementation of run-time error prevention mechanisms requires a thorough analysis of possible constraint violations (static or dynamic, timing, periodic constraints), for each of which a system able to prevent or detect that failure must be developed. Conventional processors use timer circuits as run-time modules to detect unwanted behavior.

Circuit level issues: Several design techniques, as well as rules of thumb, are widely applied to prevent unwanted perturbations or side-effects from influencing safe IC operation. Some commonly used principles are summarized below.

- Faraday shielding: unwanted noise coupling is avoided by keeping the resistors and capacitors in the middle of quiet shielding wires (see Figure 2a) [29].
- Guard rings: noise carriers are gathered to prevent them from perturbing active devices.
- Power distribution: power supplies of analog and digital, switching and quiet units are separated.

- Layout placement: device parameter mismatches due to a thermic gradient can be avoided by correct layout placement (see Figure 2b): symmetric circuit elements are placed symmetrically with respect to the heat source and hence are expected to exhibit the same electrical characteristics (resistance, for instance) [30]. Other techniques include adding dummy devices and applying common centroid localization of devices.
- Capacitor and resistor layout: resistors and capacitors follow specific design guidelines, which tend to eliminate effects of device mismatch, as well as side-effects. Figure 2c) depicts an efficient approach to the design of integrated capacitors, which are made insensitive to capacitance value variation due to possible translation mismatches during production.
- Clock feedthrough compensation: at the shut-off of a transistor, the charge in the channel is redistributed; the right-side transistor (see Figure 2 d) acts as a capacitor which absorbs this extra charge, preventing node 2 from experiencing a voltage surge [31].
- Physical separation of analog and digital units: each must have its own supply and physical localization.
- Minimization of voltage and current swings: whenever this is possible, the module will produce less noise and consume less power.
- Separation of noisy and quiet wire routing: physical separation to avoid noise coupling.

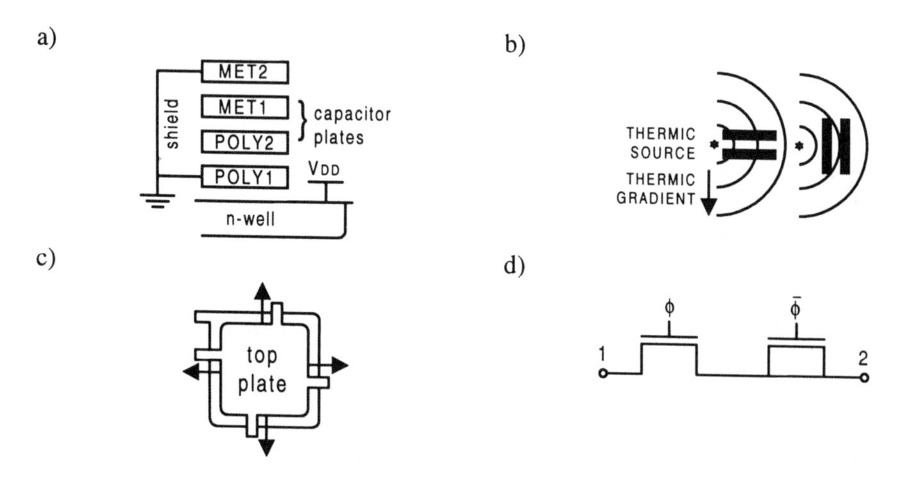

Figure 2: Circuit level common design practice.

Technology level issues: The optimal choice of an integration technology is probably the only feature that influences IC robustness at the low-end technological level. The technology itself, as well as the standard-cells involved in the automated design-flow, are characterized by the library provider, usually the foundry. However, ICs are subject to changes in some of their internal parameters, until failure occurs. Hence, an expected lifetime has to be determined, to allow preventive maintenance of safety critical systems. Mechanisms involved in IC aging, leading to premature failure, include [32]:

- Electrostatic Discharge Damage (ESD) describes a destruction of input and output MOS devices by the effect of dielectric breakdown, caused by a discharge of static electricity, which may reach several kilovolts. ESD is not a problem of concern in some applications where external conditions do not affect circuitry. However, in some medical applications, such as prostheses and portable equipment, ESD protection is vital, especially for input circuitry. The solution to this problem is basically twofold. The integration of on-chip protection consists of exploiting the nondestructive effects of diode breakdown to bypass the ESD pulse within the pads.
- Metal electromigration has been extensively studied as a cause of line rupture and shorts within ICs. Electromigration consists of the displacement of metal atoms due to the impacts of electrons, which move in an electrical field through the material, for instance, from the electrodes to the dielectric regions.
- Dielectric MOS silicon dioxide breakdown. Defects introduced during the manufacturing process of the IC may cause failure by a breakdown of the gate oxide. A common practice among memory manufacturers consists of preventively operating all ICs at elevated temperature and voltages, in order to eliminate weak devices. Improvement in the manufacturing technology of the dielectric growth helps avoid weak zones.
- Degradation and aging of the Si/SiO2 interface due to hot carrier injection causes the I-V characteristics of the MOS to change, until device failure.
- The packaging of the IC also proves to be a relevant factor of circuit reliability and cost. IC chip packaging facilitates their use on printed circuit boards (PCBs) and protects the delicate IC from the outside environment, which may be humid, ranging over a wide scale of temperature and be corrosive. Two main families of packages cohabit; the ceramic package is known to be robust and can be hermetically sealed, which qualifies it for sensitive or military applications; plastic packaging has the advantage of lower cost. Package-related issues include moisture degradation of the IC, mechanical stress affecting the IC, and IC bonding on the package substrate. Packaging and protection for implanted circuits are key factors in producing reliable circuits.

Accelerated tests are standard techniques applied in the validation of a new technology. Samples are subjected to exaggerated external conditions to accelerate their failure. A reasonable estimate of the circuit behavior under normal conditions can be derived from the test results and the testing exaggeration factors. Testing specifications for microcircuits devices have been developed under the following standard: *US MIL-STD-883: Test Methods and Procedures for Microelectronic Devices* [33]. Prediction methods based on empirical models attempt to determine the reliability of circuits use *US MIL-HDBK-217: Reliability Prediction of Electronic Components* [34], [35].

The development of safety critical electronic systems entails achieving reliability at all levels, and requires the implementation of active safety-oriented features and devices. Possible system failure situations must be identified, and emergency procedures targeted to preserve any damage to the system, the environment, and the IC itself, must be developed. Devices to be integrated are, for example, efficient failure detection and alarming, emergency stop procedures, electronic system resetting, mechanical limitations on any electronic-driven devices, and the possible disconnection of any electronic system, to be replaced by manual operation. Another aspect of safety concerns the interaction between software and hardware modules. A codesign approach, as well as formal simulations on the code, help to minimize any hardware–software hazard. Finally, the user interface, software- or hardware-based, has to be carefully developed.

2.4. Precision of computation

Mathematical continuous models assume an infinite precision of all implied values. When implemented in software, the algorithms are confronted with limited precision. The problem of data representation and data precision often becomes relevant when considering a hardware implementation of these algorithms. Fully digital realizations allow the designer to choose a desired precision. Nevertheless, this value has to be carefully specified; it has a significant influence on circuit performance, as well as on the circuit area and power consumption. Hence, software simulation has to be used to support defining the precision depth of the hardware design.

Reaching an acceptable precision in the analog domain raises the problems of development time, power consumption, and physical limitations. Once again, CAD software tools have to support the validation of any design. Nevertheless, high precision computations are usually not devoted to purely analog circuitry. Analog neural network realizations require no more than 8-bit precision [36].

2.5. Application-specific requirements

Application-specific requirements are derived from the environment in which the IC operates. Typical constraints include physical, mechanical, electro-magnetic, expected lifetime, IC pinning, and packaging. The temperature range, for instance, has a relevant impact on development, the applied CAD tools, the choice of technology, etc.

3. ARCHITECTURES APPLIED TO NEURO-FUZZY IC DESIGN

This chapter consists of an overview of architectural and design techniques which have been applied in the development of neuro-fuzzy based ICs, resulting in both application-oriented as well as research-oriented realizations. It is not possible to determine any well-established design technique, such as logic families. Rather, several implementations of identical macrofunctions have emerged, using different design techniques and various circuit-level developments. Hence, we focus on the description of the main orientations and techniques.

Most of the presented circuits make use of the CMOS design technique, as a mature technology that has several interesting properties, such as high signal-to-noise ratio, extensive CAD (Computer-Aided Design) support, and is widespread among IC foundries (thus insuring both low production cost and high degree of reliability).

3.1. Artificial Neural Network Integrated Realization

There are several classifications of artificial neural network implementations [37]. The research work in this domain has led to a wide range of solutions, each of which exhibits its own architecture and circuit technique, but without relying on any common universal structure. This difficulty in classifying ANNs is reflected by the abundance of performance criteria available, each emphasizing a specific property of the evaluated architecture [38], [39]. The taxonomy trees presented here (see Figure 3) intend to reflect the most relevant hardware implementation techniques and architectures developed so far.

The training method obviously has a relevant influence on the system to be developed. Three approaches can be distinguished (see Figure 3b). Off-chip training involves a software tool to simulate the system and set the weights as fixed values, to be hard-wired or downloaded in the hardware realization. During a chip-in-loop procedure, the intermediary ANN outputs are sampled from the hardware system to allow software computation of new weight values, which are subsequently downloaded for a new recall phase. On-chip training implies the hardware availability of specialized circuitry to reprocess the weights, as well as control the whole system sequencing.

Digital realizations

Digital implementation of ANNs benefit from the advantages of digital design, which have to be weighted against its drawbacks.

- Implementing high-precision computation is cheap and easy.
- The noise immunity is relevant.
- Interfacing with further digital hardware is simplified.
- HDL-based development allows a flexible and technology independent design-flow, with relevant reusability properties, and automatization of the design process.

On the other hand, the disadvantages are:

- The area and power consumption costs are relatively hard to constrain; they generally exceed those of the analog design.
- Interfacing with sensors and actuators requires the use of ADC/DAC systems, which increase the IC size and power consumption.
- A recursive digital structure (learning phase) may be affected by parasitic perturbation as a result of amplitude quantization; the solution to this problem requires increasing the numeric range.

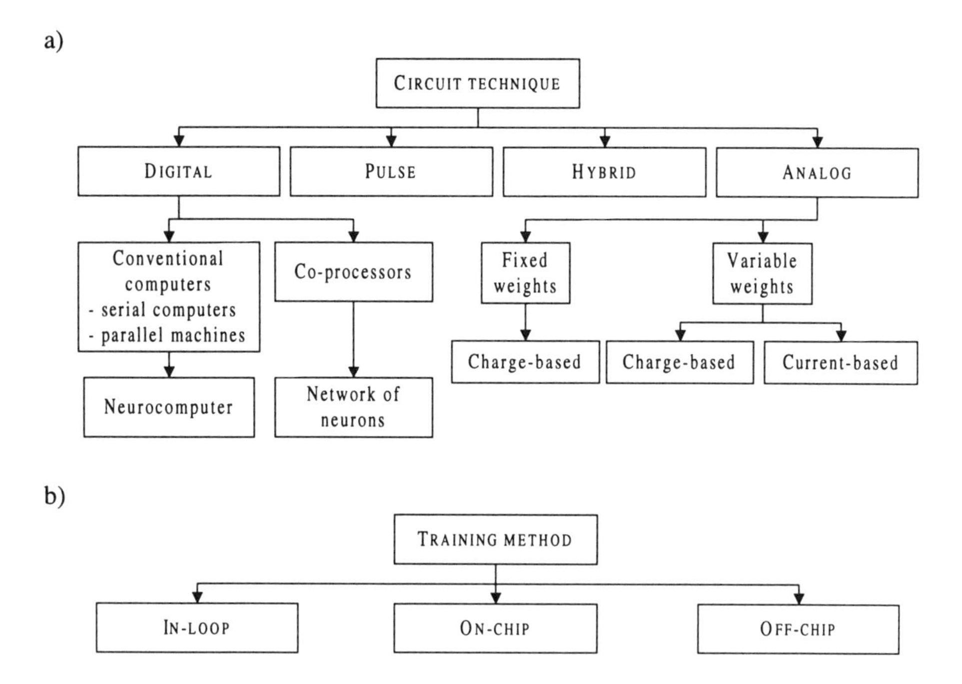

Figure 3: Taxonomic trees for artificial neural network hardware; a) circuit technique, b) training method.

Probably the most widespread technique applied to the implementation of neuro-fuzzy algorithms consists of simulating them on conventional serial computers. Given the high degree of parallelism involved in ANN algorithms, conventional parallel machines may be used to improve the overall efficiency. These techniques are referred to as software methods. The programmer and the compiler are key elements in generating a microcode that optimally exploits the processor architecture.

Co-processors are special-purpose peripheral PCB systems, dedicated to accelerating neural-based computation of standard processors by concentrating the processing power on RISC processors, DSPs, transputers, or dedicated neuro-chips. A software simulator tool is generally part of the system.

The hardware circuit technique on which this chapter focuses applies mostly to the so-called neurocomputers and networks of neurons. Neurocomputers appeared in the 1990s to satisfy the needs of mapping ANN algorithms onto targeted architectures, designed according to their specificities. A neuro-computer is defined as a dedicated processor that accelerates the computation of connectionist algorithms. The requirement of high-precision computation dictated the digital nature of all processed values. Several companies have proposed their own solution: Philips, Hitachi, Siemens, etc.

The architecture of digital neuro-computer units stems from several interconnected processing units (PUs). The topology basically defines how PUs communicate with each other within a common frame. Efficient neuro-computer architectures are expected to support scalability to achieve high density, programmability, and modularity, and should support the implementation of virtual networks for a possible simulation of large ANNs. The data-flow between PUs allows classification of the large amount of existing realizations into six architectural classes, as reported in [40]. An exhaustive description of existing realizations can be found in [37].

Networks of electronic neurons implement one neuron as a processing element, to be duplicated according to a chosen topology. Field Programmable Gate Array (FPGA) integration is well adapted to these implementations [41], [42].

Analog realizations

Analog implementations of ANNs exploit intrinsic physical properties of electronic components to efficiently implement complex functions: summation, multiplication, smashing function, memory, absolute value, square, winner-take-all (WTA), etc. The strong interaction between the application specific problem and the target architecture requires them to be developed in parallel, in order to master the following threats and trade-offs:

- Computational precision is strongly related to silicon area and power consumption.
- The storage of analog values is sensitive to the silicon area.
- The interconnection among neurons has to be minimized.
- Low-current, low-voltage and submicron feature size are design issues to be taken into account.
- Sensitivity to noise is greater than that of digital systems.
- The time of development has to be carefully considered, due to a lack of efficient analog synthesis tools.

On the other hand, analog circuit characteristics include high-speed processing, compact design technique, ease of interfacing with sensors and actuators, continuous time, and amplitude of signals, which eliminates time sampling techniques as well as ADC's extra circuitry. Many analog circuit techniques are applied to the implementation of neuro-fuzzy algorithms; here, we present the most widely applied.

Current-mode based neuron building blocks exploit Kirchhoff's laws and MOSFET internal characteristics, as well as current mirror circuit properties, as basic design techniques. Current mode electronic circuits are very popular, as they allow very compact, high-speed and low-power implementations of functions. Some function implementations such as summation and subtraction have a very simple and intuitive realization as a simple wire connection. Basic realizations of PMOS and NMOS current mirrors are shown in Figure 4. These circuits demonstrate the principle of current mirroring (reproduction) [43], but are not accurate enough for an integrated realization, since each output is modulated by the output voltage through Early conductance. Both current mirror implementations depicted in Figure 4c and d uncouple the output voltage from output current, and are thus appropriate for use as building blocks to synthesize complex functions. The precision of all mirrors highly depends on the output resistance, which is expected to be very high, and on unwanted mismatches between integrated transistors, which have to work in saturation mode. Dynamic mirrors are used to increase accuracy, but at the expense of a clocking scheme that leads to larger size and complexity. All the current mirrors depicted on Figure 4 (a through d) can have a bipolar realization as well.

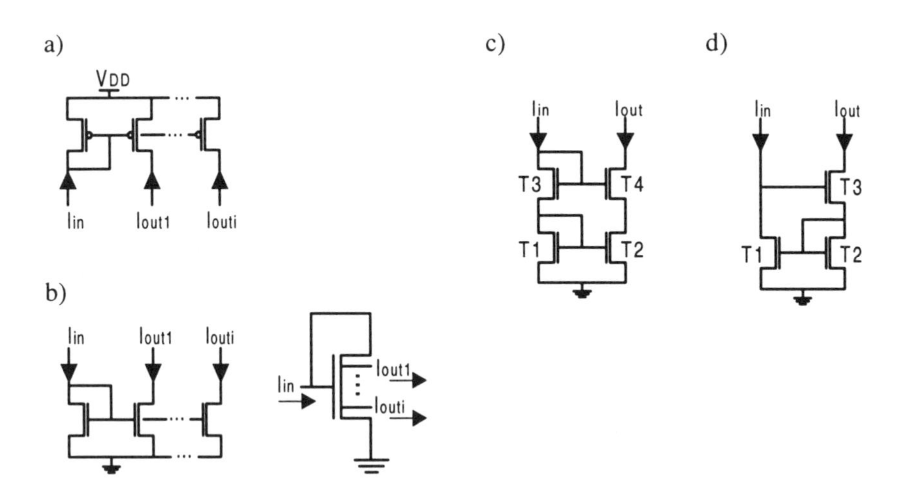

Figure 4: Conventional current mirrors: a) standard PMOS, b) standard NMOS and corresponding symbol, c) cascode, d) Wilson.

A large number of realizations have been published, as a result of the attraction of this design technique, which allows very compact realization of complex functions. The transconductance amplifier depicted in Figure 5a is used to implement multiplication. The bump circuit (c) [44] computes the similarity between two voltage inputs according to equation (d).

Charge-based circuits implement neural functions by means of charge redistribution among integrated capacitors.

- An early work in the domain of applying switched-capacitors to construct neurons is described in [45] (see Figure 6). A nonoverlaping three-clock scheme is applied to convert an incoming data into an amount of charge on an input capacitor, to reset previously stored values, and, finally, to transfer the charge toward the second capacitor. The activation function is computed by the inverter, which either implements the sum inside the brackets multiplied by β (see equation on Figure 6c or saturates.

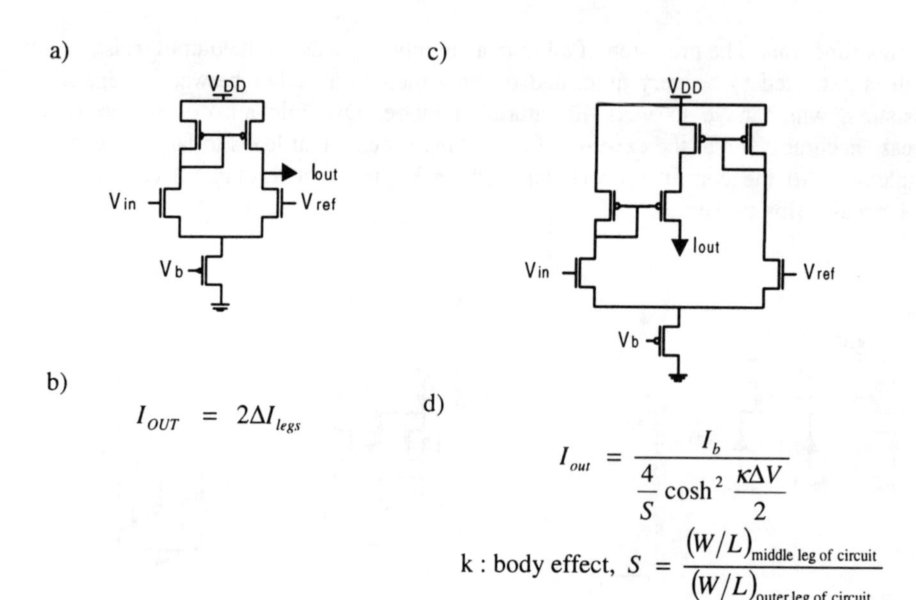

Figure 5: Conventional examples of current-based usage in ANN implementations.

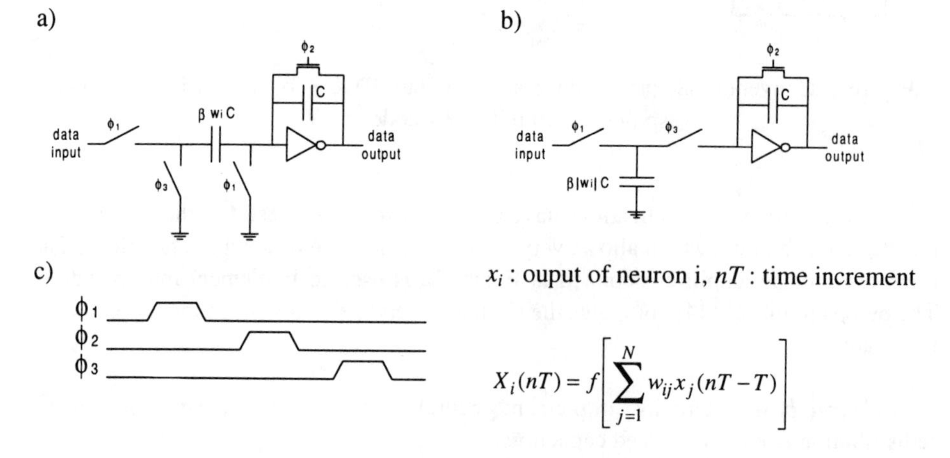

Figure 6: Switched-capacitor based neuron: a) non-inverting input, b) inverting input, c) three phase clock scheme and discrete-time equation for one neuron.

- The principle of the Neuron-MOS (νMOS) design technique [46], [47] is based on a conventional MOSFET, having its POLY1 gate floating (see Figure 7). Multiple POLY2 capacitor plates on top of this structure control the potential (ϕ_F) of this floating gate. The transistor is turned on whenever ϕ_F exceeds the threshold voltage of the transistor. The summation operation is carried out by charge sharing among several capacitors, and the thresholding is MOSFET inherent; thus all values are handled in the voltage domain of operation. This circuit technique has been applied in the development of synapses, neurons and a small feed-forward neural network, some logic functions, low-bit flash ADC, and a multiplier cell [48].

a)

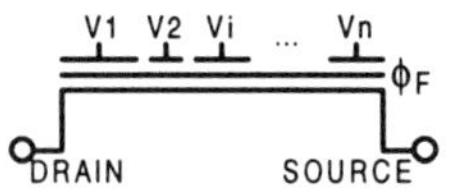

V1 V2 Vi ... Vn ΦF DRAIN SOURCE

b)

$$\phi_F = \sum C_i V_i / C_{TOT} > V_{th} \rightarrow \text{íMOS ON}$$

Figure 7: Neuron-MOS principle description: a) gate-level description, b) general circuit equation.

- A further fixed-weight charge-based realization of a Hamming ANN [49] is demonstrated in [50]. Summation and multiplication are processed by capacitive coupling (see Figure 8a and b). The circuit has a two nonoverlaping clock scheme: the dendrites and synapses are precharged during ϕ_1, and the inputs are then activated on ϕ_2 to perturb the dendritic voltage during the evaluation phase, according to equation inserted in Figure 8c. An analog charge-based winner-take-all (WTA) [51] unit selects the neuron with the smallest Hamming distance between its stored weights and the current input as the winner. The weights are realized by POLY1-POLY2 overlap capacitances, whose values have to be determined prior to circuit integration.

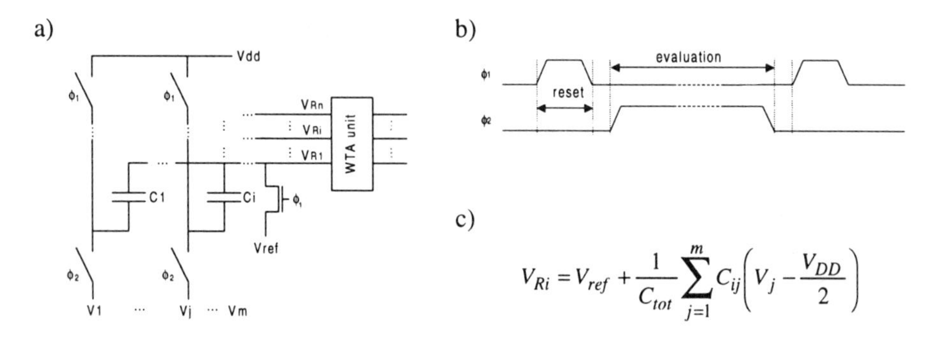

Figure 8: Charge-based Hamming ANN; a) gate-level circuit description, b) circuit clock sequencing, c) analytic expression of the dendritic voltage perturbation at the end of the evaluation phase (C_{TOT} denotes the total capacitance on a row, including parasitic).

- Charge-coupled devices (CCD) are inherently adapted to temporarily hold and process analog values as amounts of charge under the control of several electrodes that can be addressed separately. ANN functions can be synthesized by this means.

Mixed-mode realizations

Mixed analog-digital realizations attempt to combine the advantages of both, and hopefully overcome their drawbacks and weaknesses by permitting them to interact. A trade-off may lead to larger parallelism, but lower processing speed.

- *Pulse frequency or width* coding is well known in the telecommunication field. These electronic systems may be successfully adapted to satisfy the requirements of neuro-fuzzy systems integration. Rather than encoding analog information into the amplitude value, it is stored along the time axis. A pulse frequency coding technique applies the modulation of the pulses' frequency to perform the neural functions, as pulse width coding assumes digitally stored weights to modulate a chopping signal, which weights the input signals (see Figure 9) [52]. Expected advantages include high noise immunity, insensitivity to signal attenuation, low energy requirements, and an ease of interfacing with both the analog and digital world. Several types of modulation can be applied, leading to various levels of performance efficiency.

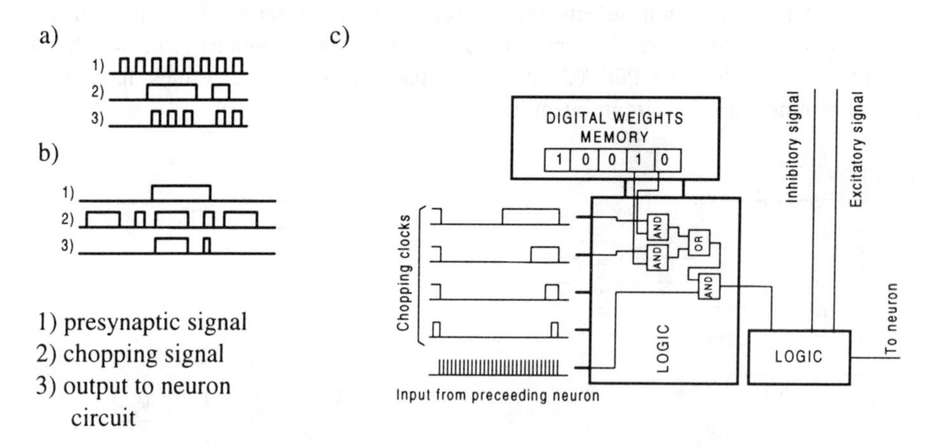

Figure 9: Pulse stream synapse circuit technique: a) weighting principle with a chopping signal slower than the presynaptic signal, b) with a chopping signal faster than the presynaptic signal, c) functional synapse description.

- A mixed analog/digital, variable-weight realization of the Hamming network is presented in [53]. This IC integrates on the same die as an analog charge-based Hamming network with a digital core implementing the on-chip error correction algorithm and chip control units (see Figure 10). The 4-bit soft-programmable capacitor values associated with each synapse are chosen as $C_i = 2^n C_u$, where n = 0, …,3 and C_u (unit capacitance) = 17 fF. The capacitors are realized by comb-shaped POLY1-POLY2 overlaps, resulting in exact integer multiples of the unit capacitance. SPICE simulations have been used to validate the full-custom designed ANN and WTA units. The learning algorithm has been implemented and functionally verified as a VHDL description, which was subsequently synthesized into a gate-level netlist and realized as a standard cell-based module using a placement and routing CAD tool. A hardware oriented adaptation of the well-known *error correction learning rule* [54] was implemented. Cell-based high-level synthesis of the learning supervision units allows considerable reduction of the development time, and offers relevant reusability properties and a technology independent development process. The development targeted was a pixel-pattern recognition application, including software C simulations to validate the hardware-friendly algorithms.

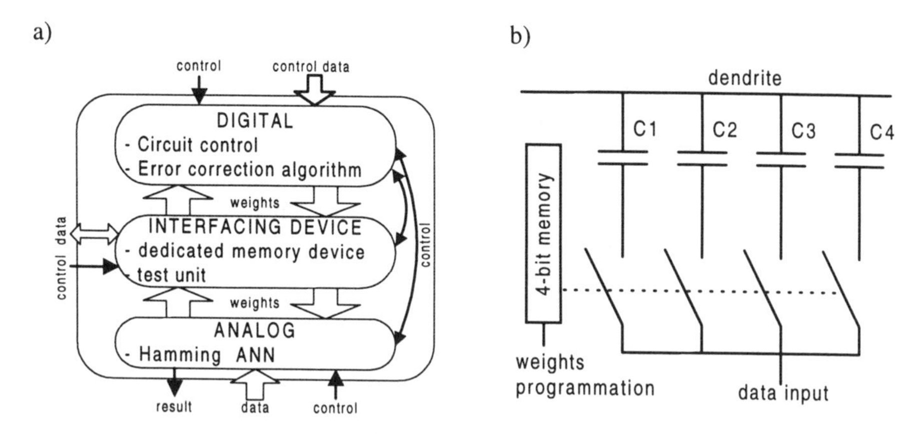

Figure 10: Hamming ANN with on-chip learning architecture: a) IC architecture, b) programmable weight synapse principle.

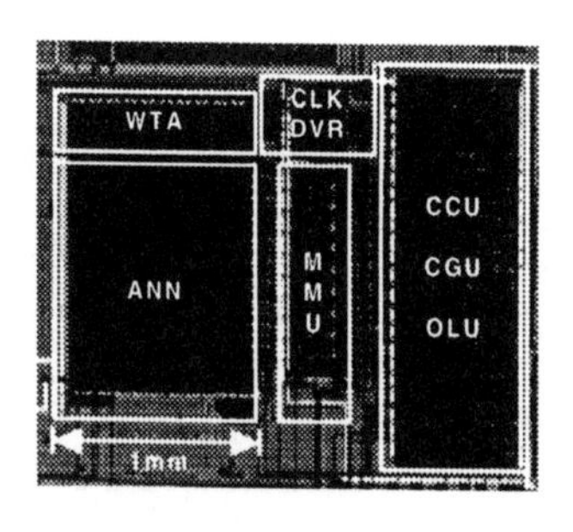

UNIT	FUNCTION	A/D
WTA	Winner-Take-All Unit	A
ANN	ANN, Hamming Capacitor Matrix	A
CCU	Chip Control Unit	D
CGU	Clock Generation Unit (weights processing)	D
OLU	On-Chip Learning Unit	D
MMU	Multipurpose Memory Unit (test purpose)	D
CLK DVR	Clock Drivers	D

Figure 11: Microphotograph of the IC to be packaged into a 100 PGA, including several test structures and test pins.

An integrated hardware implementation of the described architecture was realized using the AMS (Austria Micro Systems) CMOS 0.8 micron technology (see Figure 11).

The implemented network consists of 20 neurons, each with 10 variable-weight synapses. The die size is less than 13 mm^2 while the active chip area (including ANN, learning-supervision unit, peripheral units, and clock generators) is less than 5 mm^2. A number of test features are also included on the chip. The very small area of this implementation confirms the area efficiency of the proposed mixed analog/digital approach.

3.2. Fuzzy-Based Integrated Realization

Several ICs implementing the fuzzy paradigm as well as their related software development environment have emerged over the last decade [55] as the solution to the large amount of parallel data processing required by real-time fuzzy applications. The system architectures mostly reflect the conventional fuzzy-based strategy; that is to say, the fuzzifier interface driving the inference engine, which, in turn, drives the defuzzifier interface, to build up a structure targeted to efficiently process approximate reasoning and decision making algorithms. This modular architecture may be fully integrated on a single IC, or may be spread into different ICs. Conventional characterization of fuzzy dedicated ICs include the number and shape range of input and output variables, the number of simultaneously processed rules, the type of inference, the defuzzification method, etc., as their performance is usually evaluated in terms of fuzzy logic inferences per second (FLIPS).

Fuzzy processors have to be programmable for a wide range of real-time applications. The variable number of inputs, the membership functions' (MF) shapes, and the number and size of the rules have to be adapted to the problem, and have to undergo real-time modifications to avoid unstable behaviors. Thus extra memory and tunable elements have to be integrated.

The classification of fuzzy logic dedicated ICs presented here reflects the internal architecture and nature of processed signals (see Figure 12).

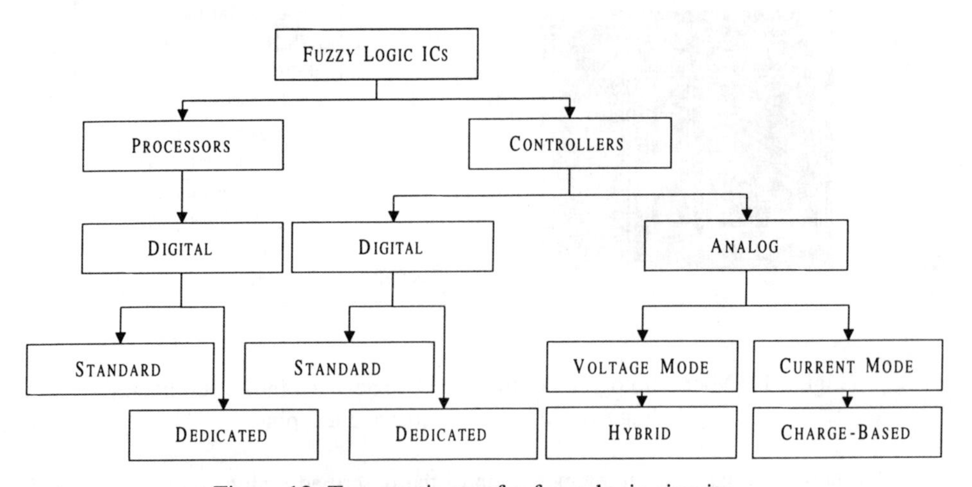

Figure 12: Taxonomic tree for fuzzy logic circuits.

Digital realizations

Digital realizations of fuzzy ICs include ADC/DAC, to adapt to the external environment, may these be on-chip or off-chip. The issue of the quantization of processed values has to be addressed by careful software simulations before integration. Four main families have emerged, taking advantage of digital implementation in the field of fuzzy ICs.

- Standard processor architectures are being used for the purpose of implementing fuzzy inferences as software (see Figure 12).
- Dedicated ICs implementing a specialized set of instructions as well as a targeted architecture aim at accelerating the processing compared to standard processor architectures. A fuzzy RISC dedicated architecture is described in [56].
- Commercial digital microcontrollers can also be used as specialized and autonomous peripheral units.
- The last family of digital implementations includes fuzzy dedicated microcontrollers.

Analog realizations

Voltage mode realizations take advantage of the easy propagation of information, as well as the very simple and efficient hardware implementation of some fuzzy operators such as MIN, MAX and truncation. On the other hand, some functions as simple as summation may lead to relatively complex developments. Difficult to integrate structures such as resistors are required to convert voltages into currents.

- Yamakawa demonstrates in [57] the circuit design and operation of several fuzzy specialized operators, applied to the inverted pendulum stabilization. The BiCMOS technology, as well as emitter-coupled fuzzy logic gates (ECFL) are combined to implement the MIN, MAX operators, as well as the membership functions (see Figure 13a, b, c, respectively).

The implementation of *current mode* circuits allows a very compact integration of several fuzzy specific functions. The bounded difference operator for instance has a simple realization as a combination of a current mirror and a CMOS transistor operated as a diode (see Figure 14). Current mirrors have a PMOS as well as an NMOS realization; this allows the development of compatible cascadable structures as depicted in Figure 14b and c. Current-mode design techniques can be used in each module in the architecture; defuzzification and normalization circuits are also reported.

Hybrid-mode realizations take advantage of both current-mode and voltage-mode designs to optimize the overall IC characteristics. Voltage-mode circuits are preferred for fuzzy IC inputs, which require propagation towards multiple rule blocks, as current mirrors have a single fan-out. Current-mode circuits efficiently implement the membership functions.

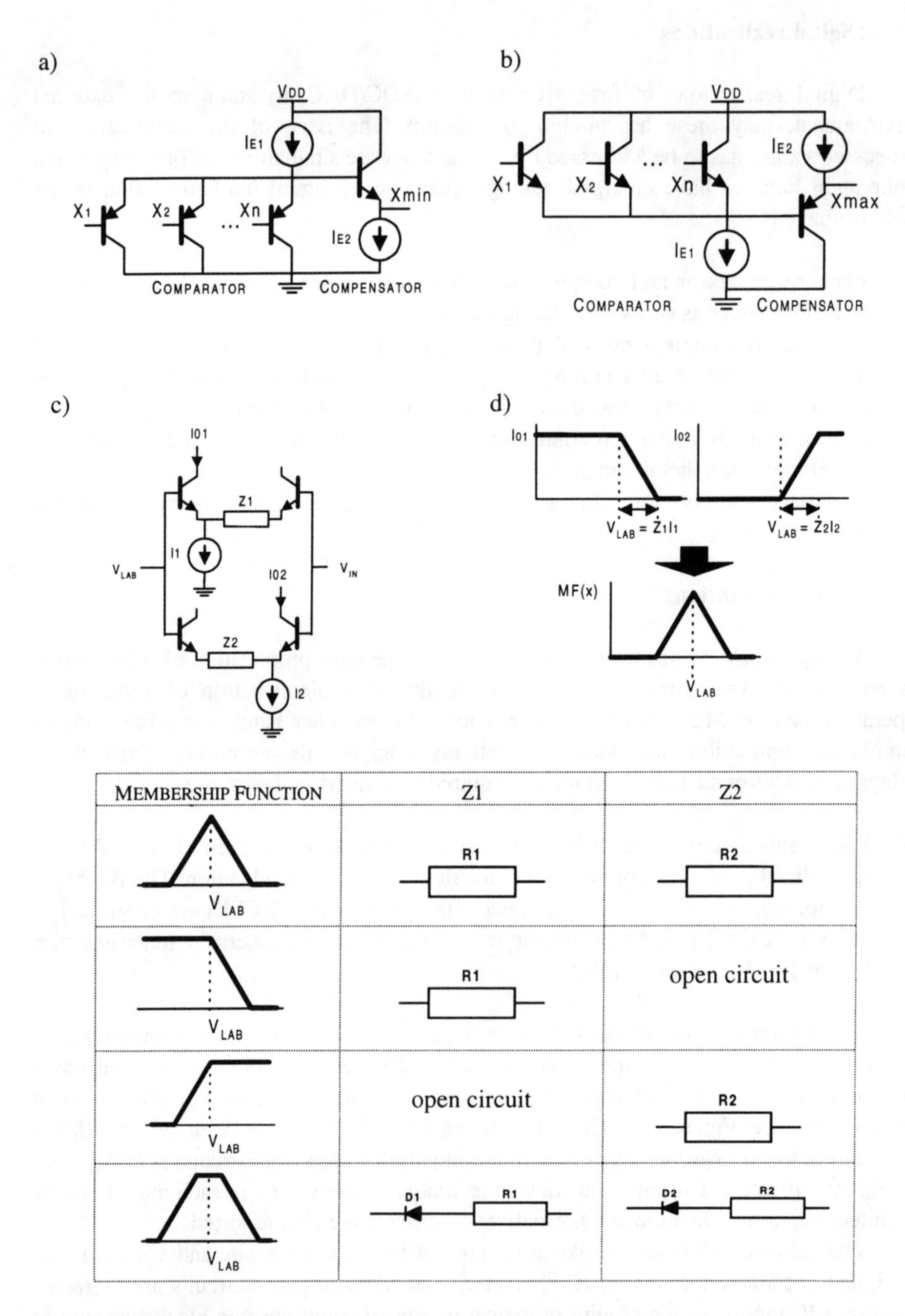

Figure 13: Voltage mode realizations of fuzzy dedicated functions: a) synthesis of the MIN function, b) synthesis of the MAX function, c) simplified gate-level description of the MF generating circuit, and table of impedances for the MF synthesis.

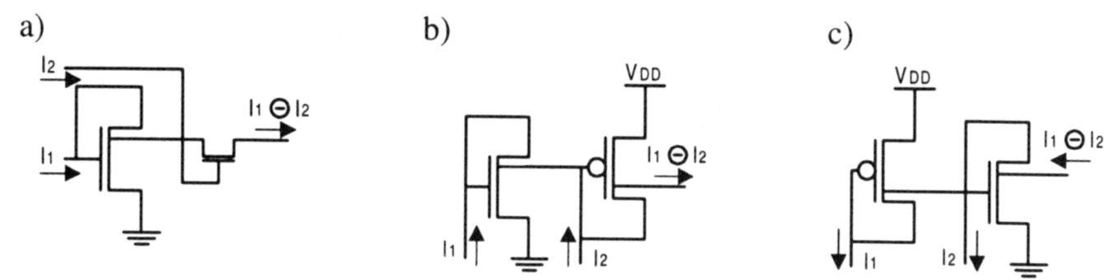

Figure 14: Current mode circuits implementing the bounded difference operation. I_1 and I_2 represent two values of membership functions μx and μy.

Charge-based realizations make use of the charge redistribution principle to implement functions.

- A fuzzifier unit is described in [58] and depicted in Figure 15. Adjustable positive and negative ramps are generated by capacitive circuitry interacting with an operational inverter (OPI) [59], which is realized by a three stage cascade of minimal geometry CMOS inverters supplied with source followers and compensation network, as a high-gain amplifier. The circuit operates in a two-phase nonoverlapping clock scheme of reset-evaluation. The inference module consists of a modified version of the WTA described in [50], which performs detection of the input with the smallest voltage. The weighted average of singletons defuzzification strategy prescribes simple addition, weighted addition, and division, each of which are computed by a dedicated unit consisting of charge-based and OPI circuit arrangements.

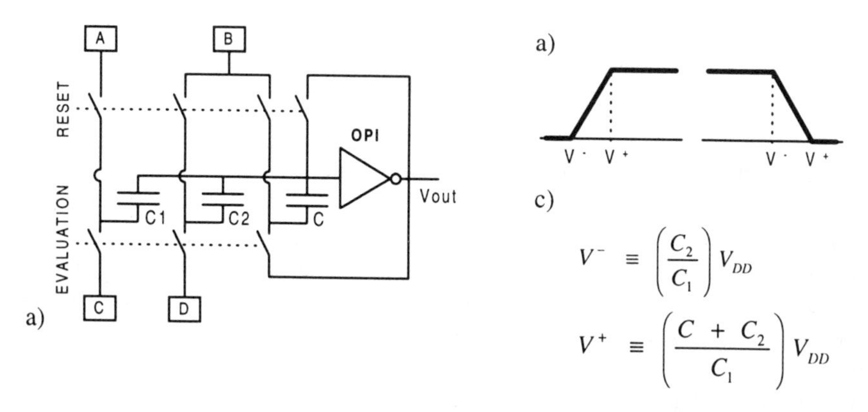

TERMINAL	POSITIVE RAMP	NEGATIVE RAMP
A	Vinput	GND
B	GND	VDD
C	GND	Vinput
D	VDD	GND

Figure 15: Charge-based fuzzifier module: a) gate-level circuit description, b) possible ramp generation, c) capacitors sizing conditions.

Hybrid analog/digital implementations of fuzzy algorithms have emerged as a way to improve the weak points of both worlds by permitting them to interact. An implementation consisting of a digital memory storage unit followed by parallel analog fuzzy inferencing in the charge-domain is described in [60].

3.3. Hybrid Integrated Realization

Neuro-fuzzy integrated algorithms are described as hybrid realizations. There are actually very few integrations of purely neuro-fuzzy paradigms. Most of them use previously described design techniques.

The IC FN305 described in [61] implements an optical character recognition task. A fuzzy neuron model is integrated on-chip in a BiCMOS 2μm technology process (see Figure 16). Scalar input signals are replaced by fuzzy input vectors, and synaptic weights are replaced by weights of fuzzy numbers characterized by a membership function. The author describes emitter-coupled fuzzy logic gates which implement the MIN and MAX operators in the analog voltage mode.

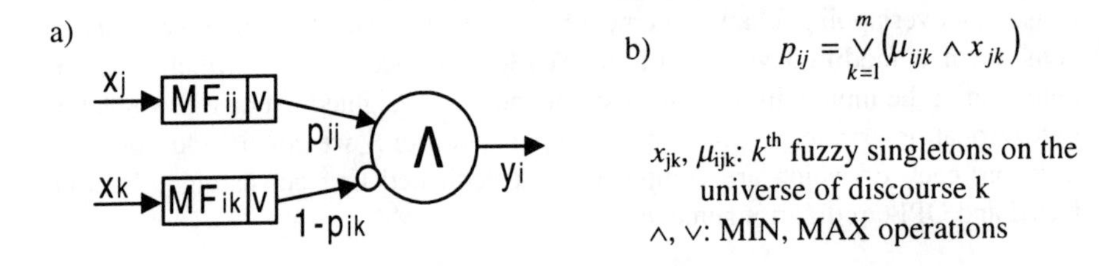

Figure 16: a) Mathematical model for the implemented fuzzy neuron, b) synaptic operation defined as fuzzy logic inner product.

3.4. An example of neuro-fuzzy realization

This section focuses on describing realizations of neuro-fuzzy ICs applied to the biomedical field. This application domain is of growing interest, but very few dedicated developments are reported. The main effort nowadays is directed toward board-level realizations. The relatively large time constants of most human body functions partially explains this.

A hybrid analog/digital realization of an ANN dedicated to intercardiac morphology classification is reported in [62] [63]. The IC is intended to detect potentially fatal cardiac arrhythmia, upon which the implanted defibrillator device administrates a therapy in the form of electrical shock. The chip includes a 10:6:3 multilayer perceptron, an input interfacing device, and an output winner-take-all unit to produce a binary decision, as well as some peripheral circuitry. The circuit was designed using a CMOS 1.2μm technology; it occupies a 2.2mm × 2.2mm area and consumes less than 200nW from a 3V supply voltage at an input sampling rate of 250Hz. The in-loop training system consists of a PC (software package), which produces ICEG signals to be fed into an ICG, which performs signal filtering and adaptation. The ANN then computes the classification, and subsequently has its 5-bit weight plus 1-bit sign modified according to the on-line learning algorithm. The ANN circuit design uses current mirror techniques. One peculiarity of this implementation lies in the smashing function, which resides at the input of synapses.

4. CONCLUDING REMARKS

The successful hardware integrated circuit implementation of safety-critical biomedical systems entails specific requirements, which increase the reliability of electronic systems and the patients' comfort and autonomy, as well as reducing the overall costs of medical treatment. Several criteria including low-power, low-area, reliable, and optimal complexity developments have to be taken into account throughout the design-flow.

Integrating neuro-fuzzy algorithms is a promising research field, which tends to provide an efficient and attractive solution to fulfillment of the severe constraints of constructing biomedical dedicated systems. The best-suited circuit architecture and circuit technique have to be chosen from the existing realizations. No universal approach is available today. An opportunistic, but intelligent development method is applied in most of the cases.

The large number of biomedical applications involving neuro-fuzzy paradigms, some of which have been described in the previous chapters of this book, offer a fascinating field of investigation for a research community involving scientists with medical and engineering education. Efficient biomedical portable systems have a brilliant future from a research and a commercial point of view. They respond to a real need that stems from doctors and patients, for both high-tech niche products as well as widespread, up-to-date, and secure medical equipment.

REFERENCES

[1] Ajluni, C., Microsensors move into biomedical applications, *Electronic Design*, May 1996, pp. 75-84.

[2] Calvez, J. P., Spécification et Conception des Systèmes, une Méthodologie, Manuels Informatiques Masson, Paris, 1990.

[3] Adams, J. K., Thomas, D. E., The Design of Mixed Hardware/Software Systems, *Proceedings of the 33rd Design Automation Conference DAC'96*, Las Vegas, NV, U.S.A., June 1996.

[4] Peterson, M. G. E., User Satisfaction Surveys, What the Engineer Should Know, *Proceedings of the Ninth IEEE Symposium on Computer-Based Medical Systems*, Ann Arbor, Michigan, June 1996, pp. 71-76.

[5] *The National Technology Roadmap for Semiconductors*, Semiconductor Industry Association, 1997 Edition.

[6] Gajski, D. D., Narayan, S., Ramachandran, L., Vahid, F., Fung, P., System Design Methodologies: Aiming at the 100 h Design Cycle, *IEEE Transactions on Very Large Scale Integration (VLSI) Systems*, Vol. 4, No. 1, March 1996, pp. 70-81.

[7] Changuel, A., Rolland, R., Jerraya, A. A., Design of an Adaptive Motors Controller based on Fuzzy Logic using Behavioural Synthesis, *Proceedings of EURO-DAC'96 with EURO-VHDL'96*, Geneva, 1996, pp. 48-52.

[8] Gonzalez, R., Torralba, A., Franquelo, L. G., AFAN, A tool for the automatic synthesis of neural and fuzzy controllers with architecture optimization, *Proceedings of the IEEE International Symposium on Circuits and Systems ISCAS'97*, Hong Kong, 1997, pp. 637-640.

[9] Pawlak, A., Bouchard, F., Bakowski, P., Survey on VHDL Modeling Guidelines, *Proceedings of the 2nd Workshop on Libraries, Component Modeling, and Quality Assurance*, Toledo, Spain, April 1997, pp. 117-128.

[10] Lemaitre, L., *Theoretical Aspects of the VLSI Implementation of Fuzzy Algorithms: Application to the Design Automation of Current Mode Fuzzy Units*, Ph.D. Dissertation No. 1226, Swiss Federal Institute of Technology of Lausanne, 1994.

[11] Sanz, A., A Unified Tool for Fuzzy/Neural Network Systems, *IEEE Micro*, March/April 1997, pp. 65-69.

[12] Manaresi, N., Rovatti, R., Franchi, E., Guerrieri, R., Baccarani, G., A Silicon Compiler of Analog Fuzzy Controllers: From Behavioral Specifications to Layout, *IEEE Transactions on Fuzzy Systems*, Vol. 4, No. 4, November 1996, pp. 418-428.

[13] Moerland, P. D., Fiesler, E., Hardware-Friendly Learning Algorithms for Neural Networks: An Overview, *Proceedings of the Fifth International Conference on Microelectronics for Neural Networks and Fuzzy Systems MicroNeuro'96*, Lausanne, February 1996, pp. 117-124.

[14] Jabri, M., Flower, B., Training Analog Neural Networks Using Weight Perturbation, *IEEE Transactions on Neural Networks*, Vol. 4, No. 1, January 1992, pp. 154-157.

[15] Marchesi, M. L., Piazza, F., Uncini, A., Backpropagation without Multiplier for Multilayer Neural Networks, *IEE Proceedings Circuits, Devices and Systems*, Vol. 143, No. 4, August 1996, pp. 229-232.

[16] Battiti, R., Tecchiolli, G., Training Neural Nets with the Reactive Tabu Search, *IEEE Transactions on Neural Networks*, Vol. 6, No. 5, 1995, pp. 1185-1200.

[17] Thiran, P., Peiris, V., Heim, P., Hochet, B., Quantization Effects in Digitally Behaving Circuits Implementations of Kohonen Networks, *IEEE Transactions on Neural Networks*, Vol. 5, No. 3, May 1994, pp. 450-458.

[18] Asanovic, K., Morgan, N., Experimental Determination of Precision Requirements for Back-Propagation Training of Artificial Neural Networks, *Proceedings of the Second International Conference on Microelectronics for Neural Networks*, Munich, October 1991, pp. 9-15.

[19] Reyneri, L. M., Filippi, E., An Analysis on the Performance of Silicon Implementations of Backpropagation Algorithms for Artificial Neural Networks,

IEEE Transactions on Computers, Vol. 40, No. 12, December 1991, pp. 1380-1389.

[20] Rumelhart, D., Hinton, G., Williams, R., Learning Internal Representations by Error Propagation, *Parallel Distributed Processing: Explorations in the Microstructure of Cognition*, Vol. 1: Foundations, MIT Press, Cambridge, Massachusetts, 1986, pp. 318-362.

[21] Khan, A. H., Feedforward Neural Networks with Constrained Weights, Ph.D. Dissertation, University of Warwick, August 1996.

[22] Satyanarayana, S., *Analog VLSI Implementation of Reconfigurable Neural Networks*, Ph.D. Dissertation, Columbia University, New York, 1991.

[23] Reed, R., Pruning Algorithms – A Survey, *IEEE Transactions on Neural Networks*, Vol. 4, No. 5, September 1993, pp. 740-747.

[24] Kerlirzin, P., Réfrégier, P., Theoretical Investigation of the Robustness of Multilayer Perceptrons: Analysis of the Linear Case and Extension to Nonlinear Networks, *IEEE Transactions on Neural Networks*, Vol. 6, No. 6, May 1995, pp. 560-571.

[25] Murray, A. F., Edwards, P. J., Enhanced MLP Performance and Fault Tolerance Resulting from Synaptic Weight Noise During Training, *IEEE Transactions on Neural Networks*, Vol. 5, No. 5, September 1994, pp. 792-802.

[26] Kission, P., Jerraya, A., Moussa, I., Hardware Reuse, *Proceedings of the 2nd Workshop on Libraries, Component Modeling, and Quality Assurance*, Toledo, Spain, April 1997, pp. 21-29.

[27] Bleeker, H., Van Den Eijnden, P., De Jong, F., *Boundary-scan Test; A Practical Approach*, Kluwer Academic Publishers, Dordrecht, The Netherlands, 1993.

[28] Vishnuvajjala, V. R., Subramanian, S., Tsai, W. T., Elliot, L., Mojdehbakhsh, R., Run-Time Assertion Schemes for Safety-Critical Systems, *Proceedings of the Ninth IEEE Symposium on Computer-Based Medical Systems*, Ann Arbor, Michigan, June 1996, pp. 18-23.

[29] Schmerbeck, T., *Mechanisms and Effects of Noise Coupling in Analog ICs*, Electronics Laboratories Advanced Engineering Course on CMOS & BiCMOS IC Design '97, Swiss Federal Institute of Technology of Lausanne, 1997.

[30] Jabri, M. A., Coggins, R. J., Flower, B. G., *Adaptative Analog VLSI Neural Systems*, Chapman & Hall, 1996.

[31] Gregorian, R., Temes, G. C., Analog MOS Integrated Circuits for Signal Processing, Wiley, 1986.

[32] Sabnis, A. G., *VLSI Reliability*, Vol. 22 in VLSI Electronics Microstructure Science, Academic Press Inc., San Diego, CA, 1990.

[33] *US MIL-STD-883: Test Methods and Procedures for Microelectronic Devices*, Available from the National Technical Information Service, Springfield, Virginia.

[34] *US MIL-HDBK-217: Reliability Prediction of Electronic Component*, Available from the National Technical Information Service, Springfield, Virginia.

[35] Pecht, M. G., Nash, F. R., Predicting the Reliability of Electronic Equipment, *Proceedings of the IEEE*, Vol. 82, No. 7, July 1994, pp. 992-1004.

[36] Ramacher, U., Rückert, U., VSLI Design of Neural Networks, Kluwer Academic, Boston, 1991.

[37] Beiu, V., Digital IC Implementations, in *Handbook of Neural Computation*, Fiesler, E., Beale, R., Editors, Institute of Physics and Oxford University Press, New York, 1996, Chapter E1.

[38] Keulen, E., Colak, S., Withagen, H., Hegt, H., Neural Network Hardware Performance Criteria, *Proceedings of the IEEE International Conference on Neural Networks*, 1994, pp. 1885-1888.

[39] Serrano-Gotarredona, T., Linares-Barranco, B., A Real-Time Clustering Microchip Neural Engine, *IEEE Transactions on Very Large Scale Integration (VLSI) Systems,* Vol. 4, No. 2, June 1996, pp. 195-209.

[40] Ienne, P., Architectures for Neuro-Computers: Review and Performance Evaluation, Microcomputing Laboratory Technical Report No. 93/21, Swiss Federal Institute of Technology of Lausanne, 1993.

[41] Salapura, V., Gschwind, M., Maischenberger, O., A Fast FPGA Implementation of a General Purpose Neuron, *Proceedings of the Fourth International Workshop on Field Programmable Logic and Applications*, Prague, Czech Republic, September 1994.

[42] Perez, A., Sanchez, E., The FAST architecture: A Neural Network with Flexible Adaptable-Size Topology, *Proceedings of the Fifth International Conference on Microelectronics for Neural Networks and Fuzzy Systems MicroNeuro'96*, Lausanne, Switzerland, February 1996, pp. 337-340.

[43] Uyemura, J. P., *Circuit Design for CMOS VLSI*, Kluwer Academic Publishers, Dordrecht, The Netherlands, 1992.

[44] Delbrück, T., Bump Circuits for Computing Similarity and Dissimilarity of Analog Voltages, CNS Memo 26, California Institute of Technology, Computation and Neural Systems Program, Pasadena, California, May 1993.

[45] Tsividis, Y. P., Anastassiou, D., Switched-Capacitor Neural Networks, *Electronic Letters*, Vol. 23, No. 18, August 1987, pp. 958-959.

[46] Shibata, T., Ohmi, T., Neuron MOS Binary-Logic Integrated Circuits – Part I: Design Fundamentals and Soft-Hardware-Logic Circuit Implementation, *IEEE Transactions on Electron Devices*, Vol. 40, No. 3, March 1993, pp. 570-576.

[47] Shibata, T., Ohmi, T., Neuron MOS Binary-Logic Integrated Circuits – Part II: Simplifying Techniques of Circuit Configuration and their Practical Applications, *IEEE Transactions on Electron Devices*, Vol. 40, No. 5, May 1993, pp. 974-979.

[48] Weber, W., Prange, S. J., Thewes, R., Wohlrab, E., Luck, A., On the Application of the Neuron MOS Transistor Principle for Modern VLSI Design, *IEEE Transactions on Electron Devices*, Vol. 43, No. 10, October 1996, pp. 1700-1708.

[49] Lippmann, R. P., An Introduction to Computing with Neural Nets, *IEEE ASSP Magazine*, April 1987, pp. 4-20.

[50] Ciligiroglu, U., A Charge-Based Neural Hamming Classifier, *IEEE Journal of Solid-State Circuits*, Vol. 28, No. 1, January 1993, pp. 59-67.

[51] Günay, Z. S., Sanchez-Sinencio, E., CMOS Winner-Take-All Circuits: A Detail Comparison, *Proceedings of the IEEE International Symposium on Circuits and Systems ISCAS'97*, Hong Kong, 1997, pp. 41-44.

[52] Murray, A. F., Smith, A. V. W., Asynchronous VLSI Neural networks Using Pulse-Stream Arithmetic, *IEEE Journal of Solid-State Circuits*, Vol. 23, No. 3, 1993, pp. 688-697.

[53] Schmid, A., Leblebici, Y., Mlynek, D., A Charge-based Artificial Neural Network with On-Chip Learning Ability, *Proceedings of the 5th European Congress on Intelligent Techniques and Soft Computing EUFIT'97*, Aachen, Germany, 1997, pp. 250-254.

[54] Haykin, S., *Neural Networks, A Comprehensive Foundation*, Macmillan College Publishing Company, New York, 1994.

[55] Kandel, A., Langholz, G., Editors, Fuzzy Hardware: Architectures and Applications, Kluwer Academic Publishers, Dordrecht, The Netherlands, 1998.

[56] Watanabe, H., RISC Approach to Design of Fuzzy Processor Architecture, *Proceedings of the IEEE International Conference on Fuzzy Systems*, San Diego, California, March 1992, pp. 431-440.

[57] Yamakawa, T., A Fuzzy Inference Engine in Nonlinear Analog Mode and Its Application to a Fuzzy Logic Control, *IEEE Transactions on Neural Networks*, Vol. 4, No. 3, May 1993, pp. 496-522.

[58] Cilingiroglu, U., Pamir, B., Günay, Z. S., Dülger, F., Sampled-Analog Implementation of Application-Specific Fuzzy Controllers, *IEEE Transactions on Fuzzy Systems*, Vol. 5, No. 3, August 1997, pp. 431-442.

[59] Pamir, B., Dülger, F., Seszin Günay, Z., Low-Voltage Rail-to-Rail Amplification with CMOS Operational Inverter, *Proceedings of the ECCTD'95 European Conference on Circuit Theory & Design*, Istanbul, Turkey, 1995, pp. 9-12.

[60] Fattaruso, J. W., Mahant-Shetti, S. S., Barton, J. B., A Fuzzy Logic Inference Processor, *IEEE Journal of Solid-State Circuits*, Vol. 29, No. 4, April 1994, pp. 397-402.

[61] Yamakawa, T., Silicon Implementation of a Fuzzy Neuron, *IEEE Transactions on Fuzzy Systems*, Vol. 4, No. 4, November 1996, pp. 488-501.

[62] Coggins, R., Jabri, M., Flower, B., Pickard, S., A Hybrid Analog and Digital VLSI Neural Network for Intercardiac Morphology Classification, *IEEE Journal of Solid-State Circuits*, Vol. 30, No. 5, May 1995, pp. 542-550.

[63] Leong, P. H. W., Jabri, M. A., A Low-Power VLSI Arrhythmia Classifier, *IEEE Transactions on Neural Networks*, Vol. 6, No. 6, November 1995, pp. 1435-1445.

Abbreviations

ADC	Analog to Digital Converter
ANN	Artificial Neural Network
ASIC	Application Specific Integrated Circuit
BiCMOS	Bipolar and CMOS integration technology
CAD	Computer Aided Design
CCD	Charge-Coupled Devices
CMOS	Complementary MOSFET design technique
DAC	Digital to Analog Converter
DSP	Digital Signal Processor
ECFL	Emitter-Coupled Fuzzy Logic
ESD	Electrostatic Discharge Damage
FLIPS	Fuzzy Logic Inferences Per Second
FPGA	Field Programmable Gate Array
HDL	Hardware Description Language
HW	Hardware
IC	Integrated Circuit
ICEG	Intercardiac Electrogram
MF	Membership Function
MOSFET	Metal-Oxyd-Semiconducor Field-Effect Transistor
OPI	Operational Inverter
PC	Personal Computer
PCB	Printed Circuit Board
PGA	Pin Grid Array
RISC	Reduced Instruction Set Computer
SIA	Semiconductor Industry Association
SW	Software
VHDL	VHSIC Description Language
VHSIC	Very High Speed Integrated Circuits
WTA	Winner-Take-All

Index of Terms